The cold spray materials deposition process

Fundamentals and applications

Edited by
Victor K. Champagne

Woodhead Publishing and Maney Publishing
on behalf of
The Institute of Materials, Minerals & Mining

CRC Press
Boca Raton Boston New York Washington, DC

WOODHEAD PUBLISHING LIMITED
Cambridge England

Woodhead Publishing Limited and Maney Publishing Limited on behalf of
The Institute of Materials, Minerals & Mining

Woodhead Publishing Limited, Abington Hall, Abington
Cambridge CB21 6AH, England
www.woodheadpublishing.com

Published in North America by CRC Press LLC, 6000 Broken Sound Parkway, NW, Suite 300, Boca Raton, FL 33487, USA

First published 2007, Woodhead Publishing Limited and CRC Press LLC

British Library Cataloguing in Publication Data
A catalogue record for this book is available from the British Library.

Library of Congress Cataloging in Publication Data
A catalog record for this book is available from the Library of Congress.

Woodhead Publishing ISBN 978-1-84569-181-3 (book)
Woodhead Publishing ISBN 978-1-84569-378-7 (e-book)
CRC Press ISBN 978-1-4200-6670-8
CRC Press order number: WP6670

The publishers' policy is to use permanent paper from mills that operate a sustainable forestry policy, and which has been manufactured from pulp which is processed using acid-free and elementary chlorine-free practices. Furthermore, the publishers ensure that the text paper and cover board used have met acceptable environmental accreditation standards.

Typeset by SNP Best-set Typesetter Ltd., Hong Kong
Printed by TJ International Ltd, Padstow, Cornwall, England

Contents

Contributor contact details

(* = main contact)

Editor, Chapter 1 and Chapter 17

V. K. Champagne
US Army Research Laboratory
Aberdeen Proving Ground
MD
USA
E-mail: vchampag@arl.army.mil

Chapter 2

A. Papyrin
Cold Spray Technology (CST)
1421 Wagon Trail Drive SE
Albuquerque
NM 87123
USA
E-mail: Apapyrin102@comcast.net
(copy to: mmszulinski@yahoo.com)

Chapter 3

M. F. Smith
Sandia National Laboratories
1515 Eubank Blvd SE
Albuquerque
NM 87123
USA
E-mail: mfsmith@sandia.gov

Chapter 4

J. Karthikeyan
ASB Industries Inc.
1031 Lambert Street
Barberton
OH 44203
USA
E-mail: karthi@asbindustries.com

Chapter 5

S. Celotto, J. Pattison*, J. S. Ho, A. N. Johnson and W. O'Neill
Innovative Manufacturing Research Centre
Cambridge University Engineering Department
Unit 26A Cambridge Science Park
Cambridge
CB4 0FP
UK
E-mail: jap54@cam.ac.uk

Chapter 6

F. J. Brodmann
F. J. Brodmann & Co.
2072 Sussex Street

Harvey
LA 70058
USA
E-mail: info@fjbco.com

Chapter 7

K. Sakaki
Department of Mechanical Systems Engineering
Faculty of Engineering
Shinshu University
4-17-1Wakasato
Nagano City 380-8553
Japan
E-mail: ksakaki@gipwc.shinshu-u.ac.jp

Chapter 8

T. Van Steenkiste
Group Leader
Advanced Materials and Processes
Advanced Thermal Components and Systems
Delphi Research Lab
51786 Shelby Parkway
Shelby Township
MI 48315
USA
E-mail: thomas.van.steenkiste@delphi.com

Chapter 9

M. Grujicic
Department of Mechanical Engineering
241 Fluor Daniel Building
Clemson University
Clemson
SC 29634-0921
USA
E-mail: gmica@clemson.edu

Chapter 10

V. F. Kosarev, S. V. Klinkov and A. N. Papyrin*
Cold Spray Technology (CST)
1421 Wagon Trail Drive SE
Albuquerque
NM 87123
USA
E-mail: Apapyrin102@comcast.net
(copy to: mmszulinski@yahoo.com)

Chapter 11

R. Maev, V. Leshchynsky and E. Strumban*
University of Windsor
Windsor
ON N9B 3P4
Canada
E-mail: emilst@tech-dm.com

Chapter 12

H. Höll
Cold Spray Technology GmbH
Wernher von Braun Str. 4
84539 Ampfing
Germany
E-mail: HHoell@crp-ag.de

Chapter 13

A. Gouldstone, W. B. Choi, W. Chi, Y. Wu and S. Sampath*
Center for Thermal Spray Research
Department of Materials Science and Engineering
State University of New York at Stony Brook
Stony Brook
NY 11794-2275
USA
E-mail: ssampath@ms.cc.sunysb.edu

Chapter 14

D. Wolfe* and T. Eden
Applied Research Laboratory
Penn State University
119 MRI Building
University Park
PA 16802
USA
E-mail: dew125@psu.edu

Chapter 15

R. C. McCune
5451 S. Piccadilly Cir.
West Bloomfield
MI 48322-1446
USA
E-mail: robert.mccune@sbcglobal.net

Chapter 16

D. J. Helfritch
US Army Research Laboratory
Aberdeen Proving Ground
MD
USA
E-mail: dhelfritch@arl.army.mil

Preface

The purpose of this book is to convey an understanding of cold gas-dynamic spray and to unravel some of the mystique that has evolved over the years about the possible uses of this process. This book is a compilation of both theory and practical knowledge that provides potential users with the information necessary to recognize the advantages as well as the limitations of cold spray. This has been accomplished by presenting data from a variety of researchers as well as from users of the technology around the globe.

From pioneering work in Russia to upwards of 30 research and industrial installations today, cold spray has evolved from a technical curiosity into an important manufacturing process. The ability of the cold spray process to produce unique coatings that are not attainable with other coating methods has advanced this technology to 'essential' status. Cold spray equipment is commercially available as stationary, robot-controlled, spray systems or as portable, hand-held, systems. The extremely dense, oxide-free, coatings available by cold spray allow this process to be used in such diverse application areas as corrosion control, electrical circuitry, and metals repair. The equipment and capabilities of the cold spray process continue to evolve, opening new application areas on a daily basis.

Cold spray is not a direct replacement for thermal spray, but it does have distinct advantages that make it the most logical choice for certain applications. There are substantial data to suggest that cold spray has a niche in the marketplace; however, it must also be realized that cold spray does have limitations that should be acknowledged in order to prevent misuse of the technology. Cold spray has sometimes been 'oversold', and acceptance of the technology has suffered as a consequence, although a steady increase in equipment sales continues through this writing.

This book is divided into the general areas of cold spray operating and performance considerations and of application demonstrations. Chapters on history, system availability, and costs are also included. The content of this book ranges from mathematically rigorous to descriptive, as is needed to convey the information. It is expected that this book will be useful to cold spray practitioners as well as to those who are considering the use of cold spray for the first time.

It is my intention that this book will provide the means to initiate interest and to spawn the development of those applications appropriate to the cold spray process. I am indebted to the contributing authors for the comprehensive cold spray analysis and for the breadth of application contained in this book.

Victor K. Champagne

1
Introduction

V. K. CHAMPAGNE, US Army Research Laboratory, USA

Cold spray is a process whereby metal powder particles are utilized to form a coating by means of ballistic impingement upon a suitable substrate.[1–3] The metal powders range in particle size from 5 to 100 μm and are accelerated by injection into a high-velocity stream of gas. The high-velocity gas stream is generated through the expansion of a pressurized, preheated gas through a converging–diverging nozzle. The pressurized gas is expanded to supersonic velocity, with an accompanying decrease in pressure and temperature.[4–6] The powder particles, initially carried by a separate gas stream, are injected into the nozzle either prior to the throat or downstream of the throat. The particles are then accelerated by the main nozzle gas flow and are impacted onto a substrate after exiting the nozzle. Upon impact, the solid particles deform and create a bond with the substrate.[7,8] As the process continues, particles continue to impact the substrate and form bonds with the deposited material, resulting in a uniform coating with very little porosity and high bond strength. The term 'cold spray' has been used to describe this process due to the relatively low temperatures (–100 to +100 °C) of the expanded gas stream that exits the nozzle.

Cold spray as a coating technology was initially developed in the mid-1980s at the Institute for Theoretical and Applied Mechanics of the Siberian Division of the Russian Academy of Science in Novosibirsk.[9] The Russian scientists successfully deposited a wide range of pure metals, metallic alloys, polymers, and composites onto a variety of substrate materials, and they demonstrated that very high coating deposition rates are attainable using the cold spray process. These experiments are described in detail in Chapter 2. Currently, a variety of cold spray research is being conducted at institutions in dozens of locations world-wide.

The temperature of the gas stream is always below the melting point of the particulate material during cold spray, and the resultant coating and/or freestanding structure is formed in the solid state. Since adhesion of the metal powder to the substrate, as well as the cohesion of the deposited material, is accomplished in the solid state, the characteristics of the cold spray deposit are quite unique in many regards. Because particle oxidation is avoided, cold spray produces coatings that are more durable with better bond strength. The exceptional adhesion of cold spray coatings is in part due to the low temperatures at which the coatings are deposited. One of the most deleterious effects of depositing coatings at high temperatures is the residual stress that develops, especially at the substrate–coating interface. These stresses often cause debonding. This problem is compounded when the substrate material is different from the coating material. This problem is minimized or eliminated when cold spray is used. In addition, interfacial instability due to differing viscosities and the resulting roll-ups and vortices promote interfacial bonding by increasing the interfacial area, giving rise to material mixing at the interface and providing mechanical interlocking between the two materials.

A key concept in cold spray operation is that of critical velocity.[10,11] The critical velocity for a given powder is the velocity that an individual particle must attain in order to deposit (or adhere) after impact with the substrate. Small particles achieve higher velocities than do larger particles, and, since powders contain a mixture of particles of various diameters, some fraction of the powder is deposited while the remainder bounces off. The weight fraction of powder that is deposited divided by the total powder used is called the deposition efficiency. High velocity is necessary for optimal deposition efficiency and packing density, and several parameters – including gas conditions, particle characteristics, and nozzle geometry – affect particle velocity. Chapters 7 through 10 consider the effects of these parameters in detail.

The two principal cold spray system configurations differ in the location of powder injection into the nozzle, and these systems are described in more detail in Chapters 11 and 12. The two configurations are depicted by Figs 1.1 and 1.2. Figure 1.1 shows a system in which the main gas stream and the powder stream are both introduced into the mixing chamber of the converging–diverging nozzle. This configuration requires that the powder feeder be capable of high gas pressure and is most often used in stationary

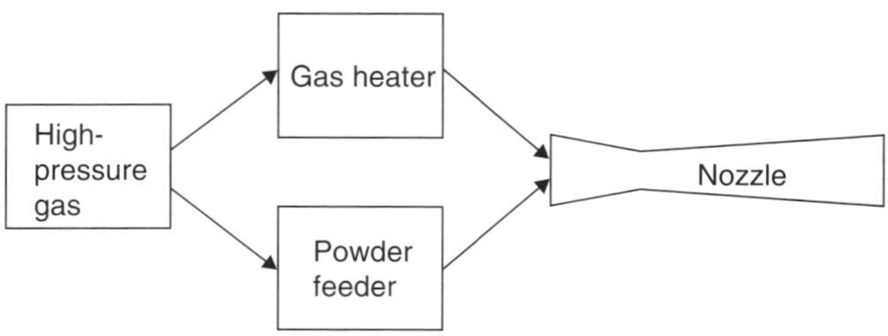

1.1 Typical configuration of the stationary system.

cold spray systems, for which the cumbersome powder feeder is acceptable. Figure 1.2 shows a system in which the powder stream is injected into the nozzle at a point downstream of the throat where the gas has expanded to low pressure. Generally atmospheric pressure air, drawn by the lower pressure nozzle injection point, is used for powder transport from the feeder. Since this system does not require a pressurized feeder, it is often used in portable cold spray systems. Figure 1.3 shows a typical stationary installation. The nozzle (pointed downward) is directed by a robot arm. The gas heater is attached to the robot arm and is coupled to the side of the nozzle

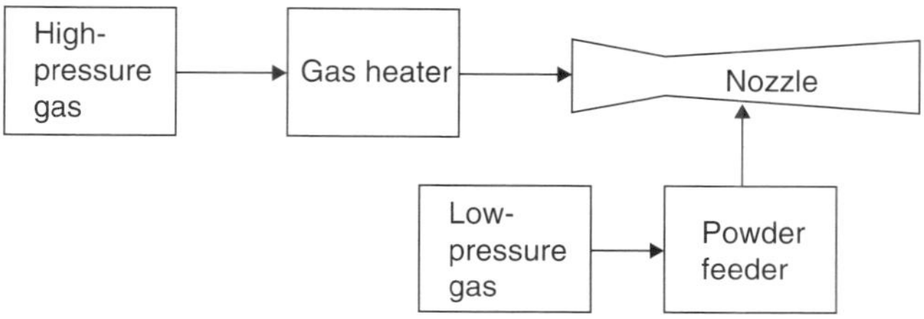

1.2 Typical configuration of the portable system.

1.3 Equipment arrangement of a stationary system.

by a flexible, high-temperature hose. The powder feed line enters the nozzle from above.

The ranges of operations for the two systems depicted in Fig. 1.1 and 1.2 are given in Table 1.1. The principal differences separating the stationary and portable systems are the gas pressure and the gas and powder flow rates, with the portable system utilizing readily available compressed air. Stationary systems utilize higher pressure gases and often have a dedicated high-pressure compressor. A low molecular weight gas, such as helium, is sometimes used as the accelerating gas when particles must be brought to very high velocity. Because of the high sonic velocity of helium, attainable particle velocities are often twice those obtained when nitrogen or air are used. The values in Table 1.1 are representative of typical values currently in use but are not necessarily limiting values for the systems.

A wide variety of metal powders can be successfully deposited on metal or ceramic substrates. Amateau and Eden[12] list 45 powders that have been successfully cold sprayed at the Advanced Research Laboratory of Penn State, USA. The manufacture and characteristics of powders that are amenable to cold spray are discussed in Chapter 6.

The deposition thickness produced by a moving nozzle can vary from 0.01 to 1.0 mm, depending on powder feed rate, nozzle sweep speed, and deposition efficiency. Multiple coating layers can result in deposits several centimeters thick. The width of a single sweep is approximately 5 mm, and large surfaces can be coated through multiple, slightly overlapping, parallel sweeps. Figure 1.4 shows a deposit of copper powder on an aluminum rod. Nitrogen was the accelerating gas, and the deposit is several layers thick. Upon closer magnification, Fig. 1.5 shows that individual particles have fused into a dense, uniform coating, which adheres tightly to the aluminum surface. The copper is seen to be mixed with the aluminum at the interface.

Table 1.1 Operating parameter ranges of the stationary and portable systems

	Stationary system	Portable system
Working gas	N_2, He, air	Air
Gas pressure (bar)	20–45	6–8
Gas preheat (°C)	20–800	20–550
Gas flow rate (m^3/h)	50–150	15–30
Nozzle exit Mach number	2–3	2–3
Powder flow rate (g/s)	0.1–1.0	0.06–0.1
Particle size (μm)	5–100	10–50

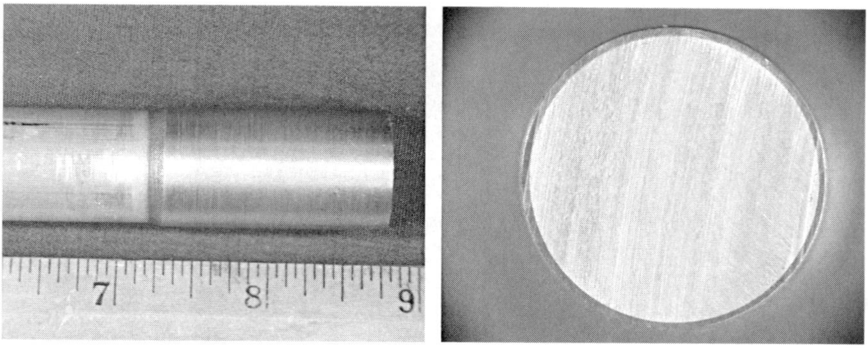

1.4 Copper deposited by cold spray onto an aluminum rod.

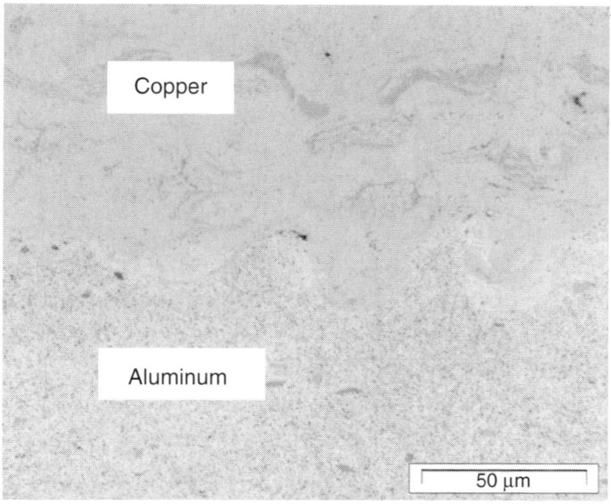

1.5 Magnified cross-section of a copper deposit on aluminum.

When a ceramic is used as a deposition substrate, results such as those shown in Fig. 1.6 can be obtained. In this case, nitrogen driving gas was used to deposit copper powder on a silicon carbide substrate. Single lines of copper are deposited as well as a uniform area coating. Ceramics such as alumina, silicon carbide, and aluminum nitride have been successfully coated.

In comparison with other thermal spray coatings, deposits produced by cold spray are characterized by being less porous, and having higher hardness and lower oxide concentration. The Young's moduli of cold sprayed deposits can be greater than 80% of bulk values.[13] The reasons for these characteristics are that cold particles are less susceptible to oxidation and

1.6 Copper deposited by cold spray onto silicon carbide.

that high-velocity impact creates dense, well-consolidated deposits. Chapters 3 and 4 compare cold spray performance with thermal spray technologies and Chapter 13 details analytical techniques for the characterization of cold sprayed coatings.

Applications for cold spray technology generally occur in situations where conventional hot metal spray technology cannot be successfully used. Chapter 14 shows that wear-resistant surfaces, such as Cr_3C_2-based powders, can be deposited by cold spray and furthermore that the coating can be tailored to the coating service requirements. Situations where conventional thermal spray is unacceptable are also found when high temperatures cannot be tolerated by the substrate, as described in Chapter 15. The superior qualities of cold sprayed deposits are often required by the application, for example the high heat transfer coefficient and electrical conductivity of cold sprayed deposits favor its use in electronic applications.[14] Chapters 15 and 17 demonstrate how good corrosion protection is achieved by dense, impermeable, cold sprayed coatings.

Chapter 5 describes how the application of cold spray systems is economically competitive with other thermal spray techniques that provide deposits of inferior quality, especially for applications with oxygen-sensitive materials such as tantalum and titanium. The chapter further describes how improvements in gas handling and powder characteristics will continue to enhance the competitiveness of cold spray.

References

1 Papyrin A, 'Cold Spray Technology', *Advanced Materials & Processes*, 2001, **160**(3), 49–51.
2 Van Steenkiste T H, Smith J R, Teets R E, *et al.*, 'Kinetic Spray Coatings', *Surface and Coatings Technology*, 1999, **111**, 62–71.
3 Stoltenhoff T, Kreye H and Richter H, 'An Analysis of the Cold Spray Process and Its Coatings', *Journal of Thermal Spray Technology*, 2002, **11**(4), 542–550.
4 Dykhuisen R and Smith M, 'Gas Dynamic Principles of Cold Spray', *Journal of Spray Technology*, 1998, **7**(2), 205–212.
5 Kosarev V F, Klinkov S V, Alkhimov A P and Papyrin A N, 'On Some Aspects of Gas Dynamics of the Cold Spray Process', *Journal of Thermal Spray Technology*, 2003, **12**(2), 265–281.
6 Grujicic M, Zhao C L, Tong C, Derosset W S and Helfritch D, 'Analysis of the Impact Velocity of Powder Particles in the Cold-Gas Dynamic-Spray Process', *Materials Science and Engineering A*, 2004, **368**, 222–230.
7 Dykhuisen R, Smith M F, Gilmore D L, *et al.*, 'Impact of High Velocity Cold Spray Particles', *Journal of Thermal Spray Technology*, 1999, **8**(4), 559–564.
8 Grujicic M, Saylor J R, Beasley D E, Derosset W S and Helfritch D, 'Computational Analysis of the Interfacial Bonding between Feed-Powder Particles and the Substrate in the Cold-Gas Dynamic-Spray Process', *Applied Surface Science*, 2003, **219**, 211–227.
9 Alkhimov A P, Papyrin A N, Kosarev V F, Nesterovich N I and Shushpanov M M, 'Gas-Dynamic Spraying Method for Applying a Coating', US Patent 5,302,414, April, 1994.
10 Gilmore D L, Dykhuizen R C, Neiser R A, Roemer T J and Smith M F, 'Particle Velocity and Deposition Efficiency in the Cold Spray Process', *Journal of Thermal Spray Technology*, 1999, **8**(4), 576–582.
11 Wu J H, Fang H Y, Lee L H, Yoon S H and Kim H J, 'Critical and Maximum Velocities in Kinetic Spraying', *Proceedings of the 2006 International Thermal Spray Conference*, Seattle, Washington, USA, 15–17 May 2006.
12 Amateau M F and Eden T J, 'High Velocity Particle Technology', *iMast Quarterly*, 2000, **2**, 3–6.
13 Mccune R C, Donlon W T, Popoola O O and Cartwright E L, 'Characterization of Copper Layers Produced by Cold Gas-Dynamic Spraying', *Journal of Thermal Spray Technology*, 2000, **9**(1), 73–82.
14 Kroemmer W and Heinrich P, 'Cold Spraying – Potential and New Application Ideas', *Proceedings of the 2006 International Thermal Spray Conference*, Seattle, Washington, USA.

Part I

General issues

2
The development of the cold spray process

A. PAPYRIN, Cold Spray Technology (CST), USA

2.1 History of cold spray

The cold spray process was originally developed in the mid-1980s at the Institute of Theoretical and Applied Mechanics of the Siberian Branch of the Russian Academy of Science (ITAM SB RAS) in Novosibirsk by Professor Anatolii Papyrin and his colleagues while studying models subjected to a supersonic two-phase flow (gas + solid particles) in a wind tunnel (Alkhimov *et al.* 1990d). They successfully deposited a wide range of pure metals, metal alloys, and composites onto a variety of substrate materials, and demonstrated the feasibility of the cold spray process for various applications. Using the results of research conducted, a number of technical solutions related to the development of cold spray equipment and technologies were suggested (patents: Alkhimov *et al.* 1990a, 1990b, 1990c, 1991, 1992, 1994a, 1994b, 1995, 1997b, 1997c. A US patent was issued in 1994 (Alkhimov *et al.* 1994a, 1997a), and the European patent was issued in 1995 (Alkhimov *et al.* 1995). At the present time, a wide spectrum of research is being conducted at many research centers and companies around the world. The evolution and current status of cold spray will be presented in Section 2.3.

2.2 Discovery of the cold spray phenomenon and its basic features

This section describes the original experiments that resulted in the discovery of the cold spray process and established the basic physical principles of the process. Some results of wind tunnel experiments that led to the creation of cold spray are presented. The interaction of solid particles with a substrate at a wide range of particle velocities is studied. It is shown that transition from erosion of the substrate to coating formation takes place at some (critical) value of particle velocity. Basic features and advantages of cold spray are formulated. Wind tunnel experiments are described in some detail because the results obtained are important in gaining a better understanding of certain aspects of the gas dynamics of the cold spray process.

2.2.1 Wind tunnel experiments that led to the creation of cold spray

One of the main tasks of the wind tunnel experiments was to study the influence of particles on flow structure and the interaction of two-phase flow with a body. To provide registration of particles in the flow and changes in wave microstructure, methods of laser diagnostics with levels of spatial and temporal resolution available at that time (1980–1985) were used. Based on such diagnostics, a wide spectrum of studies in the field of gas-dynamics of two-phase flows were conducted in the ITAM of the RAS (Alkhimov *et al.* 1981, 1982a). To study the global pattern of the two-phase flow around bodies and to clarify the main physical features of this process, the first experiments involved flow visualization with bodies of simple geometry (wedge, cylinder, and sphere) and injection of particles that differed in size and material.

The experimental setup is shown in Fig. 2.1. A supersonic gas flow was generated by a plane contoured nozzle. The air in the pre-chamber had the following parameters: pressure $p_0 = 0.85$ MPa, temperature $T_0 = 260$–280 K, and Mach number at the nozzle exit $M^* = 3$. The models to be tested were mounted in the vicinity of the nozzle exit; there were windows designed for optical measurements. A wide range of particles with disparate sizes d_p and density ρ_p were used in experiments: bronze particles 100 μm, 8.6 g/cm^3; Plexiglas particles 200 μm, 1.2 g/cm^3; aluminum particles 15 μm, 2.7 g/cm^3; lycopodium particles (plant spores) 25 μm, 0.5 g/cm^3. The particles were injected into the gas flow at a distance of 300 mm upstream of the throat section of the nozzle.

Several laser-based techniques were used to diagnose the process. The particle velocity was measured by laser Doppler velocimeter (LDV) with a direct spectral analysis for registration of the Doppler frequency shift. A pulsed Schlieren method was used to register the microstructure of the density field. A ruby laser operating with a modulated Q-factor and pulse duration of 30 ns was used as a source of light. The displacement of particles during the exposure time was less than their diameter. The time evolution of the wave structure was studied by the multi-frame Schlieren technique based on the use of a laser generating a series of pulses with a prescribed time interval between them (laser stroboscope). The trajectories of particle motion were observed by the method of laser visualization in scattered light (laser sheet). The ruby laser operated in the 'peak' generation mode. The duration of emission of a pulse series was approximately 10^{-3} s.

Some results of these studies illustrating the usefulness of these methods for two-phase flow visualization are shown in Fig. 2.2–2.5. The photographs (Fig. 2.2 and 2.3) in scattered light (laser sheet technique) show the particle trajectories; reflected particles are clearly seen; some of them have a

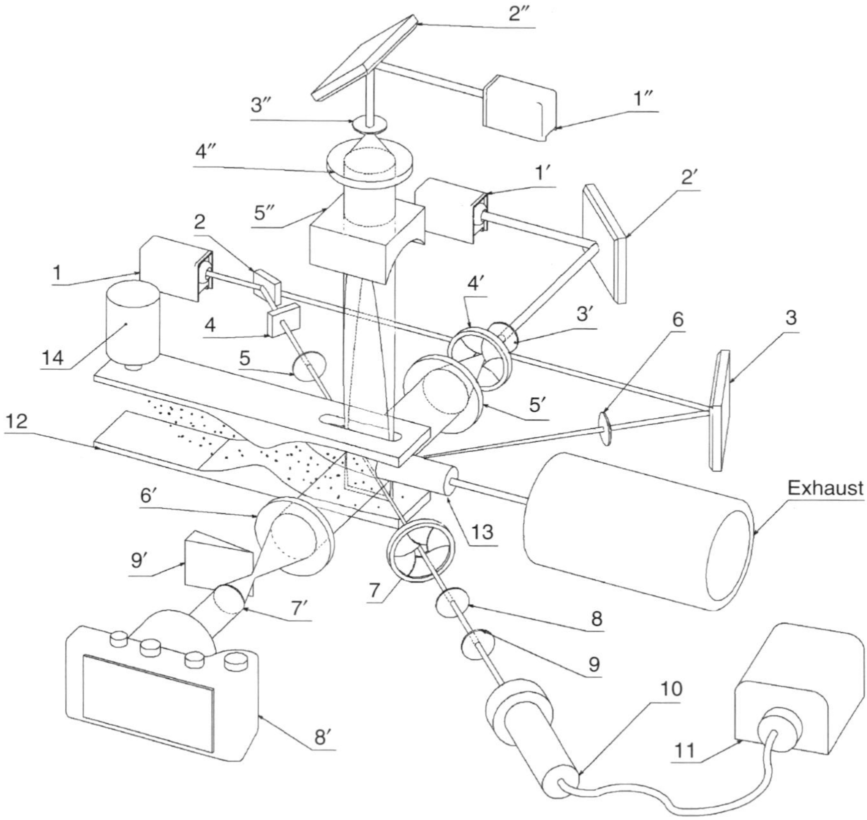

2.1 Schematic of the setup and diagnostic equipment. LDV scheme: 1, single-frequency laser; 2–9, optical elements; 10, confocal interferometer; 11, recorder. Setup elements: 12, plane supersonic nozzle; 13, model under study; 14, powder feeder. Scheme of multi-frame Schlieren technique: 1′, ruby laser with a modulator; 2′–7′, optical elements; 8′, camera, 9′, sheet. Laser sheet technique: 1″, ruby laser; 2″–5″, optical elements.

reflection angle close to the incidence angle. After reflection and subsequent deceleration by the oppositely directed flow, these particles reverse their direction and subsequently hit the body again. The presence of particles that can collide with the body several times with a gradual decrease in the rebound distance leads to their accumulation near the frontal part of the body. A zone with an elevated concentration of the disperse phase with intense interaction of incident and reflected particles is formed.

Figures 2.4 and 2.5 show typical Schlieren photographs illustrating the disturbance of the wave structure near the body owing to the presence of the disperse phase in the flow. One can clearly see local shock waves formed

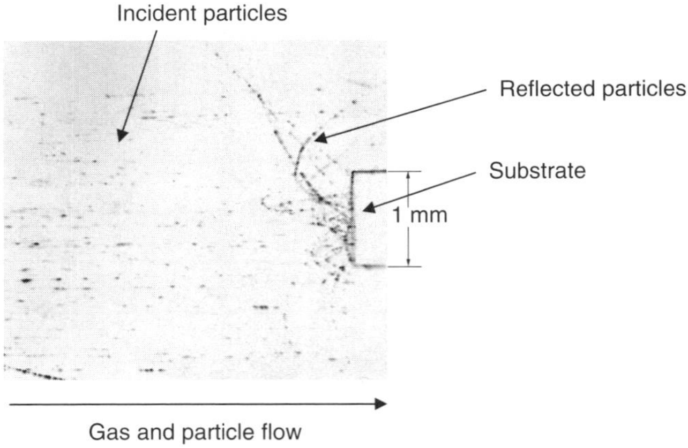

2.2 Laser sheet photograph illustrating trajectories of incident and reflected Cu particles. $d_p = 30\text{–}60\,\mu m$; $v_{pm} \approx 150\,m/s$.

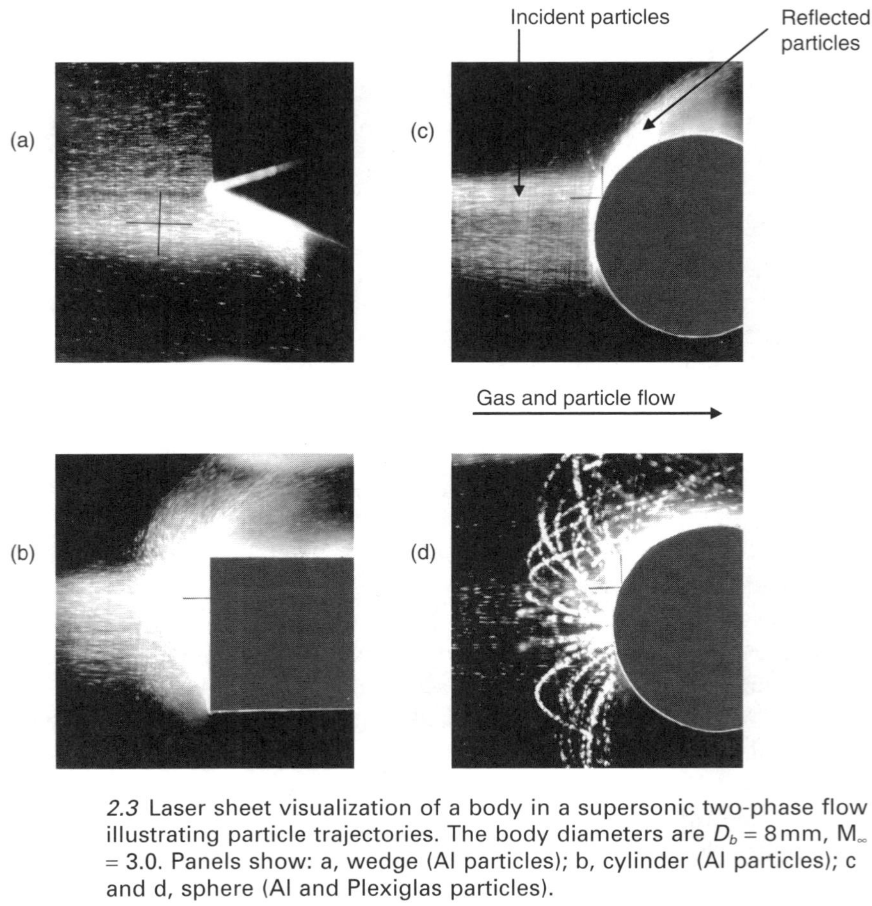

2.3 Laser sheet visualization of a body in a supersonic two-phase flow illustrating particle trajectories. The body diameters are $D_b = 8\,mm$, $M_\infty = 3.0$. Panels show: a, wedge (Al particles); b, cylinder (Al particles); c and d, sphere (Al and Plexiglas particles).

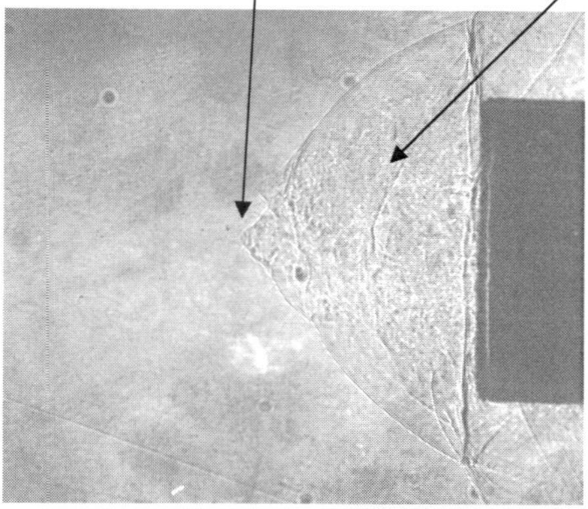

2.4 Schlieren photograph illustrating the formation of a separation region in front of the cylinder in supersonic two-phase flow by a particle with $d_p \leq 10\,\mu m$. $D_b = 11\,mm$, $M_\infty = 3.0$, $Re_D = 7.8 \times 10^5$, $t_{ex} = 30\,ns$.

by a supersonic flow around the cylinder (Fig. 2.5a) and sphere (Fig. 2.5b), and the character of changes in the bow-shock structure, which is associated with the presence of particles. Photographs of each body in a 'pure' air flow are also presented for comparison in Fig. 2.5a1 and b1.

An analysis of a large amount of experimental data revealed several features related to the influence of particles on the structure of the bow-shock front. The character of the wave structure around the body and, hence, its basic characteristics significantly depend on the body shape (blunted or sharp forebody) and on particle parameters. In the flow around bodies with a blunted forebody (sphere or stream-wise-aligned cylinder) and the injection of fine aluminum ($d_p = 1$–$40\,\mu m$) and lycopodium particles ($d_p = 25$–$28\,\mu m$) into the flow, the influence of the disperse phase on the bow-shock structure starts to manifest itself when the volume concentration of particles reaches $\varphi_p \geq 0.5$–1%. In the shadowgraphs (e.g. Fig. 2.5a4), this phenomenon is manifested as a change in the distance between the shock-wave front and the body and as a deformation of the shock-wave front shape. In the flow around blunted bodies by a gas with coarser particles of Plexiglas ($d_p = 200\,\mu m$) and bronze ($d_p = 100\,\mu m$), the changes in the wave structure acquire a different character. Strong disturbances of the bow-shock structure are observed even in the case of a low concentration of particles $\varphi_p << 1\%$. These disturbances are manifested in the form of conical shock waves (Fig. 2.5a3 and b4) with large apex angles and particles located

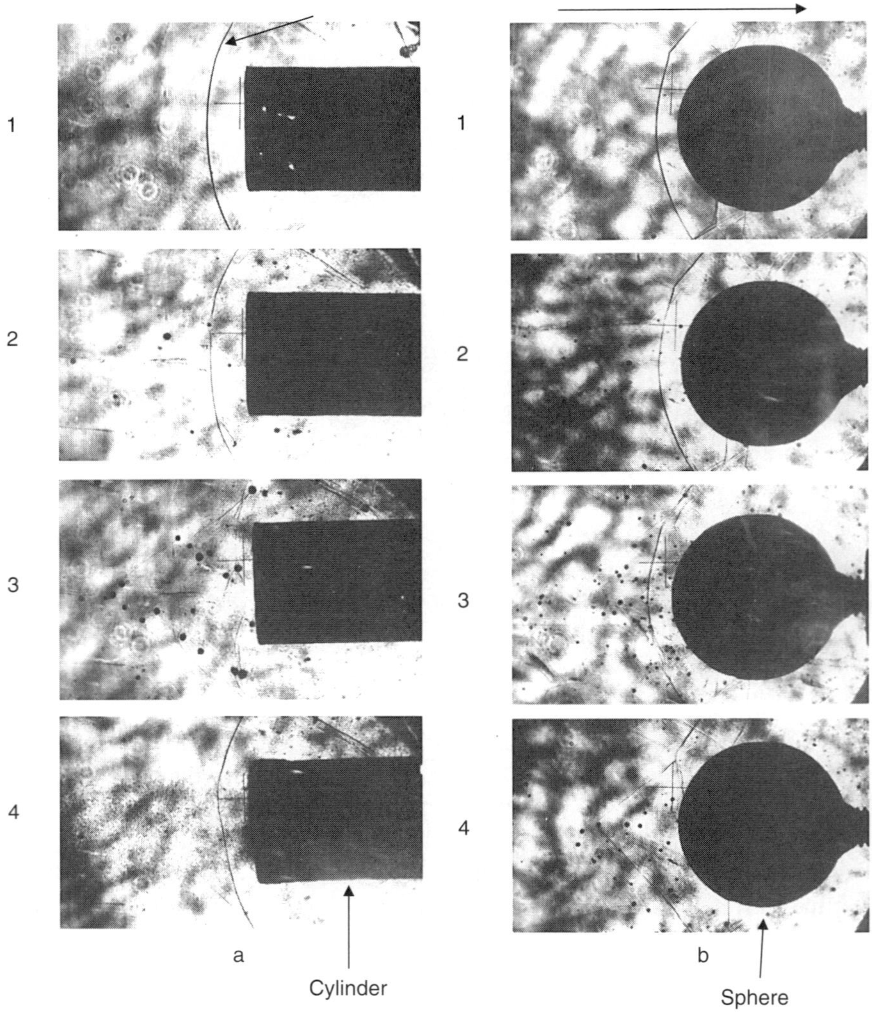

2.5 Schlieren photographs illustrating the effect of wave structure change in supersonic two-phase flow around a cylinder (a) and a sphere (b). The body diameters are $D_b = 8\,\text{mm}$, $M_\infty = 3.0$, $t_{ex} = 30\,\text{ns}$. a: 1, without particles; 2 and 3, Plexiglas particles 50–200 μm; 4, lycopodium particles. b: 1, without particles; 2, Al particles; 3 and 4, Plexiglas particles.

at the apices. This is most clearly seen in the flow around a stream-wise-mounted cylinder.

In the flow around sharp bodies (wedge, cone), the character of variation of the wave structure does not change significantly with the injection of all particles, both fine and coarse. In this case, the effects of bow-shock disturbances are similar to those observed in the flow around blunted bodies with the presence of fine particles (aluminum and lycopodium) in the flow. For $\varphi_p > 1\%$, a change in the shape and position of the bow-shock front is observed; as in the case of blunted bodies, this can be explained by the effect of 'concentration', i.e. changes in gas parameters upon its interaction with particles.

In contrast to blunted bodies, the effect of the strong disturbance of the bow shock wave in the presence of coarse particles for bodies with a sharp nose was not observed. This made it possible to assume that reflected particles moving away from the body, intersecting the bow-shock front, and entering the supersonic zone play an important role. Because of the importance of this effect in science and potential applications, the structure of the flow ahead of a blunted body, formed by single particles moving from the frontal surface of the body upward the supersonic flow, was studied in more detail.

The flight of a single particle was examined with the help of Schlieren pictures (Fig. 2.6). The experiments showed that, as the particle passes through the shock wave, an elevated pressure region is formed behind the latter, and the apex of this zone together with the particle moves upstream. The shape of this zone and the shock wave is close to conical. Depending on the distance between the reflected particle and the body (rebound distance), we can distinguish two typical modes:

(a) a regime without separation of the elevated pressure region from the particle;
(b) a regime with separation of the elevated pressure region from the particle.

The first mode is converted to the second one as the rebound distance exceeds a certain critical value.

The first mode observed, when the rebound distance is smaller than the critical value, is characterized by the absence of flow reconstruction (i.e. breakdown of the elevated pressure region) over the entire range of particle motion from the body surface to the point where it reverses direction, and the particle is 'rigidly' connected to the apex of the gas cone. The initial shape of the bow shock wave upstream of the body is reconstructed when the particle returns to the subsonic region. The second mode is observed when the maximum rebound distance is greater than the critical value at which flow reconstruction leading to recovery of a supersonic flow ahead of the cylinder begins. Figure 2.6 shows typical photographs illustrating these two modes.

Additional experiments were performed that showed that a flux of particles ejected from the frontal part of a blunted body upstream to a supersonic flow can substantially reduce the drag force of the body in a supersonic gas flow. Thus, there is a certain analogy between a spike mounted ahead

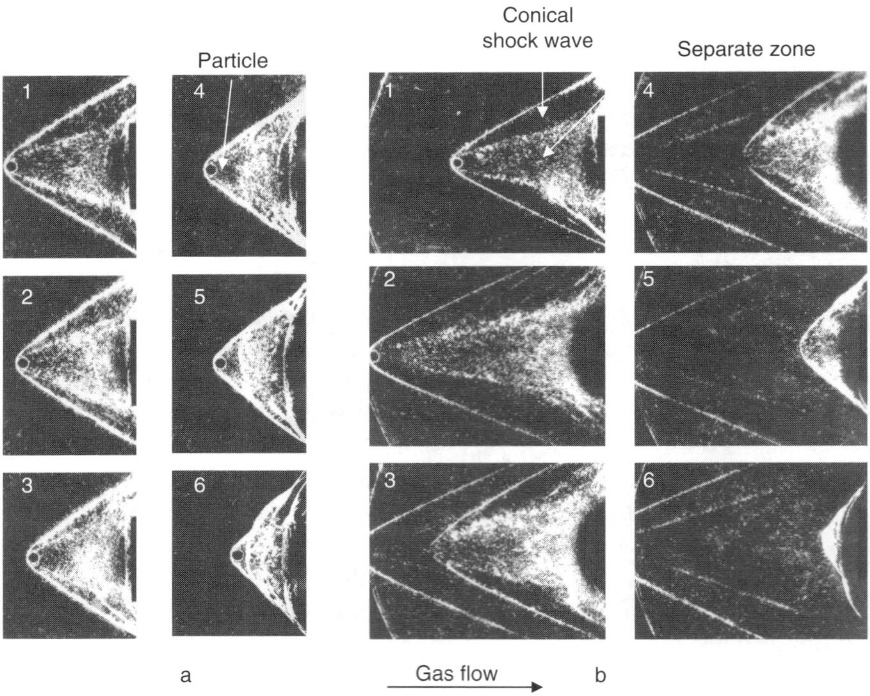

2.6 Multi-frame Schlieren photographs illustrating the development of the disturbance ahead of the cylinder under escape of 'particle' towards flow. The cylinder diameter is $D_b = 11\,\mathrm{mm}$, and the 'particle' diameter is $d_p = 1.5\,\mathrm{mm}$, $M_\infty = 3.0$, $Re_D = 7.8 \times 10^5$, $t_{ex} = 30\,\mathrm{ns}$. a, regime without flow reconstruction (the particle returns to the body), the initial flight velocity is $v_{st} = 20\,\mathrm{m/s}$. The interval between two neighboring frames is $\Delta t = 100\,\mu\mathrm{s}$. b, regime with flow reconstruction, the initial flight velocity is $v_{st} = 30\,\mathrm{m/s}$. The interval between two neighboring frames is $\Delta t = 30\,\mu\mathrm{s}$.

of the body for decreasing its drag and a single particle located at a certain distance upstream of the body. A scheme of disturbances of the wave structure in the vicinity of the body, which is formed by the reflected particle, is described in more detail in Alkhimov *et al.* (1981). The study performed made it possible to offer an explanation for the mechanism of formation and time evolution of disturbances ahead of blunted bodies with a flux of fine particles reflected or exhausted from them, and to get a patent for the method of drag reduction for such bodies in a supersonic gas flow (Alkhimov *et al.* 1986). It should be noted that the effect of the change of shock-wave structure by reflected particles in front of the body was described here in some detail because it can also have an influence on the gas dynamics of the cold spray process in the presence of reflected particles.

2.2.2 Effect of coating formation

An analysis of the body surface after its treatment by a flow of aluminum particles showed that a continuous coating of more than 200–300 μm in thickness is formed from the particle material on the frontal surface of the blunted bodies (Fig. 2.7). Coating thickness and its shape depended on particle concentration and the shape of the body. These effects were not observed in the flow with Plexiglas or bronze particles around a stream-wise-mounted cylinder. Bombardment of the target by bronze particles led to the 'reverse' phenomenon: significant erosion of the steel-target material responsible for changes in the fore-body shape.

Typical laser sheet pictures illustrating interaction of aluminum particles with the cylinder in the coating formation regime are shown in Fig. 2.8. As can be seen, the increase in powder flow rate leads to the formation of a buffer zone ahead of the body, where the concentration of particles is much higher than in the remaining free-stream region. The parameters of this zone (shape and typical size, etc.) largely depend on the body geometry and on particle concentration. The photographs shown in Fig. 2.8a and b illustrate the trajectories of aluminum particles interacting with the surface of stream-wise and cross-aligned cylinders. It can clearly be seen that in the first case the thickness of the buffer zone is less because more particles after collision with the body do not return to the body except for particles in the region of the critical point. For this reason the coating on the surface of the stream-wise-aligned cylinder at the same particle concentration was thicker and more uniform.

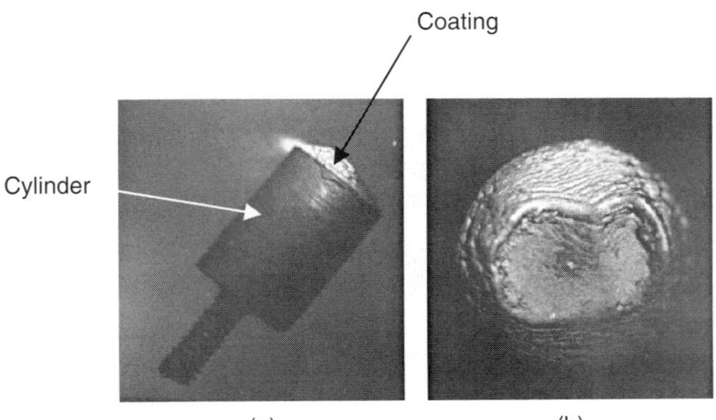

2.7 Photographs of the coating formed from aluminum particles on a cylinder, $M_\infty = 3.0$, $T_0 = 300\,K$, $d_{pm} = 20\,\mu m$.

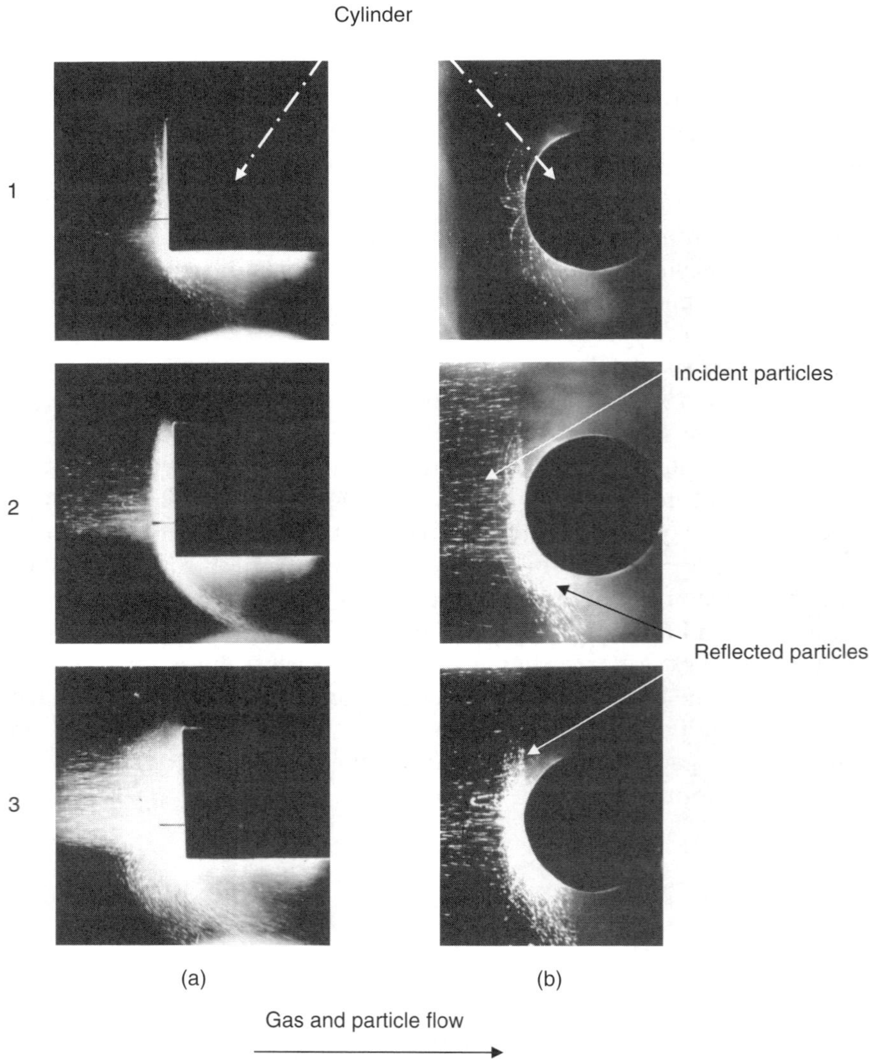

2.8 Laser sheet photographs illustrating the trajectories of particles under flow around a cylinder mounted along (a) and across (b) the flow with aluminum particles and mass flow rates of 0.5 g/s cm^2 (1a and 1b), 2 g/s cm^2 (2a and 2b), and 6 g/s cm^2 (3a and 3b). $M_\infty = 3.0$, $Re_D = 4 \times 10^5$.

Analysis of the microstructure of the sprayed layer showed that the coating consists of strongly deformed and densely packed particles uniformly covering the surface. The coating has a scaly structure with dense packing without noticeable pores and voids. Based on microscopic data, the mean strain of particles in the layer is $\varepsilon_p = 0.6$–0.8. It is important to emphasize that the flow stagnation temperature was approximately 280 K. Thus, it follows from the results presented that the effect of coating formation on the frontal surface of the body in a 'cold' ($T_0 = 280$ K), supersonic two-phase flow with aluminum-particle velocity of 400–450 m/s was obtained for the first time, and it was supposed that the principal role in coating formation belonged to particle velocity. These experiments initiated the studies of a new low-temperature method for applying coatings: cold spray. The scientific and practical importance of the above-noted effect of formation of a coating from solid particles at low temperatures (close to room temperature) of the gas and particles, stimulated experiments for a more detailed study of the phenomenon observed.

2.3 Initial studies on the development of cold spray: basic physical principles and key features

2.3.1 Spraying with a jet incoming onto target

The results described above, obtained in a wind tunnel with an external supersonic two-phase flow, demonstrated that it is possible to obtain coatings from solid particles at room stagnation temperature of the flow. It should be noted that it was commonly accepted at this time (mid-1980s) that particles should be heated to high temperatures ensuring their melting in the gas flow to obtain a coating. For instance, it was argued in Bartenev *et al.* (1982) that it is impossible to build up a coating by spraying particles in the solid state. Due to the importance of the effects observed, it was decided to perform additional studies. The objective of these studies was to find the main features of interaction of solid particles with a target at high impact velocities.

The effect of cold gas-dynamic spraying was registered in an external two-phase flow around bodies. Obviously, such a method is unsuitable for widely used applications. First, in the case of spraying in a wind tunnel, the fraction of particles hitting the body is rather low because the size of the coated body is considerably smaller than the nozzle-exit section, and the main mass of the powder does not fall onto the body surface. Second, the size of the coated body is restricted by the test-section size. It is important to emphasize that an increase in the test-section size does not allow a significant increase in the body size because the effect of particle deceleration ahead of the body becomes relevant, which can worsen the coating quality or prevent coating formation altogether. For these reasons, the cold spray method was implemented in the regime of a two-phase jet + moving target, which is typical for thermal coating techniques.

It is well known (Preece 1979) that the target material is subjected to erosion in the case of low velocities ($v_p = 10$–100 m/s) of collisions of solid particles with the target at room temperature. Therefore, it was clear that the observed effect of coating formation under conditions of an external flow is caused by the high velocity of particles (for aluminum particles, the velocity was v_p = 400–450 m/s). Therefore, the main task at the first stage was to perform experiments on the interaction of solid particles with a target at a wide range of velocities, v_p = 100–1000 m/s to register the transition from erosion of the substrate material to coating formation. It should be noted that, at that time, there were no reliable data on the collision of solid particles with a target at this range of velocities. The main data were only available for low velocities ($v_p \leq$ 100 m/s) because of the erosion problem, and for high velocities (≥1 km/s) due to military and space problems. The results of investigations in the field of thermal spraying methods were mainly obtained for interaction of particles in the melted state (Zverev *et al.* 1979; Bartenev *et al.* 1982; Kudinov *et al.* 1990).

The majority of experiments were performed with aluminum particles for the following reasons:

- It was with aluminum particles that the effect of coating formation in an external flow was detected. Therefore, it was important to continue investigations with the same particles to compare the results.
- The possibility of obtaining high particle velocities ($v_p \approx$ 1000 m/s) because of the low density of aluminum particles being accelerated in a gas flow.
- The high probability of observing a possible effect of particle-material melting at the moment of the particle–substrate collision because of the comparatively low melting point of aluminum (670 °C).

2.3.2 Description of setup: particle velocity

Investigations performed at the early stage of studying the cold spray process unambiguously showed that the velocity of the particle impact on the substrate surface plays the most important role in the application of coatings by this method. Therefore, before performing experiments in this respect, it was necessary to work out the technique for velocity control and measurement. The particle velocity was controlled by deliberate changes in the test-gas composition (the test gas was a mixture of air and helium) and was determined numerically (Alkhimov *et al.* 2001). Numerical methods were first verified by experimental results. Figure 2.9 shows the calculated velocity of aluminum particles of several different sizes for their acceleration in a rectangular nozzle with a throat size of 3 × 3 mm and an exit size of 3 × 10 mm. By changing the helium/air ratio in the mixture from 0 to 1, it was possible to change the particle velocity within the range v_p = 200–1200 m/s.

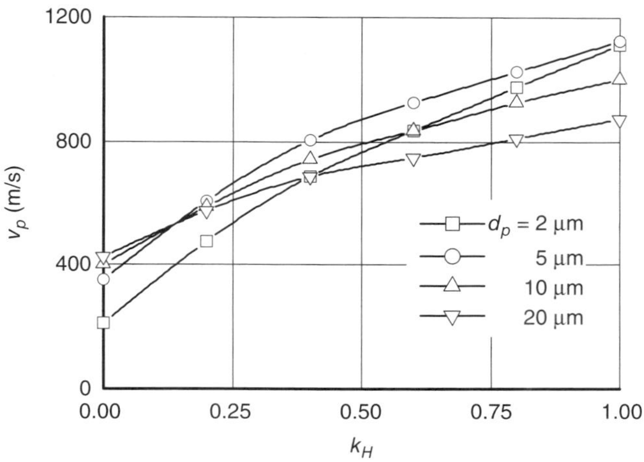

2.9 Calculated aluminum-particle velocity near the substrate surface versus helium concentration in the air–helium mixture. k_H is the helium/air flow rate ratio.

The experiments were performed in the regime of jet impact on a target (Alkhimov *et al.* 1998) normally used for coating application by thermal spray methods. The layout of this setup is shown in Fig. 2.10. The main elements are: pre-chamber and supersonic nozzle, gas heater, powder feeder, compressed air and helium, spray chamber with motion system for substrate, and exhaust system for collecting powder. The setup ensured the possibility of accelerating particles (d_p = 1–50 μm) in a supersonic nozzle up to velocities of v_p = 200–1200 m/s for different concentrations of particles. The setup was equipped with various laser-diagnostic tools (see section 2.2.1) including LDV, and Schlieren methods as well as the laser sheet technique (Alkhimov *et al.* 1978, 1982b; Alkhimov and Kosarev 1996). The particle velocity was measured by two methods: LDV and the tracking technique (Alkhimov and Kosarev 1996). Thus, the measured and calculated data made it possible to determine the particle velocity, which could be varied over a wide range: v_p = 200–1200 m/s.

The experimentally measured values of particle velocity at the nozzle exit were compared with predicted values; they are plotted in Fig. 2.11 as the ratio of particle velocity to the gas velocity versus the dimensionless quantity

$$\Omega = \sqrt{\frac{d_p}{L}\frac{\rho_p v^{*2}}{p_0}}$$

(Alkhimov *et al.* 2001). The predicted values are shown by points 1 to 11 in Fig. 2.11. Figure 2.11 also shows the results of an experiment to determine

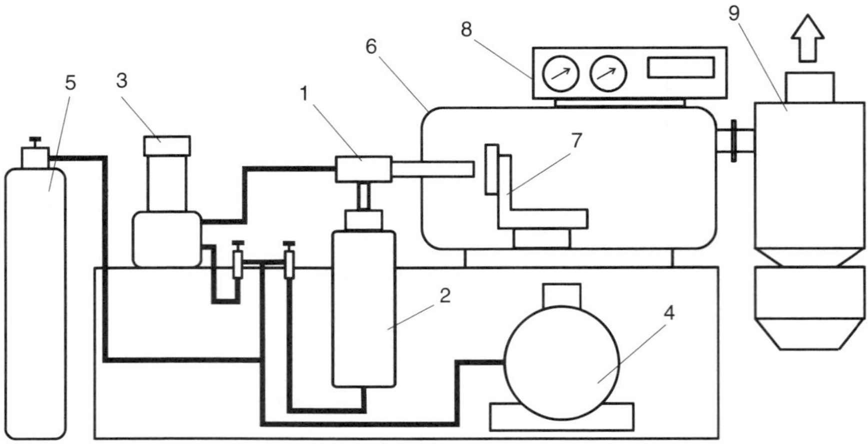

2.10 Schematic of the cold spray setup. The basic elements are: 1, the spraying unit consisting of a pre-chamber and a plane supersonic nozzle; 2, the gas heater; 3, the particle feeder; 4, the source of compressed air; 5, the source of helium; 6, the spraying chamber; 7, the traversing gear for moving the coated substrate; 8, the panel for controlling and monitoring the process parameters (gas pressure in the pre-chamber and particle dispenser, gas temperature in the pre-chamber); and 9, the particle separator.

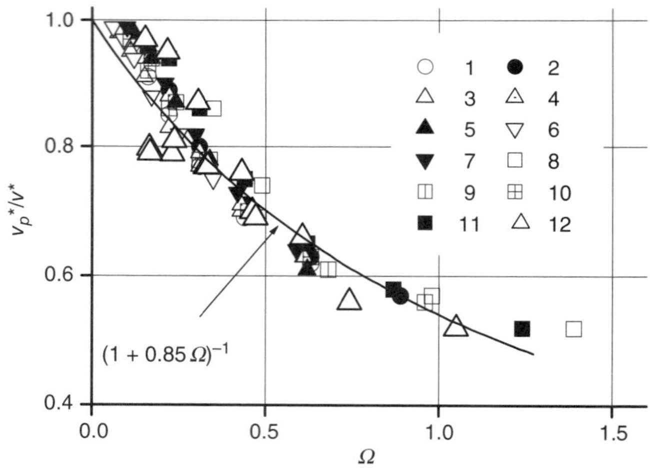

2.11 Generalized dependence of the relative velocity of particles at the exit of a plane supersonic nozzle. Data 1 to 11 are the results of calculations. All particles 1 to 7 are aluminum, $T_0 = 300\,\text{K}$: 1, $L = 50\,\text{mm}$, $p_0 = 3.0\,\text{MPa}$; 2, $L = 50\,\text{mm}$, $p_0 = 1.5\,\text{MPa}$; 3, $L = 100\,\text{mm}$, $p_0 = 3.0\,\text{MPa}$; 4, $L = 100\,\text{mm}$, $p_0 = 1.5\,\text{MPa}$; 5, $L = 100\,\text{mm}$, $p_0 = 2.0\,\text{MPa}$; 6, $L = 150\,\text{mm}$, $p_0 = 3.0\,\text{MPa}$; 7, $L = 150\,\text{mm}$, $p_0 = 1.0\,\text{MPa}$. All particles 8 to 11 are copper, $T_0 = 500\,\text{K}$: 8, $L = 50\,\text{mm}$, $p_0 = 1.0\,\text{MPa}$; 9, $L = 100\,\text{mm}$, $p_0 = 2.0\,\text{MPa}$; 10, $L = 150\,\text{mm}$, $p_0 = 3.0\,\text{MPa}$; 11, $L = 100\,\text{mm}$, $p_0 = 3\,\text{MPa}$. Data 12 are the results of an experiment, obtained under test conditions used in computations.

the particle velocity (points 12) under the test conditions used in the computations. The predicted and experimental values are in agreement and allow simple approximation convenient for the rapid evaluation of the particle velocity at the nozzle exit:

$$\frac{v_p^*}{v^*} = (1+0.85\,\Omega)^{-1} = \left(1+0.85\sqrt{\frac{d_p}{L}\frac{\rho_p v^{*2}}{p_0}}\right)^{-1} \qquad [2.1]$$

Another comparison of numerical results with the experimentally determined velocity of various particles at the nozzle exit is shown in Fig. 2.12 in the form of the particle velocity versus the particle size. The continuous curves with squares show the numerically calculated velocity of particles at the nozzle exit, accelerated by air and helium jets. The figure also illustrates the test results from determining the velocity of these particles, obtained under the conditions used in the computations. The good agreement of the numerical and experimental data allowed us to consider these data to be reliable and to use computations in further evaluations of the particle velocity during spraying.

Figure 2.12 also shows the computed velocities of aluminum and copper particles at the substrate surface. Small particles noticeably lose

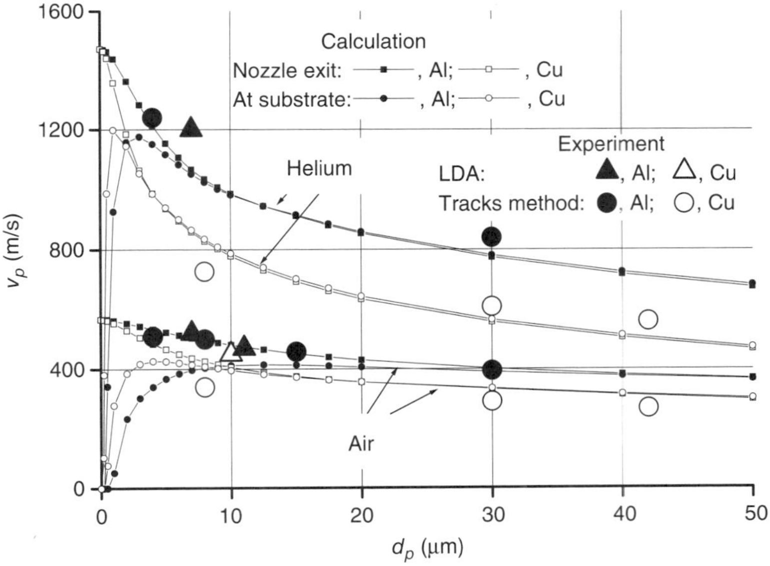

2.12 Computed velocities of aluminum and copper particles at the nozzle exit and at the substrate surface versus the particle size as compared with experimental results. L = 100 mm, p_0 = 2.0 MPa. LDA, laser Doppler anemometer.

their velocity in the frozen gas region immediately ahead of the substrate. Copper particles are less inert and have a lower velocity at the nozzle exit, but they become less decelerated behind the shock wave. As a result, both copper and aluminum particles with a size of 5–20 μm, being accelerated by an air jet, have approximately identical velocities (~400 m/s) when they hit the substrate. In a helium jet, the influence of particle inertia on their final velocity has a more pronounced effect.

Based on these results, to obtain a sufficiently high particle velocity on the substrate, one has not only to choose a nozzle with a length sufficient for particle acceleration to a high velocity but also to reduce the adverse decelerating action of the compressed gas immediately ahead of the substrate. Thus, the experiments and numerical calculations allowed for an accurate determination of particle velocity as a function of various parameters of the spraying process. This allowed us to commence experiments studying the main features of the cold spray process; in the first instance, the dependence of the cold spray process on particle velocity was investigated.

2.3.3 Interaction of individual particles with the surface of the substrate

The first task was to study the process of interaction of individual particles with the substrate at a wide range of particle velocities, in order to demonstrate the effect of transition from rebound to adhesion of 'cold' particles to the 'cold' substrate with an increase in particle velocity. The character of this interaction was considered using a moving polished substrate; the concentration of particles in the jet and the substrate velocity were chosen such that it was possible to observe individual craters and attached particles on the substrate. Aluminum particles with a mean diameter (d_{pm}) of 30.2 μm were used; they are shown in Fig. 2.13.

The particle velocity was controlled by forming different compositions of air–helium mixtures. Typical microphotographs of the substrate surface after its interaction with aluminum particles with d_{pm} = 30.2 μm and mean particle velocity v_{pm} = 730 m/s are shown in Fig. 2.14a (Papyrin *et al.* 2007). There are only some individual craters formed by particle impact on the substrate, and there are no attached particles. With the increase in particle velocity the situation is changed. Particles start adhering to the substrate and the probability of particle attachment increases with increasing particle velocity (Fig. 2.14b and c).

Thus, we can see that there are two characteristic processes in the course of the interaction of 'cold' particles with a 'cold' substrate, and these processes are separated by a certain critical velocity. If the particle velocity is low, particle rebound from the substrate occurs. With the particle velocity

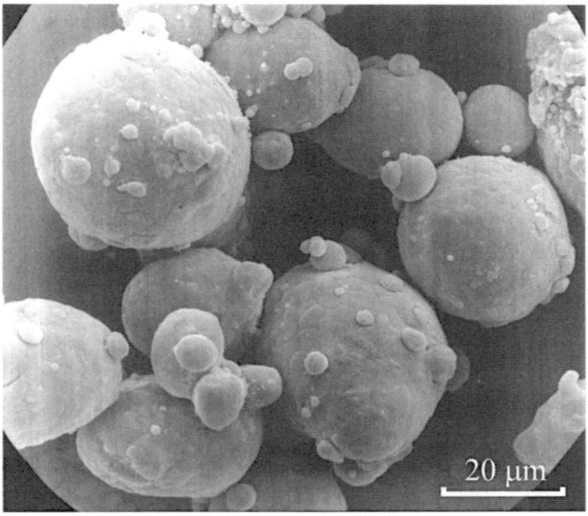

2.13 Appearance of aluminum particles.

increasing to the critical value, the process of particle adhesion to the surface begins, and the particle-attachment probability increases with an increase in particle velocity. The experiments performed showed that the critical value of particle velocity depends on many factors, including particle and substrate materials, particle temperature and size, substrate-surface state, etc. For example, Fig. 2.15 shows that the critical velocity decreases in the case of preliminary treatment of the substrate, and the probability of particle attachment increases to almost 100% for a particle velocity of 850 m/s.

The results of these experiments were of principal importance. They showed that, in terms of coating formation by 'cold' particles, a transition from the process of substrate erosion (due to particle rebound) to the process of coating formation (due to particle adhesion) should occur with an increase in particle velocity. To verify this idea, the following experiments were conducted in the regime of coating formation.

2.3.4 Transition from erosion to the coating formation process: critical velocity

The next step in understanding 'cold' spraying phenomena was to observe the process of coating formation by 'cold' particles. For this purpose, several metals were sprayed on a copper substrate with different particle velocities. The measured deposition efficiency, $k_d = \Delta m_s/M_p$ (where Δm_s is the change

(a)
(b)
(c)

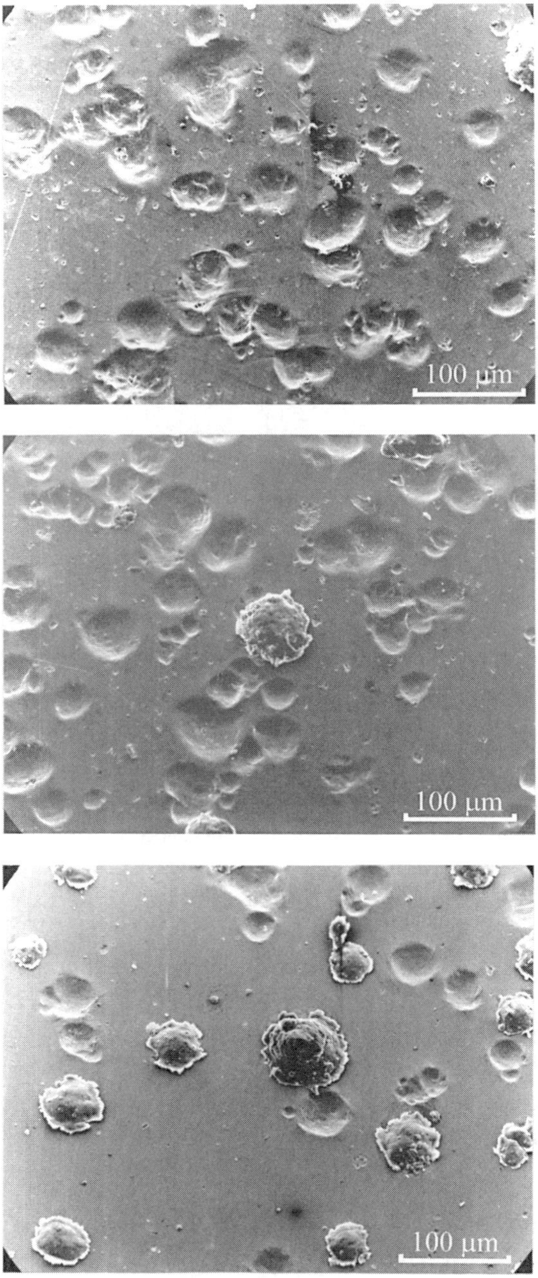

2.14 Microphotograph of a polished copper substrate after its interaction with aluminum particles with $d_{pm} = 30.2\,\mu m$. a, $v_{pm} = 730\,m/s$; b, $v_{pm} = 780\,m/s$; c, $v_{pm} = 850\,m/s$.

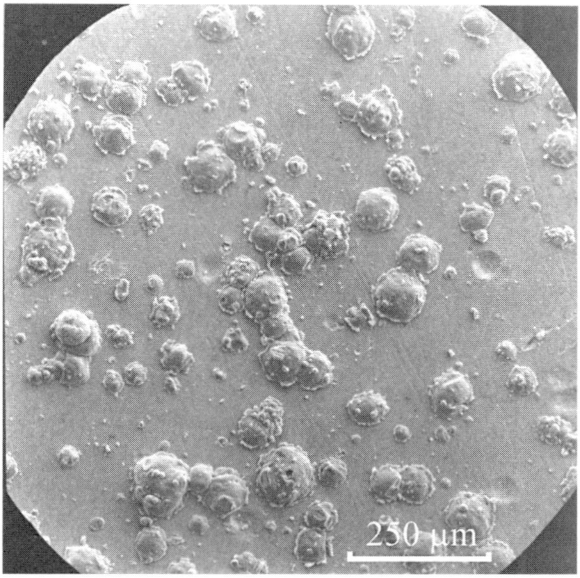

2.15 Aluminum particles attached to the preliminary treated (sand blasted) copper substrate, v_{pm} = 850 m/s.

of weight of a substrate and M_p is the weight of all particles interacting with a substrate), of various metallic particles accelerated by an air–helium mixture is shown in Fig. 2.16 (Papyrin *et al.* 2007). Figure 2.16 illustrates the fundamental concept of cold spray; namely, that the coating is formed by a high-velocity flow of 'cold' particles on a 'cold' substrate. The following results were obtained in investigations.

Two characteristic regions separated by the critical velocity v_{cr} were found. The first region ($v_p < v_{cr}$) corresponds to a well-known process of substrate erosion, which is undesirable in our case. However, as the particle velocity exceeds the critical value v_{cr}, the coating process begins. The deposition efficiency rapidly increases to 50–70% as the particle velocity significantly exceeds the critical value.

The transition from erosion to the coating formation process is illustrated by the photographs of the trajectories of incident aluminum particles and aluminum particles reflected from the substrate in Fig. 2.17. Processing of such photographs, together with studying the substrate surface, showed that all single particles with $v_p < v_{cr}$ ($v_p \approx 250$ m/s, Fig. 2.17a) are reflected. As the velocity increases within the range $v_p \geq v_{cr}$, the character of particle–substrate interaction is drastically changed: a rapidly growing coating is formed on the substrate surface ($v_p \approx 900$ m/s, Fig. 2.17b).

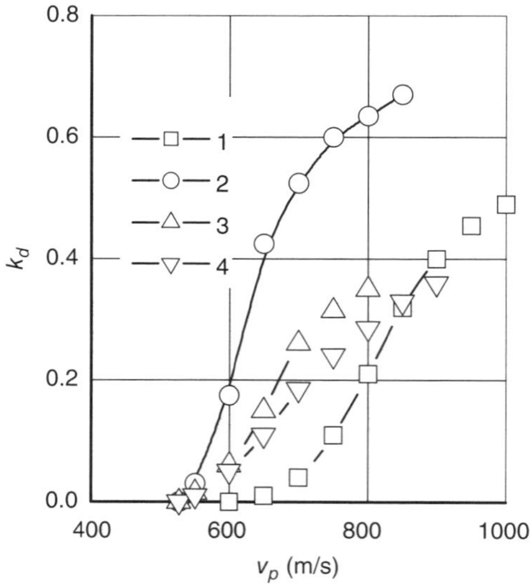

2.16 Deposition efficiency versus particle velocity for aluminum (1), copper (2), nickel (3), and zinc (4) powders accelerated by an air–helium mixture at room stagnation temperature.

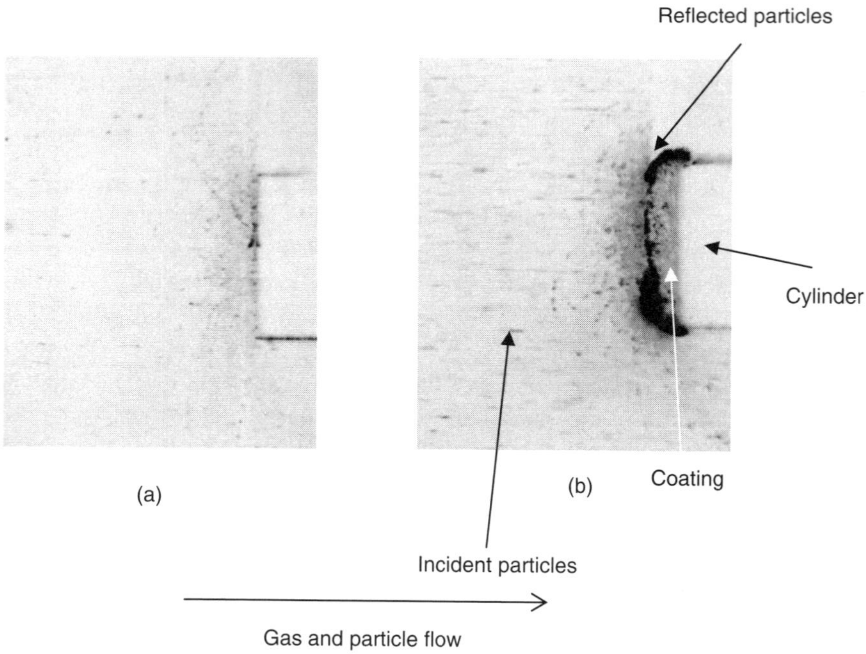

2.17 Trajectories of incident and reflected aluminum particles: a, v_{pm} = 250 m/s; b, v_{pm} = 900 m/s.

It can be seen from Fig. 2.16 that typical values of v_{cr} for various metals (Al, Cu, Ni, and Zn) are within 500–700 m/s. Various metals and alloys can be sprayed by a jet at room stagnation temperature (without any heating) if the particles reach a necessary velocity. This transition from substrate erosion to the formation of 'viable' coatings by a flow of 'cold' solid particles was the physical basis for the development of the cold spray method.

2.3.5 Effect of jet temperature on the deposition efficiency

As stated earlier, the test results plotted in Fig. 2.16 were obtained with the use of an air–helium mixture. Obviously, from a practical viewpoint, the use of expensive helium is not always justified. The use of a pure air jet at room temperature does not ensure formation of coatings for most materials. Therefore, investigations were performed with a slightly heated air jet with the objective of increasing the gas velocity and, hence, the particle velocity. It is important to emphasize that the particle temperature under such heating was always much lower than the melting point of the particle material, providing coating formation from particles in the solid state.

Figure 2.18 shows the results of measurements of the deposition efficiency for various metallic powders (aluminum, copper, and nickel) as a function of the stagnation temperature of the jet. Curves 4 to 6 in Fig. 2.19 show the data of Fig. 2.18 for the materials mentioned, but the data are plotted versus the particle velocity used in the computations (with the

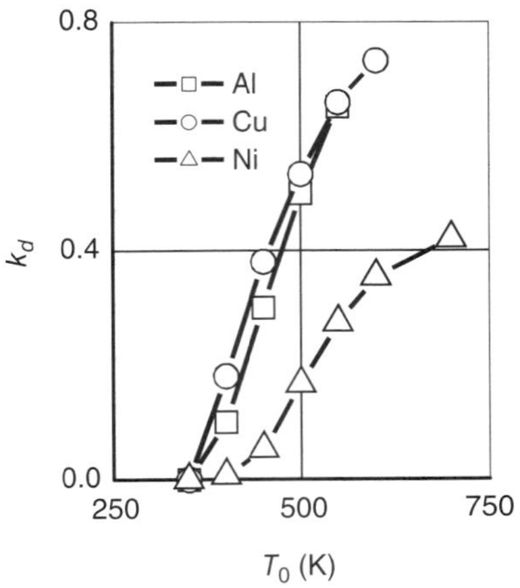

2.18 Deposition efficiency of aluminum, copper, and nickel powders sprayed on copper substrates versus the air jet stagnation temperature.

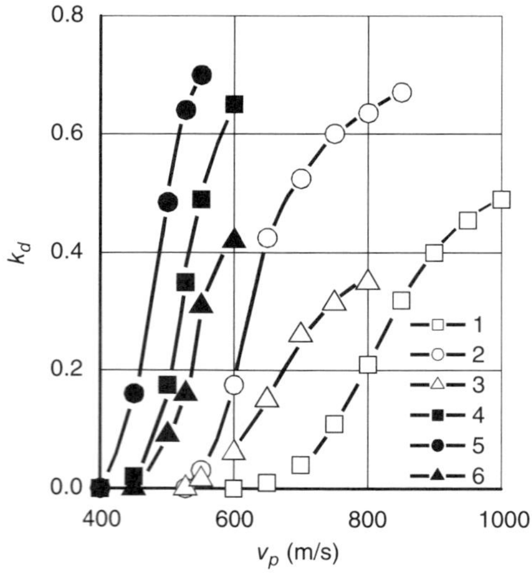

2.19 Deposition efficiency for aluminum, copper, and nickel powders accelerated by an air–helium mixture (1–3) and by heated air (4–6) versus particle velocity. The curves refer to Al (1 and 4), Cu (2 and 5), and Ni (3 and 6).

corresponding temperatures of air heating). By comparing these dependences with those obtained with the use of an air–helium mixture as a driver gas at T_0 = 300 K (curves 1 to 3), we can conclude that the particle and substrate temperatures also have a significant effect on the spraying process; otherwise these two families of curves would coincide. As the air temperature in the pre-chamber increases, both the particle velocity and the particle and substrate temperatures increase. Therefore, the drastic increase observed for the deposition efficiency can be attributed to the increase of both the velocity of sprayed particles (which increases the pressure and temperature in the contact at the impact moment) and the temperatures of the sprayed particles and the substrate. It leads to changes in their properties, an increase in temperature at the particle–substrate contact, and, hence, displacement toward lower values of the critical velocity v_{cr}. Thus, we can see that slight preheating of the jet allowed us to decrease the critical velocity and, as a consequence, to extend the range of materials that can be sprayed using an air jet.

2.3.6 Basic features and advantages of cold spray

Thus, the results presented show that the use of a supersonic jet of air (nitrogen) mixed with helium and having a stagnation temperature of

~300 K, and the use of a preheated ($\Delta T \leq$ 500–800 K) supersonic (M = 2.0–3.0) air (nitrogen) jet allowed us to obtain coatings from most metals and many alloys (Al, Cu, Ni, Zn, Pb, Sn, V, Co, Fe, Ti, bronze, brass, etc.) with particles of the size $d_p < 50\,\mu m$ onto various metallic and dielectric substrates (in particular, glass, ceramics, etc.). The deposition efficiency of the powders reaches 0.5–0.8, which is extremely important from a practical point of view in the development of particular technological processes.

A comparison of the basic parameters of the two-phase flow for cold spray with parameters typical of traditional spraying methods shows that they are significantly different (Fig. 2.20). The characteristic features of the cold spray process are the much lower temperature and the higher velocity of particles. It is important to emphasize that the key difference between cold spraying and conventional thermal spraying methods (Fig. 2.20) from the physical viewpoint is that the coating is formed from particles in the solid state. That is why the term 'cold' was introduced into the name of the process despite some jet heating. As was mentioned above, according to the concept commonly accepted in the mid-1980s (Bartenev *et al.* 1982), the incident particles should be in the melted or almost melted state in order to achieve coating formation. The above presented results in this chapter demonstrated that a high-temperature jet is not a necessary condition for many sprayed materials, and many coatings can be obtained from particles whose temperatures are substantially lower than their melting points.

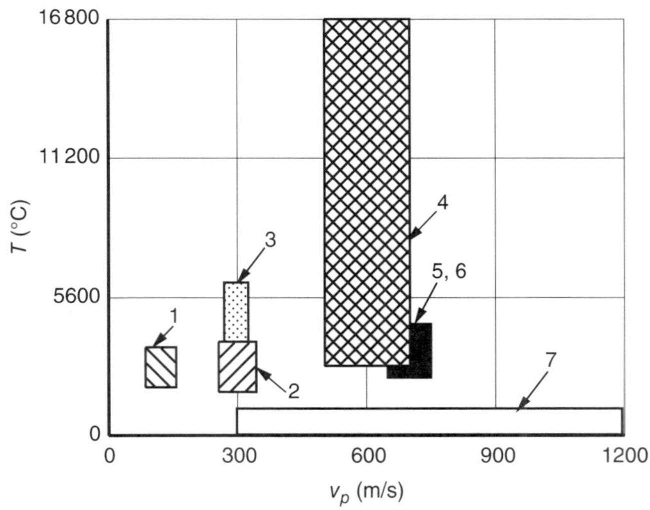

2.20 Diagram of jet temperatures (T) and particle velocities (v_p) used in different spraying methods. 1, low-velocity gas-plasma; 2, high-velocity gas-plasma; 3, electric-arc; 4, plasma; 5 and 6, detonation and high-velocity oxygen fuel; 7, cold spray.

Eliminating the harmful effects of high temperature on coatings and substrates offered significant advantages and new possibilities. These include:

- avoiding oxidation and undesirable phases;
- retaining properties of initial particle materials;
- inducing low residual stresses;
- conducting heat and electricity easily through the coatings;
- providing high-density, high-hardness, cold-worked microstructure;
- spraying thermally sensitive materials;
- spraying powders with a particle size of <5–10 μm;
- working with highly dissimilar materials;
- preparing the substrate minimally with surface preparation/masking, short standoff distance;
- feeding powder at a high rate, resulting in high productivity;
- depositing many materials at high deposition rates and efficiencies;
- collecting and re-using particles (powder utilization up to 100% with recycling);
- heating the substrate minimally;
- increasing operational safety because of the absence of high-temperature gas jets, radiation, and explosive gases.

Thus, the results presented allowed us to state that a new concept, a new approach to spraying coatings, has been developed. This offered wide possibilities for developing new technologies and new equipment for spraying coatings with any given thickness from particles in solid state including direct fabrication of parts.

2.4 Evolution of cold spray technology and current status

After the initial studies presented above (Sections 2.1–2.3), which led to the creation of the cold spray process, a wide spectrum of research has been conducted at the ITAM SB RAS (1985–2005). This research incorporated various aspects of the technology including gas dynamics, the physics of the interaction of high-speed solid particles with a substrate, as well as equipment and applications. Some of these aspects are:

- flow inside the supersonic nozzle;
- supersonic jet exhausting from the nozzle;
- gas dynamics and heat transfer under the impact of a supersonic jet on a substrate;
- optimization of the nozzle;
- deformation of particles and formation of a high-velocity flow layer in the contact zone under high-speed particle impact;

- modeling of the particle–substrate adhesive interaction;
- stationary and portable equipment and auxiliary technology development.

The results obtained are important for understanding of the cold spray process as well as for its optimization. Most of these results are published in the recently released book entitled *Cold Spray Technology*, published by Elsevier (Papyrin *et al.* 2007). A chronological timeline of some institutions that have made an important contribution to the cold spray process is presented in Fig. 2.21.

Outside Russia, the cold spray process was presented first in the United States by Dr Papyrin in 1994. In 1994–1995 Dr Papyrin, with a consortium formed under the auspices of the National Center for Manufacturing Sciences (NCMS) of Ann Arbor, Michigan, conducted the first research in the United States on the cold spray process. The consortium membership

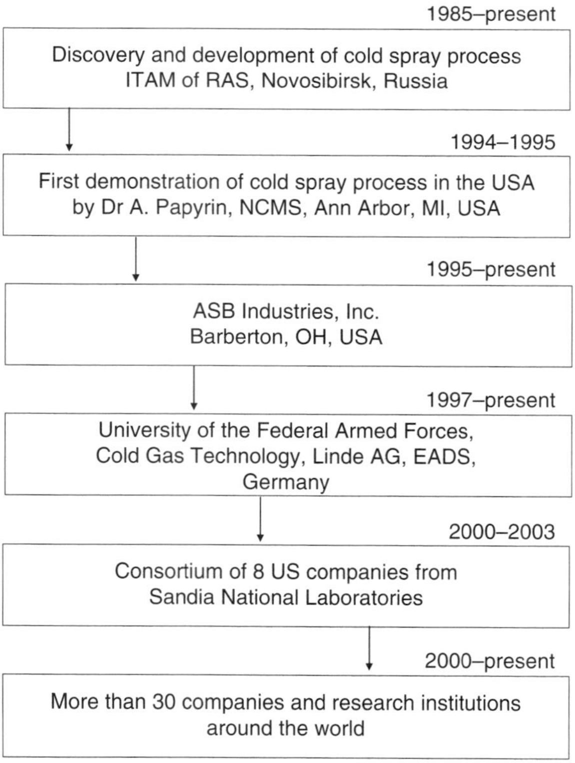

2.21 Chronological timeline of contributions to the development of the cold spray process.

included major US companies such as Ford Motor Company, General Motors, General Electric – Aircraft Engines, and Pratt & Whitney Division of United Technologies. This group established the first US cold spray capability, and published property measurements for several cold-sprayed materials.

The first US company that licensed the cold spray technology was ASB Industries, Barberton, Ohio. They started working in this field in 1995 and made an important contribution to the development of this technology.

Another consortium was formed by Sandia National Laboratories (2000–2003) to develop the process and commercial applications as well as to improve equipment. The consortium of eight companies has executed a Cooperative Research and Development Agreement (CRADA) to address pre-competitive issues leading to commercialization of the technology. The member companies included Sandia National Laboratories/DOE, Daimler Chrysler, Ford Motor Company, Jacobs Chuck Manufacturing Company, Ktech Corporation, Pratt & Whitney, Praxair, and Siemens-Westinghouse. Research and development work was conducted at Sandia's Thermal Spray Research Laboratory, providing high-level experimental studies as well as physical modeling of the process. Research in the areas of gas dynamics, bonding mechanisms, improvement of equipment, economics of the process, development of commercial applications, etc. was performed.

An important contribution in the study of the cold spray process was made by Professor H. Kreye and his colleagues at the University of the Federal Armed Forces in Hamburg, Germany. They performed a wide spectrum of systematic studies in the fields of gas dynamics, bonding mechanisms, coating properties, and applications. In cooperation with this group Mr P. Richter (Cold Gas Technology, Germany) and Mr P. Heinrich (Linde AG), along with colleagues, developed the best cold spray equipment in the world as well as many technologies for different applications.

At the present time a wide spectrum of research is being conducted at many research centers and companies around the world, including: the ITAM SB RAS; Sandia National Laboratories; Pennsylvania State University; ASB Industries Inc., Ford Motor Company, Pratt & Whitney, Dartmouth College, Rutgers University, US Army Research Laboratory, Delphi Research Laboratory (United States); the University of the Federal Armed Forces, European Aeronautic Defence and Space Company, Cold Gas Technology, Linde AG, Siemens (Germany); Cambridge University, University of Nottingham, University of Liverpool, Yasaki Europe, BOC Gases (England); Shinshu University, Plasma Gigen (Japan); National Research Council Canada, University of Windsor (Canada); Commonwealth Scientific Industrial Research Organization (CSIRO) (Australia); Mahle Metal Leve (Brazil); and companies in South Korea, China, India, and many other countries. The cold spray process has propagated around the world so fast that it is difficult to mention all the companies and institutions involved in this activity. There has been a great surge in publications on the cold spray process. Many high-level studies have been conducted at these centers and many interesting and important results for future developments and improvements of cold spray have been obtained.

In the field of gas dynamics, various models with an adequate description of the motion of the gas and particles inside the de Laval nozzle with circular and rectangular cross-sections are currently available. With allowance for the decelerating effect of the compressed layer ahead of the substrate, one can calculate the impact velocity and the temperature of particles before the impact. Still, insufficient attention is paid to the optimization of loading the gas flow with the powder and to investigations of unconventional schemes of flow formation (e.g. ejector scheme, etc.). The substrate geometry is usually plane, there are no experimental measurements of the gas and particle velocities inside the compressed layer, and unsteadiness of the bow shock wave is ignored. The influence of particles on the shock-wave structure of the flow in the vicinity of the substrate is not considered, and the substrate temperature is given little attention (in experiments as well). These and some other issues in the field of cold spray gas dynamics require additional studies.

Many studies have been performed in the field of high-speed interaction between solid particles and the substrate, and the bonding mechanism. In several papers the concept of the critical velocity is discussed; formulae for calculating the critical velocity as a function of some material properties and deposition parameters are proposed. The presence of mechanical, metallurgical, and chemical bonds is assumed. It was found that there is a viscoplastic transition and the melting point is sometimes reached in the contact, which allows us to assume that adhesion can also be ensured by micro-mixing. The significance of the impact velocity, size and temperature of the particles and substrate, heat conduction, and shock-wave loading are discussed. An analogy with explosive welding and explosive compacting of powdered materials is mentioned. It is clear that more research should be done in this field to understand and describe the nature of the bonding mechanism and the influence of various parameters on the process of coating formation.

Application research is probably the most intensely developed area of the cold spray arena. Many different technologies for several applications have been successfully developed and various coating properties have been studied. Industrial equipment is available. These studies have shown that the cold spray process allows for the application of a wide spectrum of high-quality coatings made of metals, alloys, composites, etc. The process looks promising for many applications because of important advantages that include avoiding oxidation and undesirable phases, retaining the properties of the initial particle materials, spraying thermally sensitive materials, working with highly dissimilar materials, etc. Eliminating the effects of high temperature on coatings and substrates makes cold spray promising for producing and repairing a wide range of industrial parts. Examples include: turbine blades, pistons, cylinders, valves, rings, bearing components, pump elements, sleeves, shafts, and seals for many industries. Various coatings may add strength, hardness, wear resistance, corrosion resistance, electro-magnetic conductivity, thermal conductivity, and other properties. The process is also suitable for the production of compact powder materials and for the direct fabrication of parts.

An overview of some of the studies conducted in the field of cold spray at different scientific centers around the world is presented in *Cold Spray Technology* (Papyrin *et al.* 2007). Different aspects of cold spray will be described by leading specialists in detail in the following chapters.

2.5 Some comments on the patent situation

In conclusion, we would like to make some comments related to the cold spray patent situation. It is important to emphasize that the Russian patent claims a new concept, the new approach to spraying coatings and the spray *method*. The key feature of this method is the capability to *build up* coatings from particles in a solid state with *any given thickness*. To provide such coatings, the spray parameters (particle velocity, temperature, and size) must be in a certain range. In particular, the principal concept of the cold spray process is *critical velocity*. It is on these concepts that all other patents, preceding or following the Russian patent, can be analyzed.

Certainly there is much room for improvement of this method, including equipment, specific technologies, etc. Now there are dozens of patents related to such improvements and most of the inventors recognize Russian priority. However, some of these improvements are being considered as 'new methods'. As a result, this has led to confusion in terminology (Gabel and Tapphorn 1998; Van Steenkiste *et al.* 2000). It is clear that these patents can only be considered as a possible improvement on the cold spray process. We state that this process is working at the same range of jet parameters (as we have already claimed) to make the same product: coatings (parts) from particles in the solid state. What does this mean? It means that authors of these patents are practicing the cold spray method. However, unlike authors of dozens of other patents on cold spray, they distinguished their inventions by creating a new name for the cold spray process (Kinetic Metallization, Kinetic Spray)!

An invention, patented by Thurston in 1902, is sometimes described as cold spray (Karthikeyan 2006). We should note that this patent (Thurston

1902) has been well known to specialists working in the field of cold spray for many years, and there are several references to this patent in their papers. However, for obvious reasons, nobody has claimed that this process is the cold spray process. We would also highlight a patent by Rochville (1963): this patent is exactly the same process as that of Thurston, but it contains the description of the process in more detail and at a higher level because it was published much later (1963). Below are quotations from these patents explaining the essence of these inventions:

> The method . . . which consists in forcing particles of the coating metal . . . to cause particles to became imbedded in the surface . . . (Thurston's patent).

> As a result the powder adheres to the surface of the work-piece, partly by entering the pores of the surface . . . A thin layer of around .0001 inch thickness forms on the part . . . the coating will build up over the surface of the part, but *will not build up* upon itself (Rochville's patent).

It is clear that these processes allow one to spray only a monolayer of particles (very thin coatings of around a few micron), but they can not provide cohesion and, hence, building up of coatings because of low spray parameters that are typical for the sandblasting process. The key features of 'Russian' cold spray as mentioned above are (a) the capability *to build up coatings of any thickness* including direct fabrication of parts and (b) the concept of critical velocity that allows us to understand the basic physical principles and methods of control of this process. It is also clear that the mechanism of Thurston's and Rocheville's coating formation process does not consider cohesion of the particle at all, and as such it is different from the Russian process where cohesion plays the key role. It is clear that the processes described in both of the above-mentioned patents (Thurston 1902; Rocheville 1963) can only be considered as the first attempts to spray solid particles. However, as was shown above, they do not have any relationship to the real cold spray process.

In conclusion, we would like to put forward two quotations from a paper (Browning 1997) by the world-recognized specialist James Browning who made a great contribution to spraying technologies.

1 'Until becoming aware of the "cold particle" phenomenon, we made every effort to maximize particle temperature.'
2 'Many of us wonder why decisions (we made) over 30 years to maximize jet temperature rather than jet velocity came about. Was it plasma and its newness over-awed us? At a time when combustion technology was landing men on the moon and high-velocity D-gun was so far ahead of other spraying techniques, why didn't anyone put two-and-two together?'

We would like to address these questions to those 'analysts' who 'believe' that the cold spray process was invented in years 1901–1902 or earlier. Why have they not developed the cold spray *technique*, which is very simple, if the *method* has been well known from as early as the beginning of the twentieth century?

Certain excerpts reprinted from *Cold Spray Technology*, Papyrin *et al.* (2007) with permission from Elsevier.

2.6 List of symbols

p_0	Pressure in the pre-chamber (stagnation pressure)
T_0	Temperature in the pre-chamber (stagnation temperature)
M	Mach number
M*	Mach number at the nozzle exit
M_∞	Mach number of the incident flow
d_p	Particle size
d_{pm}	Mean particle size
ρ_p	Density of the particle material
φ_p	Volume concentration of the particles
D_b	Air-flow body diameter (cylinder)
t_{ex}	Exposure time
Δt	Interval between two neighboring frames
$\mathrm{Re}_D = \frac{\rho v D_b}{\mu}$	Reynolds's number of an air-flow body
ρ	Gas density
v	Gas velocity
μ	Gas viscosity
k_H	Helium/air flow rate ratio
G_p	Particle flow rate
ε_p	Strain of the particle
v_p	Particle velocity
$v_{pm} = \frac{1}{N}\sum_{i=1}^{N} v_p^i$	Mean particle velocity
$\Omega = \sqrt{\frac{d_p}{L}\frac{\rho_p v^{*2}}{p_0}}$	Dimensionless quantity
v_p^*	Particle velocity at the nozzle exit
v^*	Gas velocity at the nozzle exit
L	Length of a supersonic part of the nozzle
v_{cr}	Critical velocity
ΔT_0	Heating of gas in pre-chamber

$k_d = \Delta m_s / M_p$	Deposition efficiency
Δm_s	Change of weight of a substrate
M_p	Weight of all particles interacting with a substrate

2.7 References

Alkhimov A P and Kosarev V F, 1996. Laser diagnostics of supersonic two-phase jets, *Proceedings of the 8th International Conference on the Methods of Aerophysical Research*, September 2–6, 1996. Novosibirsk: ITAM (Siberia), pp. 3–8.

Alkhimov A P, Boiko V M, Papyrin A N and Soloukhin R I, 1978. Diagnostics of supersonic two-phase streams from scattered laser radiation. *Journal of Applied Mechanics and Technical Physics*, **19**(2), 36–46.

Alkhimov A P, Yanenko N N, Nesterovich N I, Papyrin A N and Fomin V M, 1981. *Doklady Akademii Nauk SSSR*, **260**(4), 821–825.

Alkhimov A P, Nesterovich N I and Papyrin A N, 1982a. Experimental investigation of supersonic two-phase flow over bodies. *Journal of Applied Mechanics and Technical Physics*, **23**(2), 66–74.

Alkhimov A P, Boiko V M and Papyrin A U, 1982b. Development of laser Doppler and stroboscopic methods for study of high-speed processes, *Avtometriya*, **3**, 38–45.

Alkhimov A P, Kosarev V F, Nesterovich N I and Papyrin A N, 1986. *Method of drag reduction of a blunt-nosed body*. Russian Inventors Certificate No. 1228579, 1986. (in Russian).

Alkhimov A P, Kosarev V F, Nesterovich N I and Papyrin A N, 1990a. *Device for coating application*. Russian Federation Patent No. 1603581, July 1, 1990 (Priority of invention: October 5, 1987).

Alkhimov A P, Kosarev V F, Nesterovich N I and Papyrin A N, 1990b. *Method for coating application*. Russian Federation Patent No. 1618778, September 8, 1990 (Priority of invention: June 6, 1986).

Alkhimov A P, Kosarev V F, Nesterovich N I, Papyrin A N and Shushpanov M M, 1990c. *Device for coating application*. Russian Federation Patent No. 1618777, September 8, 1990 (Priority of invention: June 18, 1986).

Alkhimov A P, Kosarev V F and Papyrin A N, 1990d. Method of cold gas-dynamic deposition. *Doklady Akademii Nauk SSSR*, **315**, 1062–1065 (Transl: *American Institute of Physics*, 1991).

Alkhimov A P, Kosarev V F and Papyrin A N, 1991. *Device for coating application by spraying*. Russian Federation Patent No. 1674585, May 1, 1991 (Priority of invention: October 19, 1989).

Alkhimov A P, Kosarev V F, Nesterovich N I and Papyrin A N, 1992. *Method for application of metal-powder coatings*. Russian Federation Patent No. 1773072, July 1, 1992 (Priority of invention: October 5, 1987).

Alkhimov A P, Kosarev V F, Nesterovich N I, Papyrin A N and Shushpanov M M, 1994a. *Gas-dynamic spraying method for applying coatings*. US Patent No. 5,302,414, April 12, 1994.

Alkhimov A P, Kosarev V F and Papyrin A N, 1994b. *Device for coating application*. Russian Federation Patent No. 2010619. *Bulliten Izobretenii*, **7**, 32.

Alkhimov A P, Kosarev V F, Nesterovich N I, Papyrin A N and Shushpanov M M, 1995. *Method and device for coating*. European Patent No. 0484533, January 25, 1995.

Alkhimov A P, Kosarev V F, Nesterovich N I, Papyrin A N and Shushpanov M M, 1997a. *Gas-dynamic spraying method for applying coating.* Reexamination Certificate, US Patent No. 5,302,414, February 25, 1997.

Alkhimov A P, Demchuk A F, Kosarev V F and Kozhevnikov V E, 1997b. *Electrotechnical connector*. Russian Federation Patent No. 2096877. *Bulliten Izobretenii*, **32**(Part II), 376.

Alkhimov A P, Gulyaev V P, Demchuk A F, Kosarev V F, Larionov V P and Spesivtsev V P, 1997c. *Setup for coating application onto the inner surface of a pipe*. Russian Federation Patent No. 2075535. *Bulliten Izobretenii*, November 8, 184–185.

Alkhimov A P, Kosarev V F and Papyrin A N, 1998. Gas-dynamic spraying. Experimental study of the spraying process. *Journal of Applied Mechanics and Technical Physics*, **39**(2), 318–324.

Alkhimov A P, Klinkov S V and Kosarev V F, 2001. The features of cold spray nozzle design. *Journal of Thermal Spray Technology*, **10**(2), 375–381.

Bartenev S S, Fedko Y P and Grigirov A I, 1982. *Detonation Coatings in Machine Building*, Leningrad: Mashinostroenie (in Russian).

Browning J A, 1997. What if we are right? *United Thermal Spray Conference and Exposition*, September 15–18, 1997, Indiana.

Gabel H and Tapphorn R, 1998. *Coating or ablation applicator with debris recovery attachment.* US Patent No. 5,795,626, August 18, 1998.

Karthikeyan J, 2006. Evolution of cold spray technology. *International Thermal Spray & Surface Engineering, the TSS Newsletter*, **1**(1), 12–13.

Kudinov V V, Pekshev P YU, Belashchenko V E, Solonenko O P and Safiulin V A, 1990. *Plasma-Induced Application of Coatings*, Moscow: Nauka (in Russian).

Papyrin A N, Kosarev V F, Klinkov S V, Alkhimov A P and Fomin V M, 2007. *Cold Spray Technology*, A.N. Papyrin, ed., Amsterdam: Elsevier.

Preece C (ed.), 1979. *Erosion*, New York: Academic Press.

Rocheville C F, 1963. *Device for treating the surface of a workpiece.* US Patent No. 3,100,724, August 13, 1963.

Thurston S H, 1902. *Method of impacting one metal upon another.* US Patent No. 706,701, August 12, 1902.

Van Steenkiste T, Smith J, Teets R, Moleski J and Gorkiewicz D, 2000. *Kinetic spray coating method and apparatus.* US Patent No. 6, 139, 913, October 31, 2000.

Zverev A I, Sharivker S YU and Astakhov E A, 1979. *Detonation-Induced Application of Coatings*, Leningrad: Sudostroenie (in Russian).

3
Comparing cold spray with thermal spray coating technologies

M. F. SMITH, Sandia National Laboratories*, USA

3.1 Introduction

Cold spray has many similarities to traditional thermal spray processes. However, there are also important fundamental differences that make the cold spray process and cold-sprayed materials highly unique. In some applications, the unique features of cold spray offer significant advantages, overcoming inherent limitations of traditional thermal spray processes. In other applications, more traditional processes may offer superior capability and performance. To identify the best potential applications for cold spray, it is important to understand the similarities and differences of cold spray as compared with traditional thermal spray alternatives. This chapter provides a high-level overview of spray process technologies and discusses important similarities and differences between cold spray and traditional thermal spray. Some of the unique features of cold spray that will be introduced in this high-level process comparison are discussed in greater detail in other chapters of this book.

3.2 Traditional thermal spray technologies

The term 'thermal spray' denotes a broad family of spray process technologies, such as arc spray, plasma spray, flame spray, and high-velocity oxy-fuel spray (HVOF). Although the specific details of different spray processes vary, the fundamental principles for all traditional thermal spray processes are essentially the same, derived from a landmark patent recorded by M. A. Schoop and co-workers in 1911 (Schoop, 1911).

* Sandia National Laboratories is a multiprogram laboratory operated by Sandia Corporation, a Lockheed Martin Company, for the United States Department of Energy under contract DE-AC04-94AL85000.

3.2.1 Generic thermal spray process

In a generic thermal spray process (Fig. 3.1), electrical or chemical energy is used to create small molten or semi-molten droplets (typically 10–100 microns in diameter) from powder, wire, or rod feedstock. These droplets are propelled onto a workpiece surface by a subsonic or supersonic gas stream at velocities that range from a few tens of meters per second, up to roughly 1000 m/s (3280 ft/s).

Upon impact, each droplet spreads out and quickly solidifies at cooling rates of the order of 10^4–10^8 K/s forming the basic microstructural unit of spray-deposited material, called a 'splat'. Due to the very high cooling rates, an individual splat typically solidifies before the next molten droplet impacts on top of it. Splats randomly stack up on each other (much like randomly stacking snowflakes or playing cards) forming the layered or lamellar microstructure that is characteristic of most thermal spray-deposited materials. The metallographic cross-section in Fig. 3.2 illustrates the lamellar microstructure of a typical thermal-sprayed coating. The quality of the interfaces between the solidified splats, called 'splat boundaries', strongly influences the physical and mechanical properties of the spray-deposited material, much like grain boundaries influence the properties of wrought or cast materials. Deposition rates vary for different processes and materials, but they can range up to many kilograms per hour.

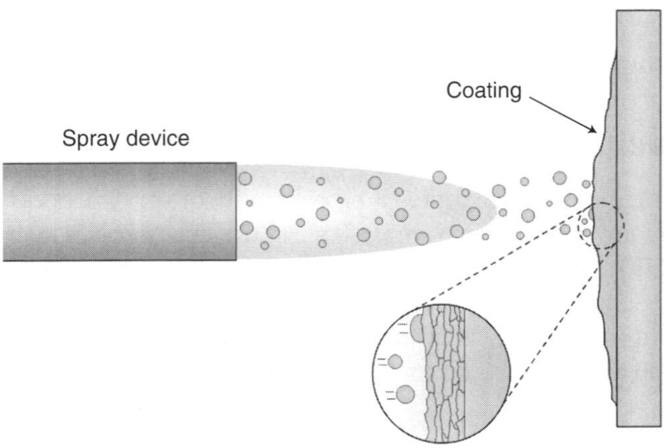

3.1 Generic thermal spray process. Molten or semi-molten droplets are sprayed onto a target surface where they 'splat' cool and form a layered microstructure.

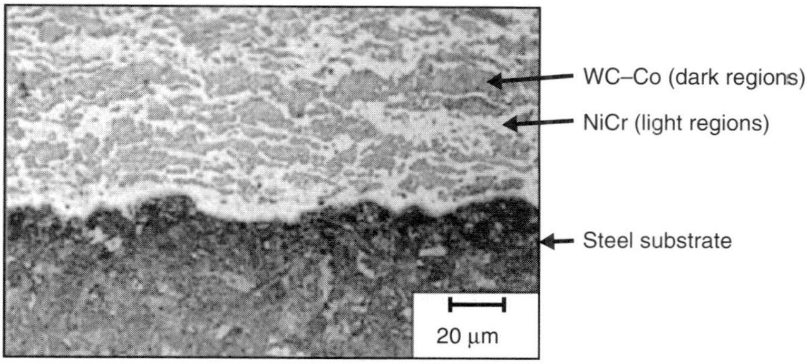

3.2 Composite cemented-carbide/nickel-chromium wear coating plasma sprayed onto a steel substrate. (WC–Co, tungsten carbide + cobalt; NiCr = 80% nickel, 20% chromium). Micrograph courtesy of Thermal Spray Technologies.

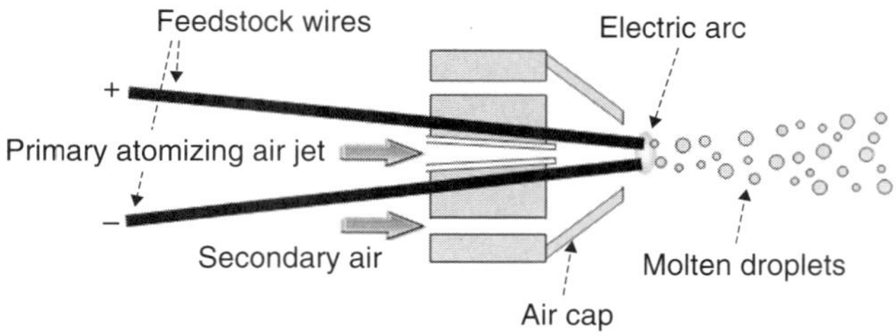

3.3 Schematic diagram of a twin-wire arc spray gun. Electrically isolated feedstock wires are continuously fed into the spray gun and serve as consumable electrodes.

3.2.2 Commercially important thermal spray processes

Major commercial thermal spray process technologies can be broadly grouped into two basic categories: those that use electrical energy and those that use chemical energy. Electrical processes typically use either an electric arc or an electrically driven plasma to heat and melt the spray material. For example, in the twin-wire arc spray process (Fig. 3.3) the tips of two metal wires are brought together at a slight angle and an electric arc is struck between them, melting the material at the wire tips. A jet of compressed air or inert gas is directed across the wire tips, atomizing the molten metal and creating a spray stream of molten droplets. As the tips melt, additional wire is fed into the arc, making this a continuous process, with the two wires serving both as feedstock material and as consumable electrodes. An obvious limitation of this process is that it can only be used to deposit materials that can be made into conductive metal wires. Yet, this simple and inexpensive process is in wide commercial use for applications such as spraying zinc- or aluminum-based anti-corrosion coatings onto bridges, ship decks, large metal tanks, etc.

Plasma spray is another widely used electrical thermal spray process. In this process, a partially ionized conductive gas, known as 'plasma', is used to melt and propel powdered feedstock material onto the substrate. A typical plasma spray gun (Fig. 3.4) consists of two electrodes, a roughly 'bullet shaped' tungsten cathode and a hollow, cylindrical, copper anode that surrounds the cathode. The cathode and anode are both water cooled and electrically isolated from one another by a suitable insulating material. The copper anode also serves as a nozzle to direct the flow of the hot plasma jet. To create the plasma jet, inert plasma-forming gas, usually argon or nitrogen with minor additions of helium or hydrogen, is injected into the annular space between the two cylindrical electrodes, and a high amperage direct current (DC) arc is then struck between the electrodes. The arc partially ionizes the gas to form a high-temperature, electrically conductive plasma, which expands and escapes through the open end of the anode to form a very hot, high-velocity, plasma jet. Flow velocities of the plasma jet

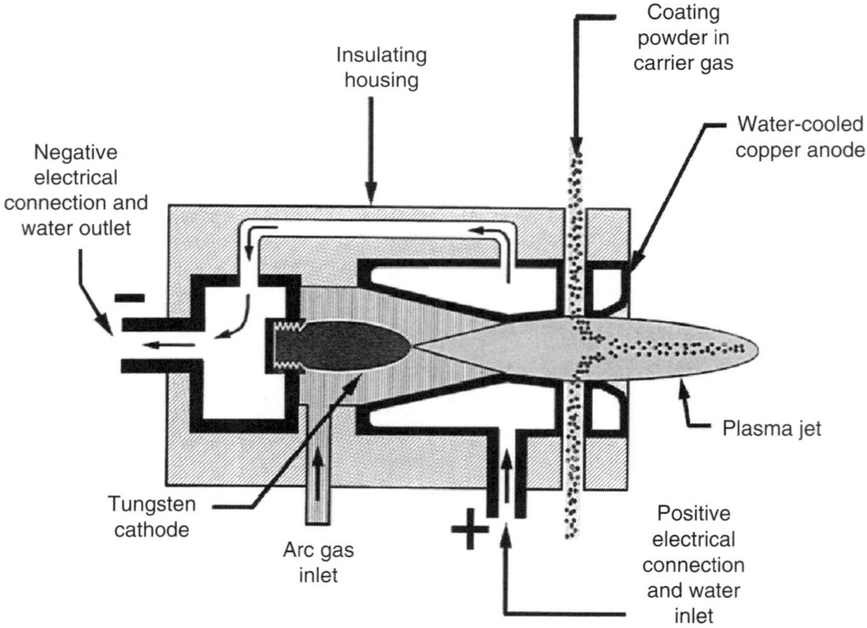

3.4 Schematic diagram of a plasma spray gun.

as it exits the torch may be either subsonic or supersonic, depending upon the design of the anode (nozzle) and the operating conditions. Powder of the material to be deposited is injected into the plasma jet, where the powder melts and is propelled toward the substrate.

Core temperatures in the plasma jet can exceed 15 000 K, hot enough to melt even the most refractory metals or ceramics. For this reason, the plasma spray process is one of the most versatile of all spray processes, able to deposit an exceptionally wide range of materials including metals, many ceramics, glasses, and even some polymers, such as nylon. By simultaneously feeding more than one material into the spray gun, highly unique composite materials can be deposited. For example, aluminum and polyester are co-deposited to create a novel aluminum–polyester abradable seal material that is used to surface stationary shroud seal segments that surround rotating fan blades in some turbine aircraft engines. This permits a design where the rotating blade tips travel very close to the surrounding stationary shroud to improve engine efficiency, but if a blade tip should occasionally rub against the shroud, the soft composite seal material is easily scraped away without damaging the blade or causing a catastrophic engine failure.

An important variant of plasma spray is a process variously known as vacuum plasma spray (VPS) or low-pressure plasma spray (LPPS). Both terms denote plasma spray deposition carried out inside a vacuum chamber at a relatively low dynamic pressure of inert gas, i.e. gas in the chamber is continuously pumped out to balance the inflow of plasma-forming gas and powder feed carrier gas, with an operating pressure in the chamber typically above 6 MPa (60 bar or 45 torr). The process chamber is initially sealed and pumped down as far as possible with the spray gun off; it is then back-filled to the desired operating pressure with an inert gas, such as argon, before coating deposition begins. This process can produce sprayed coatings of exceptionally high quality with very low oxide content and very little porosity, i.e. near theoretical density. However, the equipment to produce VPS/LPPS coatings is complex and expensive, so this process is used primarily for high-value-added applications, such as coatings for biomedical or aerospace hardware.

Moving on to chemical energy thermal spray processes, flame spray is an oxy-fuel process that is widely used in industry. Burning oxygen with various fuel gases – such as acetylene, propane, or propylene – can produce flame temperatures sufficient to melt many metals and polymers as well as some ceramics. Different flame spray devices are designed to spray powder, wire, or ceramic rod feedstock. A typical flame spray device (Fig. 3.5) utilizes a jet of compressed air to accelerate the molten droplets toward the substrate, creating droplet velocities of the order of 50–300 m/s (160–980 ft/s), depending upon the design of the device, the material being sprayed, and the operating conditions. Flame spray operations in the lower part of this

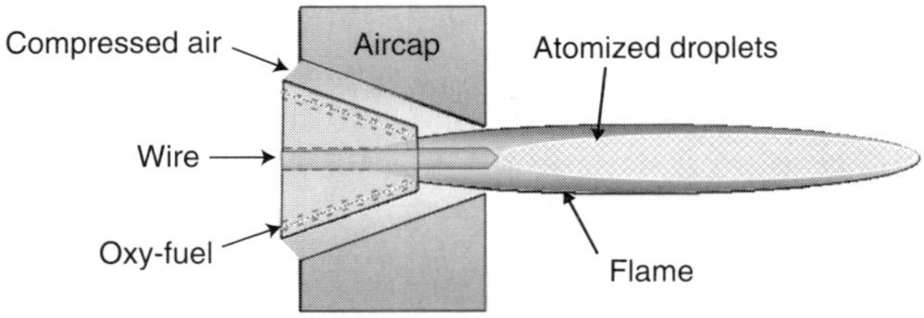

3.5 Schematic diagram of a wire flame spray gun. A powder flame spray gun is similar, but powder suspended in a carrier gas is injected along the central axis of the gun where the wire is shown in this diagram.

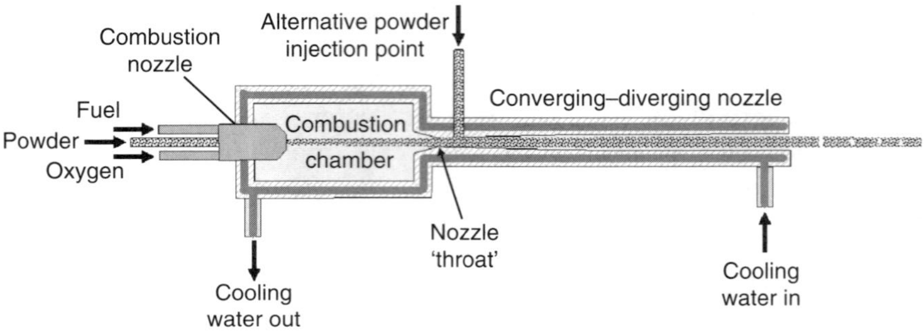

3.6 Schematic diagram of a high-velocity oxy-fuel (HVOF) spray gun.

velocity range tend to produce coatings that are more porous (less dense) than those produced by more energetic spray conditions with higher impact velocities. Flame-sprayed metal deposits may also contain several percent metal oxide, due to reaction of molten metal with oxygen in the flame and in the ambient air atmosphere. Nevertheless, the properties of flame-sprayed coatings are well suited to many applications, and the relatively low process cost makes flame spray a preferred process for many commercial purposes.

The HVOF process is similar to flame spray in the respect that a fuel gas or liquid fuel is first mixed with oxygen and then burned in the combustion chamber of the torch (Fig. 3.6). However, HVOF combustion occurs at much higher combustion chamber pressures, and the resulting heated gas exits the torch through a converging–diverging nozzle, which generates a supersonic gas jet. As the combustion gas is accelerated to supersonic velocities, the jet expands and cools as thermal energy is converted to kinetic energy. Powder particles entrained in this high-velocity gas jet are accelerated up to velocities of the order of 450–1000 m/s (1475–3280 ft/s), depending upon the spray device, spray material, and operating conditions. The 'peening' effect of these high-velocity particles impacting the work-piece surface can produce coatings with very little porosity (near theoretical density) and also create a more favorable residual stress state in the spray-deposited material. The lower jet temperatures and higher particle velocities of HVOF make it a preferred process for applications such as spraying cemented carbide wear-resistant coatings, where the lower process temperatures prevent de-carburization and the high impact velocities produce a relatively pore-free coating with a favorable residual stress state that makes it less prone to cracking.

The Detonation Gun™ (D-Gun™) is a thermal spray process technology similar in concept and attributes to HVOF; however, with the D-Gun™ combustion occurs as a rapid series of individual detonation events, much like firing a machine gun, rather than the continuous combustion process employed in HVOF. Among the traditional thermal spray processes, HVOF and the D-Gun™ are the processes that most closely resemble cold spray due to their relatively high particle velocities and somewhat more moderate heating of the spray particles.

There are other thermal spray process technologies, but the processes just described account for the vast majority of commercial thermal spraying today and serve to illustrate the key attributes of traditional thermal spray technologies as a basis for comparison with cold spray. For those interested in more detailed information on traditional thermal spray processes, additional information can be found in the *Handbook of Thermal Spray Technology*, published by the Thermal Spray Society of ASM International (Davis, 2004), or in Chapter 11 'Thermal Spraying and Cold Spraying' in volume 3, part 1, of the 9th edition of the *Welding Handbook*, published by the American Welding Society (Smith *et al.*, 2007).

3.2.3 Advantages of traditional thermal spray

A principal advantage of traditional thermal spray processes is the ability to deposit an extremely wide range of materials. Virtually any material that has a stable molten phase can be deposited with one or more traditional thermal spray processes. This includes virtually all metals and metal alloys, many ceramics and glasses, as well as some polymers. Even some materials that do not melt, such as graphite and many carbide or boride ceramics, can often be co-deposited with another sprayable material to create a composite coating material. Indeed, the ability to co-deposit highly dissimilar raw materials to create unique composite materials with mixed, layered, or gradational microstructures is another important advantage of thermal spray. Such composites often have unique properties that can provide a materials engineer or component designer with important performance advantages and valuable new design options.

Another advantage of thermal spray is that the range of suitable substrate materials is even greater that the range of sprayable materials. In addition to metals, ceramics, glasses, and polymers, thermal spray coatings have been successfully applied to many other substrate materials, including wood, concrete, and even paper. It is often possible to apply a high-melting-temperature material, such as a metal, to a temperature-sensitive substrate, such as wood or a polymer composite. This is possible because the thermal energy in a single droplet of molten spray material is quite limited, and if proper thermal management is used in the coating process, excessive heat build-up in the substrate can usually be avoided.

For many applications, traditional thermal spray also offers the advantage of very high deposition rates. Because the coating material is spray deposited as 10–100 micron molten droplets, thermal spray deposition rates are typically orders of magnitude higher than the deposition rates of most alternative coating technologies, such as plating or vapor deposition, where deposition occurs at the atomic or molecular level. This affords the opportunity to rapidly and economically coat very large surfaces – such as the bridges, ship decks, or large metal tanks mentioned earlier – and to build up very thick deposits, ranging up to several millimeters and even many centimeters in thicknesses for some applications. When spray coating objects that are very large or otherwise difficult to move, the ability to apply thermal spray coatings *in situ* by taking portable spray equipment to the worksite is also an important advantage.

Compared with most alternative coating technologies, the effluent from thermal spray processes is also relatively easy to control and dispose of in an environmentally friendly manner. The major effluent consists of macroscopic solid particles, primarily overspray, which can be readily trapped with standard air filtration equipment. In some instances, it is even economical to recycle the trapped overspray by sending it to a suitable re-processing plant.

3.2.4 Limitations of traditional thermal spray

Two important limitations of traditional thermal spray processes are the presence of porosity and oxides in the spray-deposited material. These imperfections in the sprayed coating microstructure can significantly degrade the mechanical, electrical, and thermal properties of a spray-deposited material as compared with the same material in conventionally processed form. Excessive porosity can also be an issue if the coating is intended to protect the underlying substrate material from exposure to liquid or gaseous species that may cause corrosion or other problems. Conversely, high levels of porosity can actually be an advantage in some applications. For example, intentional high levels of porosity in plasma-sprayed zirconia thermal barrier coatings reduce the thermal conductivity of the coating and also enhance the resistance of the coating to cracking or spalling.

Porosity varies greatly according to the specific spray process, material, and deposition conditions. In general, low-velocity processes, such as flame spray and arc spray, tend to have higher levels of porosity, typically in the range of 5–15% by volume (vol.%). Higher velocity processes, such as plasma spray, can produce coatings with less porosity, typically of the order of 3–8 vol.%, although more porous coatings can be produced if desired for specific applications, such as the thermal barrier coatings noted earlier. Vacuum plasma spray, HVOF, and the D-Gun™, which have relatively high spray particle impact velocities, tend to produce coatings with the lowest levels of porosity. With these higher velocity processes, coatings with near theoretical density (<1 vol.% porosity) can be achieved under some conditions.

When spraying metals, reaction of the hot molten metal with oxygen in the ambient atmosphere results in the formation of metal oxide impurities in the coating. Excessive heating of the spray material can also result in preferential vaporization of more volatile species in a complex metal alloy, causing a shift in the chemical composition of the coating material as compared with the feedstock alloy. As with porosity, oxide levels vary greatly according to the process, material, and spray conditions. Typical oxide levels in sprayed coatings are of the order of a few percent by weight (wt.%). Processes that minimize heating of the spray material, such as HVOF, typically result in lower oxide concentrations and minimal changes in alloy chemistry. Of course, the controlled inert atmosphere of vacuum plasma spray creates very little or no oxide during the deposition process; however, some changes in alloy chemistry may still occur due to the relatively high temperatures in the plasma jet and the reduced ambient pressure in the spray chamber.

Another limitation of traditional thermal spray processes is residual stress. As each molten droplet solidifies and then cools to room temperature, it undergoes thermal contraction in direct proportion to the temperature change (ΔT) and the thermal coefficient of expansion (α) for that material. Since underlying, already solidified material is typically at a lower temperature, the net result is that each successive layer of material is left in a state of residual tensile stress, and the overall tensile residual stress in the coating steadily increases as subsequent layers of material are added. Indeed, it is the steadily increasing residual tensile stress that limits the maximum thickness of thermal spray coatings in many instances. As the coating thickness increases, the residual tensile stress in the coating eventually rises to the point that the coating will either crack or separate from the substrate. In general, high-ductility metals with low to moderate melting temperatures can often be sprayed to greater thicknesses than high-melting-temperature brittle materials, such as ceramics. This is because there can be some reduction of residual stress by localized yielding of the ductile metal as it cools. Also, the lower melting/solidification temperatures of such metals reduce thermal contraction ($\Delta T\alpha$) as newly solidified splats cool to the ambient room temperature. Coatings sprayed with HVOF and D-Gun™ processes generally tend to have less residual tensile stress, and such coatings may even be in a state of residual compressive stress. The reason for this is that HVOF and D-Gun™ spray particles, some of which may not be fully melted, impact at very high velocities and thus tend to 'peen' the underlying material, compressing it and pushing it toward a more compressive residual stress state.

A final noteworthy limitation of virtually all thermal spray processes is the simple fact that deposition is limited to surfaces that are directly accessible to the spray stream. Put another way, only those surfaces with a direct line-of-sight to the spray gun can be coated. This limitation can be also be turned to an advantage in some applications because simple masking techniques, such as special heat-resistant masking tapes, rubber or metal masks cut to specific shapes, etc., can be used to limit deposition to selected areas of the workpiece surface. However, the cost of labor and materials to create special masks, apply them, remove them after coating and, in some cases, clean them for re-use may add significantly to the total cost of a commercial thermal spray coating operation.

3.3 Comparison of cold spray with traditional thermal spray

In many respects, a generic cold spray gun (Fig. 3.7) looks very similar to some of the traditional thermal spray devices described earlier in this chapter. Pressurized gas is heated, usually with electrical energy, to temperatures generally in the range of 573–1073 K (300–800 °C) and then passed through a converging–diverging nozzle to create a supersonic gas jet. However, unlike traditional thermal spray processes, the reason for heating the cold spray process gas is not to melt the spray material. The gas is heated primarily to increase the sonic velocity of the gas in the 'throat' (point of smallest diameter) of the converging–diverging nozzle, which creates a higher spray jet velocity, while also reducing process gas consumption. Because expansion of the gas in the diverging portion of the nozzle rapidly cools the gas, it exits the spray gun nozzle at a much lower temperature, in some cases even below room temperature. Hence, the name 'cold' spray. Figure 3.8 shows a comparison of approximate process temperatures

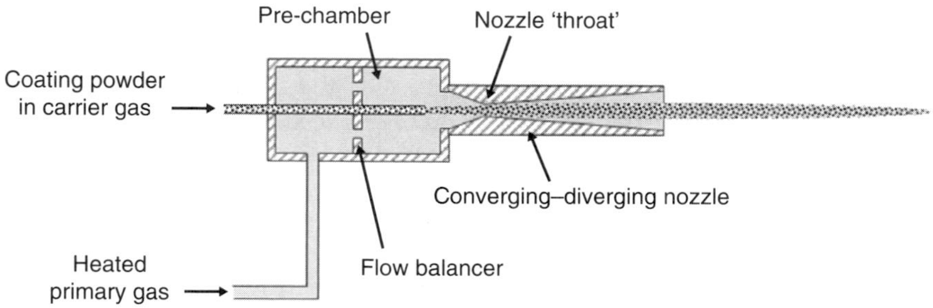

3.7 Schematic diagram of a cold spray gun.

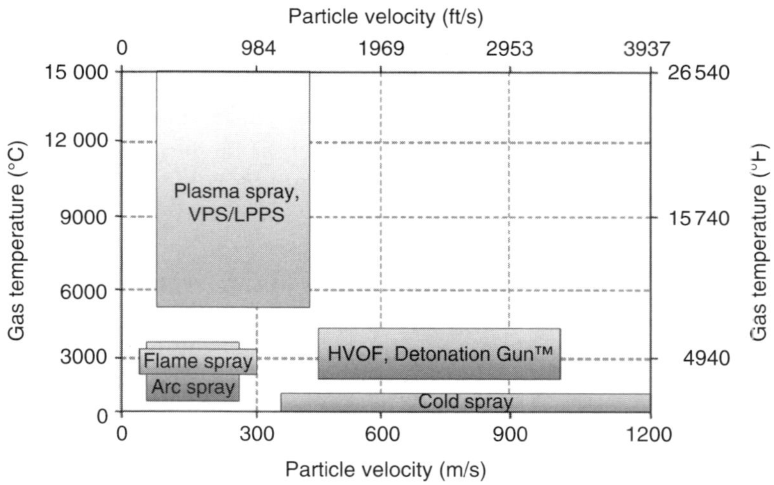

3.8 Comparison of approximate process gas temperature and particle velocity ranges for several common thermal spray processes and cold spray. VPS, vacuum plasma spray; LPPS, low-pressure plasma spray; these are two common names for plasma spray operations inside a vacuum chamber at a sub-atmospheric pressure of inert gas.

and particle velocities for cold spray and for several traditional thermal spray processes. It is clear from this comparison that cold spray occupies a relatively unique position with respect to other spray technologies, offering exceptionally low process temperatures combined with very high particle velocities.

Powdered feedstock material, typically 5–50 microns in diameter, is injected along the central axis of the cold spray gun (Fig. 3.7). The powder particles are exposed to the hot process gas for only a short while inside the spray gun, after they are injected and before the gas jet is expanded in the diverging portion of the nozzle. Hence, the powder particles arrive at the workpiece surface in the solid state, usually far below their melting point. The particles are accelerated to velocities of the order of 500–1200 m/s (1640–3940 ft/s) before they impact the workpiece surface. If the velocity is sufficient and the material properties are correct, the solid particles plastically deform and flow out upon impact, creating a hydrodynamic flow instability at the interface between the incoming particle and the underlying material, which results in bonding at that interface by a process similar in concept to explosive welding, but on a micro-scale. A more detailed discussion of this bonding process is presented in Chapter 9 of this book.

3.3.1 Advantages of cold spray

Because cold spray is a relatively low-temperature process with deposition carried out in the solid state, it is fundamentally different from traditional thermal spray technologies. As discussed earlier in this chapter, porosity and metal oxide impurities are two important limitations of traditional thermal spray processes. In some applications, cold spray can offer significant advantages in these two areas. For example, Fig. 3.9 shows a comparison of two copper coatings deposited using the same feedstock powder, but two different spray processes – plasma spray and cold spray. The ~5% porosity in the plasma-sprayed coating of Fig. 3.9(a) results from splashing of molten droplets upon impact and failure of the molten metal to completely fill surface irregularities during deposition. The greatly reduced level of porosity (<1%) in the cold-sprayed material of Fig. 3.9(b) results from

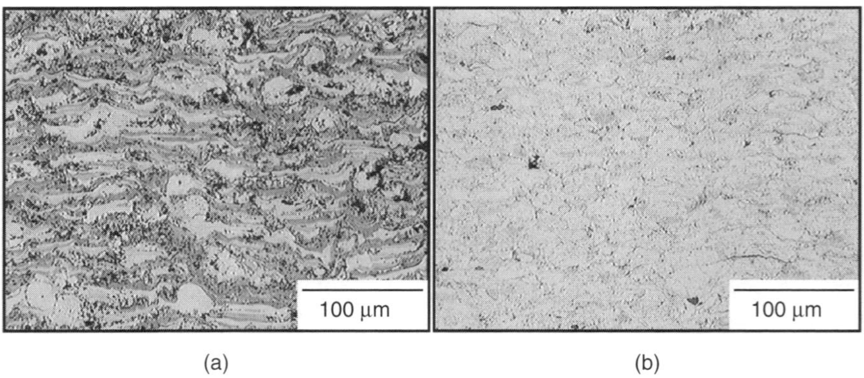

3.9 Comparison of two copper coatings produced from the same feedstock powder: (a) copper plasma sprayed in ambient air showing ~5% porosity (black areas) and 1.7 wt.% oxide (dark-gray bands); (b) copper cold sprayed in ambient air with <1% porosity and only 0.3 wt.% oxide (same oxide content as the feedstock powder).

the fact that cold spray is a solid-state process, so there is no splashing. In addition, the peening effect of incoming high-velocity solid particles in the cold spray process deforms underlying, previously deposited material, which tends to close any small pores or gaps in the underlying material.

In a similar comparison, the plasma-sprayed coating of Fig. 3.9(a) contains 1.7 wt.% oxide, almost six times more oxide than the 0.3 wt.% oxide present in the feedstock copper powder. The high process temperatures of plasma spray and consequent melting of the copper powder in an air environment resulted in significant oxidation of hot metal during the spray process. However, within measurement uncertainty, the measured oxide content of the cold-sprayed coating in Fig. 3.9(b) is exactly the same as that of the feedstock powder. Since cold spray is carried out at much lower process temperatures, reaction of metals with oxygen in the ambient spray environment is greatly reduced or eliminated.

The presence of porosity and oxides in most traditional thermal spray coatings is important because these microstructural defects can significantly degrade the mechanical, electrical, and thermal properties of sprayed materials. For example, the conductivity of the plasma-sprayed copper in Fig. 3.9(a) is only 15% of the conductivity of oxygen-free high-conductivity (OFHC) copper. In contrast, the relatively defect-free cold-sprayed copper in Fig. 3.9(b) has a conductivity that is 85% of the conductivity of OFHC copper. Since the conductivity of copper is very sensitive to oxygen content, a significant portion of the 15% reduction in the conductivity of the cold-sprayed copper as compared with OFHC can be reasonably attributed to the 0.3 wt.% oxide in the initial feedstock powder. An example commercial application where the superior thermal conductivity and lower oxide content of cold-sprayed copper has been used to great advantage is cold spray deposition of a thin layer of copper onto the bonding surface of finned aluminum heat sinks that are used to cool computer chips. The cold-sprayed copper layer makes it easy to solder the aluminum heat sink to other components in the heat sink assembly (it is very difficult to solder to aluminum, due to its very stable surface oxide). In contrast, the relatively high porosity and oxide levels inherent in most thermally sprayed copper would degrade the thermal conductivity and solder wetability of such copper coatings to the point that they would not be suitable for this application.

Porosity and metal oxides also have a profound effect on the mechanical properties of sprayed materials. Many metal coatings deposited in an ambient air environment with traditional thermal spray processes have poor ductility and, for this reason, they are sometimes difficult to machine with conventional methods. In the as-deposited condition, cold-sprayed metals also tend to have low ductility due to the extensive plastic deformation inherent in the deposition process and consequent work hardening of the material. However, since the cold-sprayed coatings are largely free of porosity and oxide defects, a simple post-deposition heat treatment to anneal the material can dramatically improve the ductility of a cold-sprayed metal. For example, in one study, modified ASTM E8 tensile test bars were machined from thick deposits of cold-sprayed commercially pure aluminum (Hall *et al.*, 2006). The bars were then pulled to failure in a standard tensile testing machine. The ductility of the cold-sprayed aluminum in the as-deposited condition was less than 0.5% elongation-to-failure, with a brittle inter-splat fracture mode shown in Fig. 3.10(a) that exhibits no evidence of any plastic deformation on the failure surface. However, after a simple annealing heat treatment of 22 hours at 573 K (300 °C), the cold spray aluminum shown in Fig. 3.10(b) exhibited more that 10% elongation-to-failure with clear evidence of extensive plastic deformation and trans-splat fracture on the failure surface. Indeed, the mechanical properties of the annealed cold-sprayed aluminum were comparable with the properties of conventional wrought 1100 aluminum in a partially strain hardened (H14) condition. With traditional thermal spray, it would be nearly impossible to achieve such properties for a highly reactive metal, such as aluminum, without the considerable expense of a controlled atmosphere process, such as vacuum plasma spray, to reduce porosity and oxide in the sprayed coating material.

Another potential advantage of solid-state deposition in the cold spray process is that the chemistry, phase composition, and crystal (grain)

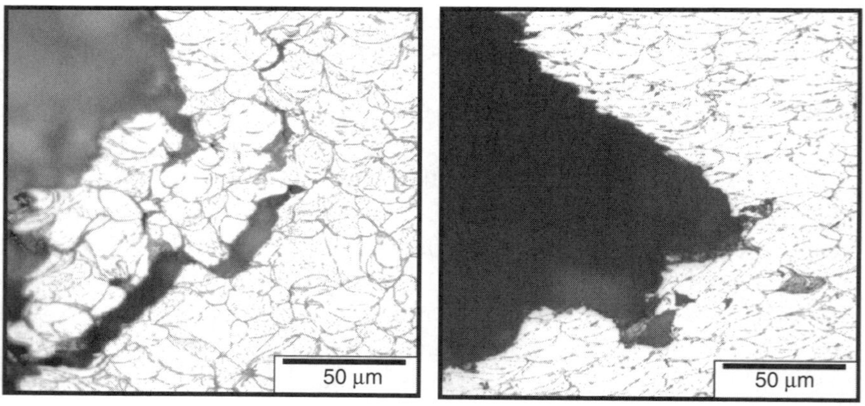

(a) (b)

3.10 Comparison of tensile test fracture modes in cold-sprayed aluminum in the as-deposited and annealed conditions: (a) the as-deposited condition shows a brittle failure mode with no splat deformation and fracture along splat boundaries; (b) after an anneal of 22 hours at 573 K (300 °C) a ductile failure mode is observed with significant splat elongation and fractures across individual splats.

structure of the feedstock powder are preserved in the coating. Metal alloys that contain multiple elements may experience some shift in chemistry in traditional thermal spray deposition due to preferential vaporization of more volatile elements as the alloy is exposed to high temperatures in the deposition process. In addition, the melting/solidification process of traditional thermal spray can dramatically alter the phase composition and crystal structure of the sprayed coating material as compared with the initial feedstock material. Conversely, the solid spray particles in cold spray experience minimal heating, so there in no shift in the chemistry of a complex alloy, and the initial phase composition and crystal structure are preserved, which may afford significant advantages in some applications. For example, nanocrystalline materials have offered great promise as a means to obtain exceptional mechanical properties. However, the challenge has always been how to consolidate nanocrystalline powders to create a structure of useful size without heating the material to the point that grain growth occurs. Cold spray may offer a practical route to consolidate nanocrystalline powders into structures of useful size without destroying the fine grain size that imparts the unique properties of these materials. Nanocrystalline powders have already been successfully cold spray deposited with no measurable change in grain size distribution or chemistry.

Another limitation that was mentioned for most traditional thermal spray processes is tensile residual stress. It was noted that high tensile residual stress resulting from thermal contraction as a thermal spray coating cools from the solidification temperature can sometimes lead to cracking or coating separation. Also, tensile residual stress often limits the maximum coating thickness that can be achieved with traditional thermal spray processes. It was further noted that HVOF and D-Gun™ coatings tend to have more favorable residual stress states due to the peening effect of very-high-velocity spray particles. Cold spray has particle impact velocities that are comparable with those of HVOF and D-Gun™, and the solid particles of cold spray are very effective in peening the underlying material. In addition, since the cold spray particles are deposited at relatively low temperatures, there is little temperature-driven dimensional change ($\Delta T\alpha$ thermal contraction) as the cold-sprayed material equilibrates with the ambient room temperature. The net result is that cold-sprayed coatings, unlike most traditional thermal spray coatings, are typically in a state of *compressive* residual stress. Figure 3.11 shows an experimentally measured residual stress distribution for a steel coating cold sprayed onto an aluminum substrate (McCune *et al.*, 1996). This profile confirms that the primary residual stresses in this coating are compressive, with a stress profile that is essentially a mirror image of the typical residual stress profile in most traditional thermal spray coatings. Since cold-sprayed coatings generally have no tensile residual stress to drive the opening or extension of cracks in the

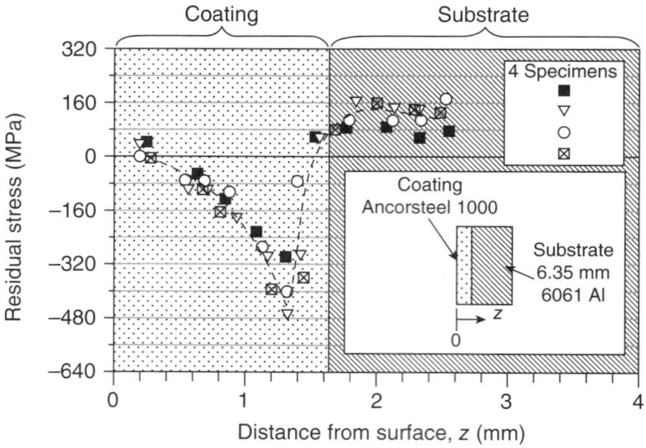

3.11 Experimentally measured residual stress distribution in cold-sprayed steel on an aluminum substrate (McCune *et al.*, 1996). Negative values indicate compressive residual stress.

3.12 Cold-sprayed copper deposited onto smooth and rough aluminum and aluminum oxide (Al_2O_3) substrates without masking. The tall 'spike' on the lower-left sample was built by simply holding the spray gun stationary over the substrate for a few seconds.

coating material, most ductile metals can be cold spray deposited to almost any desired thickness. For example, copper has been cold sprayed to a thickness of more than 10 cm (4 inches) and could be sprayed to an even greater thickness if necessary.

The possible added expense of masking and de-masking operations for some applications was another potential limitation noted previously for traditional thermal spray processes. The spray pattern of a cold spray gun is inherently very highly focused, and it is possible to spray highly localized areas with little or no masking. Figure 3.12 illustrates some examples of copper deposits cold sprayed onto a variety of smooth and roughened, metal and ceramic substrates without masking. The coating strips in this photograph were sprayed using a rectangular spray nozzle with an opening at the nozzle exit that was 2 mm (0.08 inch) wide and 1 cm (0.39 inch) long. The resulting coating strips in Fig. 3.12 were approximately 1 cm wide when sprayed with the 1 cm nozzle dimension oriented normal to the direction of spray gun motion and approximately 3 mm (0.12 inch) wide when the 1 cm nozzle dimension was oriented parallel to the direction of the spray gun motion. The edges of the coating strips are sharply defined, even though no masking was used. In some applications, the ability to deposit coating material selectively in a highly localized area without the need for masking operations could be a significant advantage.

3.3.2 Limitations of cold spray

Although cold spray can offer important advantages over traditional thermal spray processes in selected applications, cold spray also has its own inherent limitations. The fact that the spray particles in the cold spray process are not melted has profound implications with regard to the range of materials that can be cold sprayed and also the range of process-compatible substrates. Unlike traditional thermal spray processes that can deposit an exceptionally wide range of metals, ceramics, glasses, and polymers onto almost any substrate material, the cold spray process is essentially limited to depositing ductile metals – such as aluminum, copper, steels, nickel-base alloys, etc. – onto metal, ceramic, or other relatively hard substrate materials. This limitation arises from the fact that the incoming solid particles must plastically deform upon impact in order to create the hydrodynamic shear instability that bonds the incoming particles to the underlying material. For this to occur, the spray particle must have sufficient ductility to plastically deform and flow, and the substrate material must be hard enough to cause the incoming particle to plastically deform. For this reason, brittle spray materials, such as ceramics or brittle inter-metallic metal alloys, and softer substrate materials – such as plastic, fiber-reinforced composites, etc. – are generally not compatible with the cold spray process. In some cases, it is possible to embed a layer of a cold spray coating material into the surface of a softer substrate and then build a deposit on top of the embedded material. However, this is difficult at best, and the adhesion of the coating may be relatively poor in comparison to more suitable substrate materials.

Another potential issue is that the cold spray process uses much larger quantities of process gas than traditional thermal spray processes, typically of the order of 1–2 m^3/min (35–70 ft^3/min). In applications where a relatively inexpensive process gas, such as compressed air or nitrogen, is suitable, this may not be a serious issue. However, with some materials and applications, it may be necessary to use helium as the primary cold spray process gas in order to achieve the necessary impact velocities and coating quality. Since helium is relatively expensive, the process cost may increase dramatically for applications where helium is required. Some preliminary studies indicate that it may be feasible to address this issue by capturing and recycling helium in selected cold spray production applications. However, this greatly increases the cost and complexity of the complete cold spray system. Fortunately, the latest generation of cold spray equipment can operate at higher process temperatures, up to ~1073 K (800 °C) with nitrogen. The higher gas temperature at the inlet of the spray gun provides higher sonic velocities in the throat of the cold spray nozzle, so many materials that previously could be sprayed only with helium can now be successfully sprayed with less expensive nitrogen. It is important to note that although the inlet gas temperature may be relatively high when operating in this mode, the gas does cool significantly as it expands in the diverging section of the nozzle. So although the particle and substrate temperatures will increase somewhat as the inlet gas temperature is raised, the increase in spray particle and substrate temperatures is not as large as might be expected.

Another limitation of cold spray for some applications is the fact that its well-defined spray plume is relatively small, typically less than 1 cm (0.39 inch) in diameter. Therefore, the cold spray process is not as well suited for coating very large surface areas as compared with some of the traditional thermal spray processes, such as wire arc spray, where the spray plume is substantially larger. In addition, the extensive plastic deformation inherent in the deposition of cold-sprayed metals work hardens the sprayed material and results in very low ductility of the coating in the as-sprayed condition. While this may be an issue for some coating applications, in others it is not a significant drawback. In still other applications, the slightly higher hardness of the work-hardened material may actually be an advantage.

While it is true that most metals can be cold spray deposited in an ambient air environment with little or no oxidation during the spray process, some highly reactive metals, such as titanium, may still react with atmospheric oxygen even though deposited at relatively low temperatures. If this is of concern for a specific application, it may be necessary to take special measures to limit the oxygen level in the area where the metal is being deposited (e.g. inert glove box, spray shroud, etc.).

Finally, as with traditional thermal spray processes, cold spray deposition is limited to surfaces that are in a direct line-of-sight with the spray gun. In addition, cold spray is a relatively new process in comparison to the traditional thermal spray technologies discussed earlier in this chapter. For this reason, there is substantially less published information on the properties and performance of cold-sprayed materials and coatings in real-world applications. The applications and available data for cold spray are growing, but at this point in the evolution of the technology, the lack of previously demonstrated commercial success in specific applications means that some additional testing may be required when evaluating this technology for a new application.

3.4 Conclusions

In conclusion, cold spray is still an emerging, but very promising, technology. In some applications, it offers a way to overcome important limitations of traditional thermal spray technologies, such as high levels of porosity and oxide in the sprayed material. However, it is also more limited in the range of materials and substrates that are process compatible, and there are currently very limited application and property data for cold-sprayed materials as compared with traditional thermal spray processes. For this and other reasons, cold spray will not replace traditional thermal spray processes. Nevertheless, it represents an important new process capability that may replace more traditional spray processes in selected applications, and it is likely that it will lead to new applications that are not suited to traditional thermal spray technologies. Cold spray offers the potentially valuable possibility of new design, manufacturing, and component reclamation alternatives for those willing to explore the intriguing new opportunities that it provides. It is likely that some of the best applications for cold spray have yet to be discovered!

3.5 References

Davis J R (ed.) (2004), *Handbook of Thermal Spray Technology*, Materials Park, Ohio, USA, ASM International.

Hall A C, Cook D J, Neiser R A, Roemer T J and Hirschfeld D A (2006), 'The Effect of a Simple Annealing Heat Treatment on the Mechanical Properties of Cold-Sprayed Aluminum', *J Thermal Spray Tech*, **15**(2), 233–238.

Mccune R C, Donlon W T, Cartwright E L, Papyrin A N, Rybicki E F and Shadley J R (1996), 'Characterization of Copper and Steel Coatings Made by the Cold Gas-Dynamic Spray Method', in Berndt C C (ed.), *Thermal Spray: Practical Solutions for Engineering Problems*, Materials Park, Ohio, USA, ASM International.

Schoop M U (1911), *An Improved Process of Applying Deposits of Metallic Compounds to Surfaces*, United Kingdom Patent 21,066.

Smith M F, Berndt C C, Crawmer D E, Kay A, Lenling W J, Miller R A, Moreau C, Neiser R A, Russo L D and Zammit P F (2007), 'Thermal and Cold Spray', Chapter 11 in O'Brien A and Guzman C (eds), *Welding Handbook*, 9th edition, Volume 3, Part 1, Miami, Florida, USA, American Welding Society.

4 The advantages and disadvantages of the cold spray coating process

J. KARTHIKEYAN, ASB Industries Inc., USA

4.1 Introduction

As with any other materials processing technique, the cold spray process has its own advantages and disadvantages. The main advantage of the cold spray process is that it is a solid-state process, which results in many unique coating characteristics. The main disadvantage arises due to the plastic deformation process, which leads to a loss of ductility of the coating.

In the thermal spray processing technique, electrical, chemical or mechanical energy is used to produce a high velocity–high temperature gas jet.[1] The coating material, usually in powder form, is introduced into this gas jet. In processes such as plasma and flame spraying, these particles are molten and the liquid droplets impinge on the substrate surface. Liquid flow and quenching leads to the formation of splats, which interlock with each other to form the coating. Processes such as the detonation gun and high velocity oxy-fuel spray (HVOF) heat the particles close to their melting temperature, but do not fully melt them. These processes accelerate the particles to a greater extent than the plasma and flame spray processes. On impact onto the substrate surface, the enhanced kinetic energy is converted to thermal energy, resulting in particle fusion/melting and leading to the formation of splats.

Since the powder particles are heated, molten and quenched, many undesirable high temperature reactions – such as phase changes, decomposition, preferential evaporation, etc. – occur during conventional thermal spray processing. Typical examples include the formation of metastable γ alumina in alumina coating, decarburization of tungsten carbide in WC–Co coating, and preferential depletion of carbon and boron in self-fluxing alloys.[2,3]

The cold spray process takes another step along the direction of *higher velocity at lower temperature*.[4] By increasing the particle velocities to supersonic values and reducing its temperature to near ambient at or near the point of impact, the cold spray process leads to the formation of a splat-like morphology by pure solid-state plastic deformation of the particles. Since the particle does not experience high temperatures, the above-noted deleterious high temperature reactions are avoided and this leads to many advantages.

4.2 Advantages of the cold spray process

Some advantages of the cold spray process are discussed below.

High deposition efficiency

Very high deposition efficiency (DE) values have been achieved with the cold spray process for most metals, alloys and composites. For instance, DE values in excess of 95% have been achieved for aluminum, copper and their alloys. Experiments have shown that, by (a) minimizing the oxygen content of the feedstock material; (b) stress relieving the powder particles; (c) optimizing the particle size distribution; and (d) optimizing the spray parameters, it is possible to obtain very high DE values for most materials including refractory metals (such as tantalum, niobium, etc.), high strength materials (such as Inconels, Ti6Al4V, etc.), MCrAlYs and others.

High deposition rate

The spray footprint of the cold spray beam is narrow and well defined. A typical spray beam has about a 10 mm diameter with sharp edges. A narrow spray beam with appreciable powder feed rate translates into very high deposition rate (DR) values. For most materials, it is possible to produce a 1–2 mm thick coating in a single pass. Unlike other coating processes, cold spray produces these ultra-thick layers with high bond strength. This aspect has been exploited in developing cold spray as a tool for the fabrication of near-net-shape articles and also for additive manufacturing of features.

No or little masking

As noted above, the particle beam in the cold spray process is narrow and well defined. Although a typical spray beam has a diameter of about 10 mm, it is possible to design special rectangular nozzles and obtain spray beams as small as 1–2 mm wide with sharp edges. Hence, masking of regions where overspray can not be permitted is not necessary in many applications.

No grit blast

The cold spray process can be viewed as a triplex (grit blast–spray coat–shot peen) process. As is to be expected, the velocity distribution across the spray beam shows Gaussian distribution. When the system is running with optimized parameters, particles in the core have higher than the critical velocity and the particles at the rim have lower (than critical) velocity. When the spray beam is scanned, the particles at the leading edge impact on the substrate surface with less than the critical deposition velocity (V_c) which leads to solid particle erosion of the surface, resulting in *in-situ* micro-grit blasting. Particles in the core impact on the substrate surface with higher than V_c, plastically deform over the freshly cleaned rough surface and form a coating. Particles in the trailing edge, which again impact with lower than V_c, not only sputter off any loosely bound particles but also 'shot peen' freshly formed coating. Experiments have shown that conventional grit blasting is not required for many substrate materials including aluminum, copper, titanium, etc.

High density

In the cold spray process, only particles impacting with higher than V_c plastically deform and deposit. During the deformation process, high strain rate deformation leads to additional thermal energy added onto the conversion of kinetic energy to thermal energy on impact. These microscopic processes lead to the formation of hydrodynamic instabilities and produce a metal vapor jet.[5] This jet in effect produces 'vapor deposition' of material at the interparticulate interfaces and fills any pores and cracks that exist there. Thus, cold spray can be viewed as a combination of particulate and microscopic vapor deposition processes.

Moreover, any loosely bound particles are sputtered away, and deposited splats are micro-peened by the trailing edge particles of the spray beam. Finally, every subsequent pass of the spray beam 'shot peens' the underlying layer and thereby increases its density. Due to a combination of all these phenomena, cold spray produces near-theoretical density coatings. Figure 4.1 gives the microstructure of typical cold-sprayed coatings.

Flexibility in substrate-coating selection

Many engineering applications require substrate-coating couples with dissimilar materials, and these are difficult to produce by conventional processes. Even some of the widely used materials such as aluminum over copper or copper over aluminum can produce unacceptable interfaces. For instance, when a molten droplet of copper impacts over aluminum substrate, a brittle Cu–Al intermetallic phase is formed, which can lead to failure of the coated structure during service. Hence, it is necessary to study the phase diagrams carefully and select the substrate-coating couples to ensure that intermetallic phases or metastable phases, which can lead to

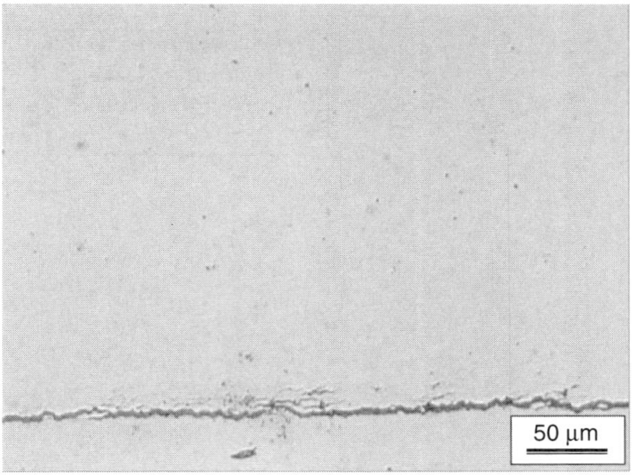

4.1 Microstructure of cold-sprayed copper coating.

premature failure of the coated system, are not formed during the high temperature processing. Since cold spray does not heat and melt the coating material, the formation of weak interfaces is avoided. Hence, the engineer has the flexibility to select the materials on the basis of design requirements.

Minimum thermal input to the substrate

In conventional thermal spray processing, the substrate is heated by the flame to varying degrees. Hence, it is difficult to process temperature-sensitive materials such as magnesium, polymers, etc. Moreover temperature-induced warping of the substrate is of concern, particularly when the thickness of the specimen is small. In the cold spray process, there is no high temperature jet to heat the substrate, and the substrate receives only the stagnation enthalpy of the impacting particles. Hence, cold spray can be used to repair parts and components made of temperature-sensitive materials.

High bond strength

Cold-sprayed coatings exhibit very high bond strength over many substrate materials including metals (aluminum, copper, titanium, nickel, etc.), alloys (Inconels, steels, etc.) and composites (metal matrix composites, carbon composites, etc.). Even glass can be coated with some materials, such as aluminum. Figure 4.2 presents the variation of bond strength with coating thickness of cold sprayed aluminum coating over 6061 aluminum alloy

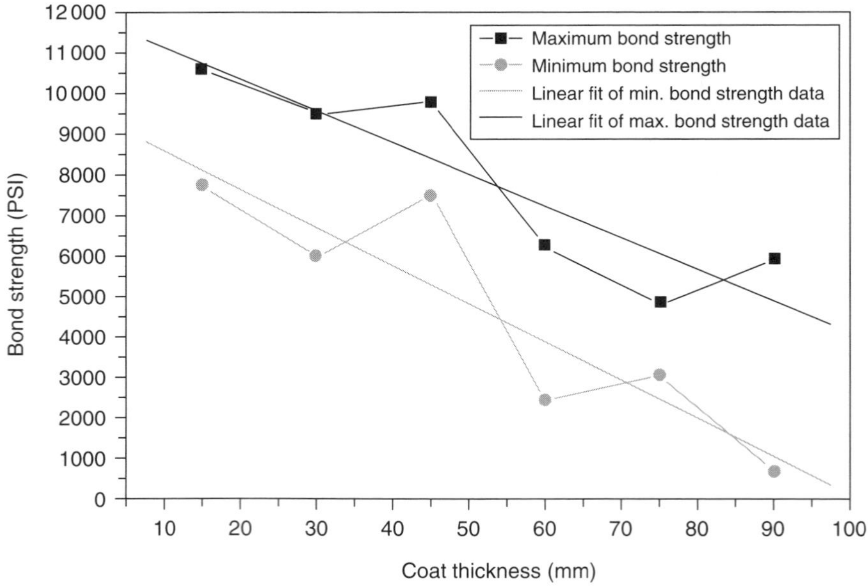

4.2 Variation of bond strength of cold-sprayed coating with thickness.

substrate. As can be seen, even ultra-thick coatings show sufficiently high bond strength values.

Compressive residual stresses

Most coating processes (thermal spray, electroplating, etc.) produce deposits with tensile residual stresses. In addition to bond failure, these stresses lead to micro-cracking on the top surface of the coating. This leads to lowering of the fatigue properties of the coating and results in deterioration of the performance characteristics of the coating in the actual application environment. In the cold spray process, the coating is generated by plastic deformation in the solid state and hence the residual stress across the entire coating thickness is compressive in nature. In many high technology industries such as gas turbines, coatings are required to have compressive residual stresses, as better fatigue performances are achieved with such a stress state. In some critical applications, post-spray shot peening is carried out to achieve better fatigue performance. Since the cold spray process itself produces compressively stressed coatings, these coatings show better fatigue properties and the additional shot peening step can be avoided.

Ultra-thick coatings

As noted above, most processes produce tensile stresses during coating generation. As the coating thickness is increased, these stresses pile up at the substrate–coating interface, leading to a decrease in the bond strength. As the coating thickness reaches a critical value, the coating spontaneously spalls from the substrate. Since a cold-sprayed coating is compressively stressed, ultra-thick layers can be produced over many substrates without bond failure. This aspect has been used to advantage in some applications such as replacement of thick electroplated layers.

Figure 4.3 shows the production of a 2.5 mm thick copper coating over a stainless steel manifold used in rocket engines. In order to produce such thick coatings by the electrolytic process, engineers used to immerse the substrate for 7–10 days in electrolytic tanks and continuously monitor and control the plating process. Even then, the plated layers could not withstand the subsequent machining, brazing and testing operations. Cold spray can produce the coating in under an hour of spray operation, and the sprayed coating can successfully meet all the requirements of subsequent machining operations, brazing cycles, cryo shock and other test procedures as well as the subsequent operating conditions.[6]

The possibility of producing well-bonded, ultra-thick coatings, even over dissimilar materials, has been exploited in producing high thermal mass layers in heat sinks, additive manufacturing of large features in difficult-to-machine parts, etc.

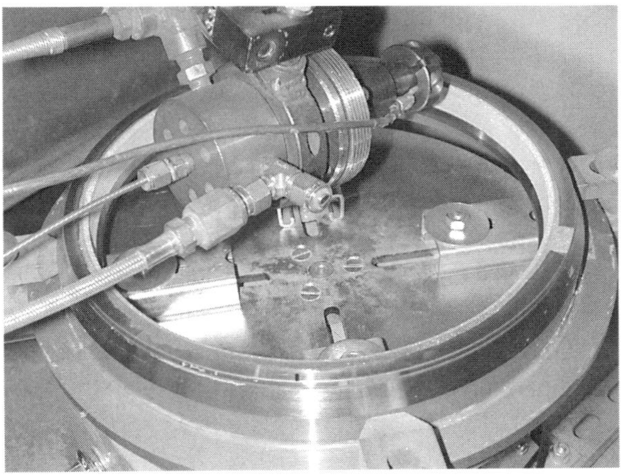

4.3 Cold spray processing of a rocket engine manifold.

No phase changes

As is to be expected, oxidation, decomposition, formation of metastable phases, preferential loss of some constituent, etc. occur during any high temperature processing. In the cold spray process, the particles remain near ambient and hence no phase changes occur. Figure 4.4 shows the X-ray diffraction (XRD) patterns of WC–Co powder and cold-sprayed coating. As can be seen, the phase structure of the feedstock is maintained in the coating and neither decarburization nor formation of η phases is observed.[7]

No oxidation

Thermal spray processing of oxygen-sensitive materials – such as aluminum, copper, magnesium, titanium and their alloys – is difficult as increasing the temperature exponentially accelerates the oxidation. Cold spray does not heat the material to high temperature. Moreover, an inert gas is used to propel the particles, which effectively shields the splats that are formed over the substrate. Hence, oxidation of the particle is almost completely avoided in the cold spray process. In fact, it has been observed that cold-sprayed coatings have a slightly *lower* oxygen content than the starting material.[8] This deoxygenation occurs as follows. Each fine particle in the feedstock material has a thin oxide layer. When the particle impacts on the substrate surface, the brittle oxide skin shatters and is swept away by the gas jet, while the nascent metal surface bonds to the underlying surface.

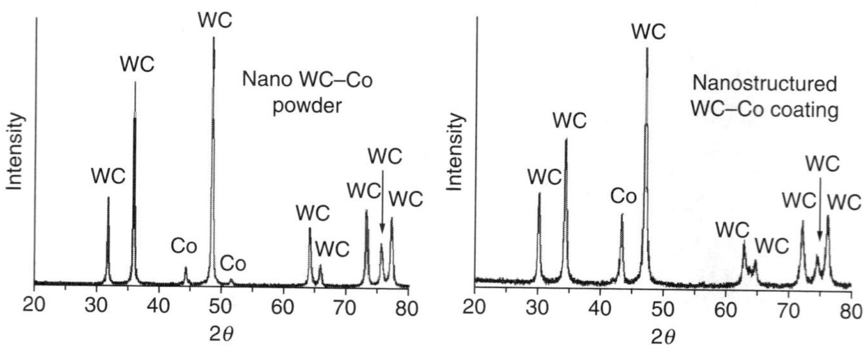

4.4 X-ray diffraction of WC–Co powder and cold-sprayed coating, showing that there are no phase changes during the cold spray process; 2θ, diffraction angle.

No grain growth

Cold spray is a low temperature, solid-state consolidation technique and can be used to process temperature-sensitive materials such as nanomaterials and amorphous materials. Unlike other powder consolidation techniques – such as pressing and sintering, thermal spray, etc. – which lead to unacceptable grain growth, cold spray produces nanostructured coatings without any appreciable grain growth. Experiments have shown that unique properties such as very high hardness can be achieved by the cold spray consolidation of nano WC powders.[9] Similarly, retention of amorphicity has been recorded while amorphous alloys are cold spray processed.[10]

High conductivity

Cold spray produces high density coatings with fairly high interparticulate bonding. Moreover, coatings exhibit phase purity, and inclusions – such as pores, oxides, etc. – are minimal or even absent. This leads to excellent thermal and electrical conductivity values of cold sprayed coatings. Experiments have shown that a typical cold-sprayed coating shows bulk conductivity values in excess of 92%, which can be compared with about 40–63% for thermal-sprayed coatings.[8]

High corrosion resistance

High density, phase purity and homogeneous microstructure of cold-sprayed coatings combine to yield exceptional corrosion characteristics. In fact, cold spray-produced aluminum and other coatings have even shown corrosion resistances that are superior to those of the bulk materials.[11]

High strength and hardness

Mechanical property evaluation has shown that the ultimate tensile strength (UTS) of cold-sprayed coatings is always higher than the bulk values. Moreover, coatings exhibit higher hardness than the bulk values, due to the high degree of plastic deformation of the particles.[12]

In summary, the cold spray process has many advantages. Many of the deleterious high temperature reactions such as oxidation, decomposition, phase transformation, grain growth, etc., are avoided. Phase-pure coatings are produced with compressive rather than tensile stresses, with wrought-like microstructure and near-theoretical density, with high electrical and thermal conductivities and excellent corrosion characteristics. Oxygen- and temperature-sensitive substrates can be spray coated to produce protective coatings or repair/refurbish the parts. Ultra-thick (several centimeters thick) coatings with high bond strength to many substrates can be produced. Cold spray does not require grit blasting for many substrate materials. The footprint of the cold spray beam is very narrow and well defined, which results in high growth rate of coating thickness with better control over the shape of the coating, without masking requirements. This can be exploited in producing freeforms and features on parts. Since cold spray leads to almost no grain growth, it can be used for consolidation of nanomaterials into coatings and structures.

4.3 Disadvantages of the cold spray process

The cold spray process has some disadvantages, as given below.

- Although mechanical property values equivalent to bulk values can be obtained by post-spray heat treatment, the cold-sprayed coating has near-zero ductility in the as-sprayed condition.
- Although composites can be sprayed, pure ceramics and some alloys (such as work-hardening alloys) can not be processed.
- Cold spray requires at least limited ductility of the substrate to produce well-bonded coatings. Hence, cold-sprayed coatings over ceramic substrates show only limited bond strength.
- At present, high quality coatings such as MCrAlYs, Inconels, etc., are in general produced with expensive helium as the processing gas, in order to achieve the velocities necessary for deposition.
- As with many other coating techniques, cold spray is also a line-of-sight process and hence spraying complex shapes and internal surfaces is difficult.

4.4 References

1 Davis J R (Ed), *Handbook of Thermal Spray Technology*, ASM International, Materials Park, Ohio, 2004.

2 Mcpherson R, Formation of Metastable Phases in Flame and Plasma Sprayed Alumina, *J. Materials Sci.*, **8**, 851–858, 1973.

3 Lovelock H L de Villers, Powder/Processing/Structure Relationships in WC–Co Thermal Spray Coatings: A Review of the Published Literature, *J. Thermal Spray Technol.*, **7**(3), 357–373, 1998.

4 Karthikeyan J, Process Technologies: Cold Spray, *Adv. Mater. Processes*, **163**(3), 33–35, 2005.

5 Assadi H, Gärtner F, Stoltenhoff T and Kreye H, Bonding Mechanism in Cold Gas Spraying, *Acta Materialia*, **51**, 4379–4394, 2003.

6 Haynes J and Karthikeyan J, Cold Spray Copper Application for Upper Stage Rocket Engine Design, *Thermal Spray 2003: Advancing the Science and Applying the Technology*, (Ed.) C. Moreau and B. Maple, pp. 79–84, ASM International, Materials Park, Ohio, 2003.

7 Lima R S, Karthikeyan J, Kay C M, Lindemann J and Berndt C C, Microstructural Characteristics of Cold Sprayed Nanostructured WC–Co Coatings, *Thin Solid Films*, **416**, 129–135, 2002.
8 Borchers C, Gärtner F, Stoltenhoff T, Assadi H and Kreye H, Microstructural and Macroscopic Properties of Cold Sprayed Copper Coatings, *J. Appl. Physics, B.*, **93**, 10064–10070, 2003.
9 Kim H J, Lee C H and Huang S Y, Superhard Nano WC–12%Co Coating by Cold Spray Deposition, *Mater. Sci. Eng. A*, **391**, 243–248, 2003.
10 Ajdelsztajn L, Lavernia E J, Jodin B, Richter P and Sansoucy E, Cold Gas Dynamic Spraying of Fe-Based Amorphous Alloy, Paper s2__5-11745, *Proceedings of ITSC* 2006, Seattle, ASM International, Materials Park, Ohio, 2006.
11 Karthikeyan T, Laha K, Balani A, Agarwal A and Munroe N, Microstructural and Electrochemical Characterization of Cold-Sprayed 1100 Aluminum Coating, Paper 017, pp. 22–27, *Proceedings of ITSC-2004*, Osaka, Japan, ASM International, Materials Park, Ohio, 2004.
12 Haynes J, Pandey A, Karthikeyan J and Kay A, Cold Sprayed Discontinuously Reinforced Aluminum (DRA), Paper s2__6-12091, *Proceedings of ITSC-2006*, Seattle, ASM International, Materials Park, Ohio, 2006.

5
The economics of the cold spray process

S. CELOTTO, J. PATTISON, J. S. HO, A. N. JOHNSON and W. O'NEILL, University of Cambridge, UK

5.1 Introduction

This chapter discusses the technical aspects of cold spray from an economic viewpoint. There are several different ways in which the commercial potential of cold spray can be fulfilled. Firstly, the process can add value to existing products by enhancing performance and improving functionality. This can enable a product to gain market advantage and be sold at a premium price. For example, the low porosity and oxide content of cold-sprayed coatings provide superior thermal and electrical conductivity, as well as corrosion resistance, compared with their thermally sprayed counterparts.[1] These enhanced coatings are especially useful where pure metals and oxide-free alloys are required. In addition, the process's non-melt approach to material deposition – with low thermal input to the substrate – is opening up new applications using oxygen-sensitive materials[2] and high melting point refractory metals.[3] An extension of this would be the production of unique products that until the development of cold spray could not be realised because of technical or economic obstacles. The process can enable entry into established markets by reducing production costs through lower material waste and by reducing the number of manufacturing steps required for part production.[4,5] Another prospect is using cold spray to repair existing products, potentially *in situ*, through additive fabrication, and thus extending service lifetimes and reducing the need for replacement products. Finally, cold spray can offer a more energy- and material-efficient manufacturing route that provides environmental benefits such as the elimination of harmful liquid waste.

This chapter gives a breakdown of the necessary capital equipment investment and consumable costs for various cold spray system configurations. Process gas selection is discussed with regard to different deposition materials and applications where deposition efficiency is an important factor; here helium is considered with and without recycling. It will be shown how the use of expensive powders that are difficult to deposit can justify the use of helium as the process gas. Also discussed are the cost comparisons between cold spray and other thermal spray processes and manufacturing methods, in particular some common additive fabrication techniques.

There are three key difficulties in presenting the economics of processes that are heavily reliant on proprietary equipment and raw materials. The first is that the prices of these items are generally not released into the public domain. Even if such information were released, it would rapidly become redundant because of the second problem, which is how quickly such costs vary. One reason for this comes from the worldwide variability and temporal volatility of energy and commodity prices that determine the day-to-day running costs. To highlight this, between 2005 and 2006, rising commodity prices have increased the cost of raw materials for powders by approximately 100%. We have tried to circumvent these issues by providing cost comparisons in the form of normalised ratios. The other cost that will undoubtedly change in the future is the cost of cold spray equipment, as competition increases in this developing sector. The third reason for the difficulty in stating specific costs is that they are application-specific. The size, complexity, number and variability of the products determine the basic choices of infrastructure and equipment investment. The infrastructure and labour costs will be the same as for other thermal spray techniques so we have concentrated on the aspects that set cold spray apart.

5.2 Cold spray equipment and infrastructure

Setting up a cold spray facility can require extensive capital equipment investment in such things as gas supply, spray apparatus and waste powder extraction. Capital costs of auxiliary equipment are similar to those in other thermal spray processes, with actuators, spray booths and jigging depending on the type, size, number and variability of the desired product. Figure 5.1 shows a schematic diagram of this equipment and infrastructure, along with a list of major consumables.

An example of a cold spray system that has low floor space requirements is the portable apparatus available from Supersonic Spray Technologies (SST),[6] which uses compressed air and has a built-in gas heater and extraction system. Such portable systems are discussed in Chapter 11. The main concern of this chapter is the economics of larger systems that use nitrogen or helium process gases.

5.2.1 Equipment

The cost of cold spray systems depends on the level of control and scale required for the gas parameters (pressure, temperature, etc.) and nozzle–

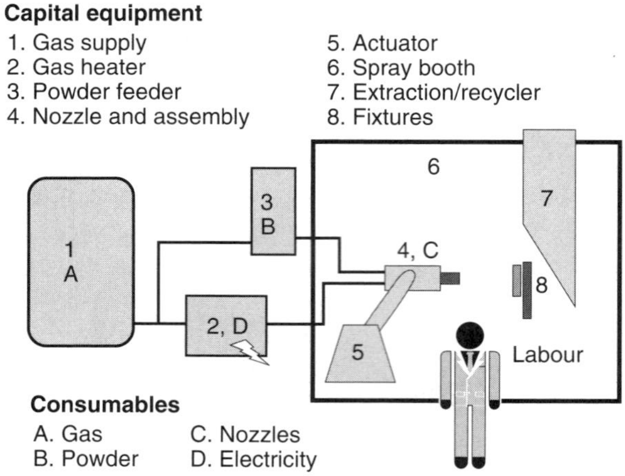

5.1 Schematic diagram of a cold spray system highlighting cost contributions from consumables, capital equipment, infrastructure and labour.

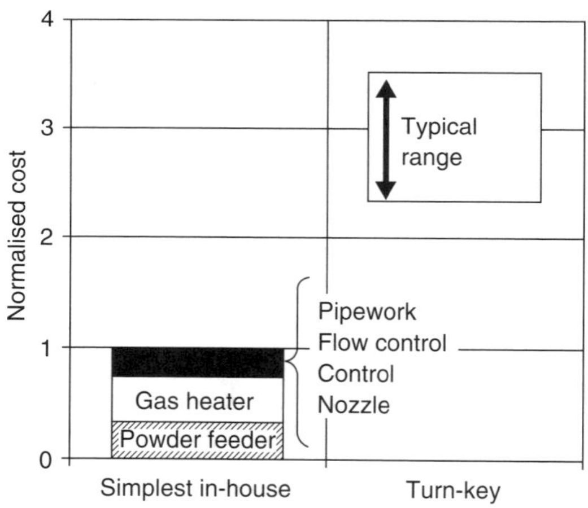

5.2 Normalised costs of cold spray equipment for different levels of complexity.

substrate positioning. Since the process is relatively simple, especially without gas heating, most of the apparatus can be constructed in-house. A cheap and simple system may be achieved with a single gas cylinder, a high-pressure regulator and a high-pressure hose to a nozzle; although to ensure steady flow conditions, feedback control is required. As can be seen in Fig. 5.2, the powder feeder and gas heater make up the greater part of the total cost. This is because they are certified pressure vessels and sold in relatively low numbers; the gas heater has the added complexity of operating at high temperatures. The other specialised component is the convergent–divergent nozzle. A simple nozzle with a linear profile can be manufactured using standard milling techniques and cheap materials. However, the performance of such a nozzle may be poor compared with more technically advanced nozzles that have optimised de Laval profiles and polished surfaces with low wearing and anti-clogging properties. One further consideration is that of a basic positioning system such as an H-frame gantry. However, such simple and rudimentary systems are only useful for basic studies and would not achieve state-of-the-art results consistently.

Investment in a turnkey solution that is fully controlled and optimised would be more appropriate for a high-volume user. These systems consistently deliver the best performance and are the only viable solution for a commercial application. At present, the only commercial systems available are from TevTech (USA)[7] and Cold Gas Technology (CGT; Germany).[8,9] Systems from both suppliers are modular and consist of a powder feeder, gas heater, gas distribution and control system, supersonic nozzle and related assembly. All of the equipment is specific to cold spray and comes with a license to the patent. Such units vary greatly in price depending on the configuration and level of performance required. In the case of the CGT Kinetic 4000 system, the price will depend on the number and power rating of the gas heaters.

5.2.2 Infrastructure

Many of the fixed infrastructure costs related to cold spray are equivalent to those of similar thermal spray processes. The floor space requirements include gas supply, powder feeder and control unit, and depend on the configuration of the gas heater and spray booth. If spray runs of several hours at a time are required, then the floor space may be dominated by the gas supply vessels or recycling system; both of which are described in later sections. Note that because gas storage containers have large volumes, they are usually stored in an external gas storage compound for ease of delivery and collection. These can be constructed without the additional precautions and expense associated with the safe storage of combustible gases used in other thermal spray processes. Acoustic booths or spray booths are often used for cold spray and thermal spray processes in order to contain any powder that is not deposited and the high noise levels that arise (~100 dB). The amount of floor space required for a spray booth can be as little as 2 m^2.

5.3 Consumables

The primary consumables in the cold spray process are gases and powders. Cold spray consumes substantially more gas than any other thermal spray process, as well as many other manufacturing methods. The flow rates are in the range of 60–120 m^3/hour and depend on the gas composition, nozzle throat diameter, gas pressure and gas temperature. The cost of powder can be one of the deciding factors in choosing which gas species to use, either helium or nitrogen. Hence, in this section the powder feedstock is discussed before the gas type, followed by the relatively minor costs of electricity and nozzles.

5.3.1 Powder feedstock

In cold spray, the feedstock material must be in fine-powder form. This is by no means a limitation, as there are a wide range of pure metals and alloy systems available in powdered form, some of which can not be manufactured into wire feedstock. Novel powder blends can be sprayed that comprise different metallic materials,[10] or metallic and non-metallic materials,[11–14] that would be impossible to deposit individually because of their lack of ductility. This approach can also add value through the application of functionally graded coatings.[15,16]

In general, metallic powders are inherently more expensive than bulk monolithic forms such as cast or extruded rod, bar or sheet. This is because of the additional processing and handling required. In addition to the primary processing needed to produce the bulk material (e.g. mineral refining, smelting, etc.), secondary processes such as gas or water atomisation also consume energy and considerable volumes of gas. The powder feedstock is therefore a highly value-added product compared with the bulk forms, which is reflected by its elevated cost. Figure 5.3 shows the difference in costs between the typical base metals of aluminium, copper and titanium, as well as comparisons with the high-value powder forms. In this graphic, the bulk form for titanium is replaced with sponge powder, which is in effect a by-product from the production of titanium. Here prices are shown for spherical powders with typical cold spray size distributions. Whereas aluminium and copper powders are produced by gas atomisation, commercial titanium powder is produced by the reduction of the precursor mineral.[17] The intrinsic cost of titanium still makes it more expensive than other base metals. Spherical titanium powder is manufactured by transforming titanium sponge into a monolithic form before it is inert-gas (argon) atomised like other spherical powders; the additional processing stage makes it even more expensive. Aside from the other processing costs, a product fabricated by cold spray can be a factor of 10 times more expensive than bulk material on a unit mass basis.

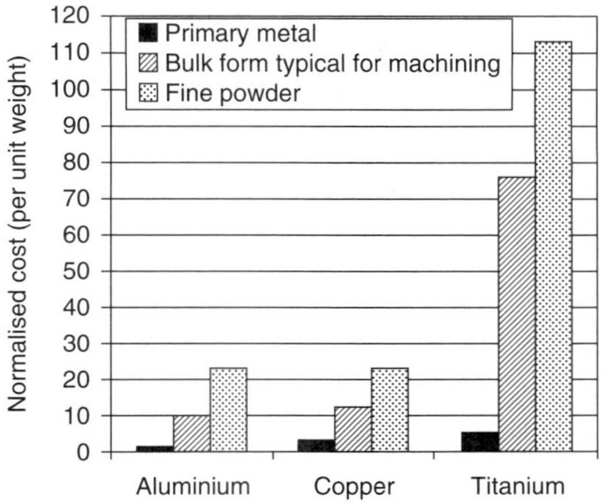

5.3 The comparative costs per unit mass of raw metals in the bulk and powder form (spherical, size classified) for aluminium, copper and titanium. Here the primary metal is that which has come direct from smelting, while the bulk metal includes the costs of casting/forming. In the case of titanium, non-size classified (<100 μm) sponge powder is substituted for the bulk form. The costs are normalised to bulk aluminium.

Data used in this chart were sourced as follows: bulk metal costs for aluminium and copper are based on a 6-month average between April and September 2006 from the London Metals Exchange. The cost of titanium is for bulk sponge in August 2005. The costs for bulk material forms for aluminium, copper and titanium are an average value for thick-section extruded and rolled products taken from a survey of suppliers. The costs of powders are from a range of suppliers. The power characteristics are spherical, inert-gas atomised and classified to be <50 μm.

The basic requirement for powder in thermal spraying is that it is flowable; this is so that it can be fed through a powder feeder. Generally, spherical powders are considered the most flowable, although irregular-shaped particles are still practicable.[18,19] Atomisation processes using air, water or inert gases are used to break up a molten metal stream to form fine particulates. Atomisation using air or water is used to produce relatively cheap powder consisting of irregular-shaped particles with high oxide contents. Using inert gases like nitrogen, argon and helium produces the most spherical metal particles with the least oxide content. However, since argon and helium are expensive, powders such as titanium and its alloys produced in this way are also expensive. Low oxide content powders are needed to achieve high deposition efficiencies and coating quality. It has been shown that copper powders with the lowest oxide contents produce the best coatings.[20] This is also likely to be the case for titanium powders, whose hardness increases and ductility decreases with increasing amounts of oxygen or nitrogen in solid solution.[21]

Another constraint on the process is particle size.[22] In order for the process to work efficiently, powders in the range of 5–50 μm are needed. Particles have poor flowability below 5 μm, making them difficult to feed. Furthermore, such fine powders have been reported to lose velocity downstream of the bow shock.[23] Above 50 μm, particles generally have too much inertia to be accelerated above the critical velocity. The powder size distribution for some thermal spray processes can be quite flexible, with average sizes ranging between 2 and 200 μm and with spreads of several tens of microns. However, there are reports suggesting that the best cold spray results are obtained with powders having spreads of less than 20 μm.[24] The tighter the range, the more expensive the material becomes, because of the lower production yield of material in the desired size range. Sizing powders below 50 μm is not possible with mechanical sieves as they clog; instead, these are classified with gas centrifuge systems. This additional processing increases the cost of the feedstock material.

Overall, the requirements for cold spray powders are more restrictive than for many other thermal spray powders, which leads to cost increases. The powder is generally finer with an average size below 50 μm, a size range of tens of microns and higher metallurgical quality in terms of oxide content (internal and surface). Spherical powders give better flowability and deform more uniformly upon impact;[18] however, irregular particles are purported to achieve greater velocities.[25] Therefore, powders produced by air or water atomisation may give lower deposition efficiency compared with inert-gas atomised powder due to their irregular-shaped, high oxide content particles.

Figure 5.4 shows the cost comparisons of several powders with different characteristics. It can be seen that alloying additions, narrower size ranges and atomisation type increase the cost of powder. Powder prices vary greatly from supplier to supplier and are material-dependent. Of course, with economies of scale the price per kilogram of the feedstock material will drop.

5.3.2 Gases

A key consideration for the choice of process gas is the quality of the resulting coatings. Here bond strength, coating density, and internal structural integrity and microstructure can all be influenced by gas type. Another consideration is productivity, which is determined by deposition rates and deposition efficiency. These parameters are material-dependent, with some gases giving superior performance compared with others. Cost and safety issues can also affect the choice of process gas.

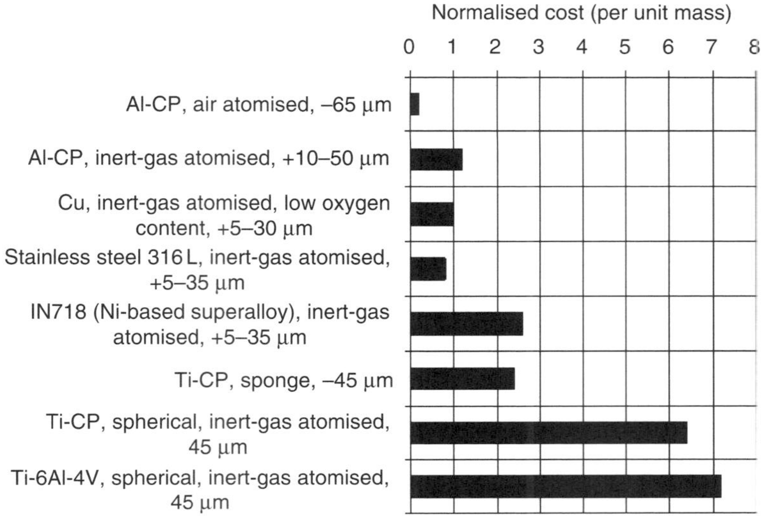

5.4 Normalised cost of different types of metallic powders used for thermal spraying and cold spray in particular. Comparisons can be made for powders of different size ranges, oxide contents, types of atomisation and alloy contents. Values are averages taken from a survey of powder suppliers. CP, commercially pure.

Choice of gas

The fundamental resource in cold spray is the process gas. This gas forms the supersonic jet stream that is used to accelerate the powder particles above the critical velocity for deposition and consolidation. Ideally, the gas needs to have a high sonic velocity and mass in order to transfer momentum to the powder. It is also important that the gas is inert, so that it is safe for fine metal powders – which are potentially explosive – and does not change the properties of the deposit. It is also preferable that the gas is cheap because of the high consumption rates involved.

A list of candidate process gases that fulfil some of these criteria is given in Table 5.1. While hydrogen has the highest sonic velocity due to its low molecular mass, its use is essentially prohibited due to its inherent risk of explosion, the prevention of which requires significant investment in intrinsically safe equipment and safety procedures. Hydrogen can spontaneously combust upon contact with air at high velocities, burning with an invisible flame. On top of this the unit cost is relatively high compared with nitrogen. Argon is inert and has a high mass, but its sonic velocity is low; it is also expensive.

The two main gases used in cold spray are helium and nitrogen, both of which are inert. Initially, cold spray was developed with helium, principally

Table 5.1 The physical properties and qualitative costs of candidate process gases for cold spray

Gas	Sonic velocity (m/s)	Density at s.t.p. (kg/m^3)	Inert or reactive	Cost
Air (80% N_2 + 20% O_2)	343	1.225	Reactive	Cheap
Nitrogen (N_2)	349	1.185	Partially inert	Cheap
Hydrogen (H_2)	1303	0.085	Reactive	Expensive
Argon (Ar)	319	1.69	Inert	Expensive
Helium (He)	989	0.168	Inert	Very expensive

s.t.p., Standard temperature and pressure; s.t.p. is 101 325 Pa and 15 °C

because of its high sonic velocity, which is approximately three times that of nitrogen; however, there is a significant cost penalty associated with its use. This penalty can be partially overcome by using recycling, which can make it economically feasible to use helium for certain materials. On the other hand, nitrogen is cost-effective for most materials. Unfortunately, when spraying materials that have high critical velocities, nitrogen cannot always produce high-quality coatings at high deposition efficiencies, even with high degrees of gas heating. There is also the issue that with ever-increasing levels of gas heating, nitrogen can no longer be considered inert, particularly with titanium and its alloys.

The sonic velocity of air is a little less than that of pure nitrogen, but it is the cheapest process gas available. However, air is a gas mixture containing inert nitrogen but reactive oxygen that is potentially explosive with metal powders. Consequently, there are deposit quality and safety issues that need consideration, especially when using gas heating and reactive powders. Nevertheless, the portable version of cold spray (i.e. SST) uses a compressed air supply.

The final consideration for gas selection is gas purity, which is also relevant to recycling and cost. Helium and nitrogen can be manufactured off-site and delivered at various purity levels. On-site 'point-of-use' nitrogen generators are available, which have capital costs and running costs that depend on the flow rates and purities required. However, the impurity content, in particular oxygen, may have serious adverse effects on the lifetime of equipment (gas heaters, nozzles) at elevated temperatures, as well as detrimental consequences to coating quality. The cost of spraying with helium is significantly reduced if greater impurity levels are allowed in the gas product after recycling. Therefore, it may be necessary to determine the maximum impurity levels that can be tolerated while cold spray remains successful, and to what degree this changes for different applications.

Helium versus nitrogen

Cold spraying with helium gives significantly better performance than with nitrogen, even without gas heating.[26] However, helium is inherently more expensive than nitrogen because it is a finite and scarce resource. Helium is extracted as a secondary product by fractional distillation from certain natural gas reservoirs in the United States, Algeria, Poland, Russia and Qatar.[27] The greatest supply of helium exists in the United States due to stockpiling that was sanctioned by the government in the 1950s.

The cost of helium can vary significantly around the world and has an inherent influence on the cost of cold spray. Key determinants of local helium pricing include its wholesale price, the size of the regional market and distribution costs. Prices in the past have been volatile, but since the US government entered the market in 2001 to liquidate its reserves, wholesale pricing has become much more stable. The stability of helium prices is expected to continue with new sources balancing increasing demand. While the United States is the main consumer of helium, its stockpiles are sufficiently great that it remains the dominant world supplier with over two-thirds of world supply, even with new sources in Algeria and Qatar coming online. As such, helium pricing has typically been lower in the United States than in the rest of the world.

In contrast to helium, nitrogen is abundant, making up 78% of the Earth's atmosphere, and it is relatively easy and cheap to extract. The main cost in the manufacture of nitrogen is electrical power and the capital equipment needed to extract it. In some instances, nitrogen is the secondary product from oxygen generation used for other industries such as iron and steel production. Hence, there is a loose correlation for the cost of nitrogen with local energy costs and available volumes. In rapidly developing economies, nitrogen demand can exceed supply, pushing market prices up. While helium can only be supplied from industrial gas companies, nitrogen has the added flexibility that it can be delivered in gaseous or liquid form; or alternatively can be generated on-site.

End-user gas pricing is again not uniform around the world, as costs depend on market availability, its distribution and storage on-site. The volume consumed on-site is a major factor, as this will affect storage method and delivery frequency, which will have a significant effect on overall cost. For example, one volume of liquid nitrogen gives 651 volumes of gaseous nitrogen under normal conditions. These factors – i.e. gas consumption, gas supply and storage options – are the next subjects of discussion.

Gas consumption

Per unit volume, helium will always be more expensive than nitrogen. However, the unit cost of each process gas is dependent on the type of supply, which in turn should be dependent on consumption. In cold spray, consumption can vary widely depending on the hours of operation and the system parameters. The usage rates may vary from only a few hours per week to more than a hundred for continuous operation.

Gas consumption rate is dependent on the process variables of gas type, operating pressure and temperature, as well as nozzle diameter. This is shown in Fig. 5.5 for different cold spray system configurations. Increasing the operating pressure and nozzle throat diameter increases consumption, while increasing the gas temperature decreases consumption, this is an additional benefit on top of improved coating quality.

For a given system configuration, the flow rate of helium is 4.2 times greater than that of nitrogen, due to the large difference in sonic velocity. In practice, however, the system and process parameters are usually different for helium and nitrogen. Helium systems generally use lower pressures (15–20 bar) and smaller nozzles (2 mm). Gas heating is not always necessary when using helium, but it can still significantly enhance process performance when spraying materials that are difficult to deposit. Nitrogen systems require higher gas pressures (30–40 bar), larger nozzle throats (2–3 mm) and gas heating typically in the range 300–500 °C, but increasing up to 800 °C. When using typical pressures and nozzle diameters for both gases, while employing the same degree of gas heating, the flow rates equalise at around 90 m^3/hour. In general, gas flow rates can vary from 50 to 200 m^3/hour.

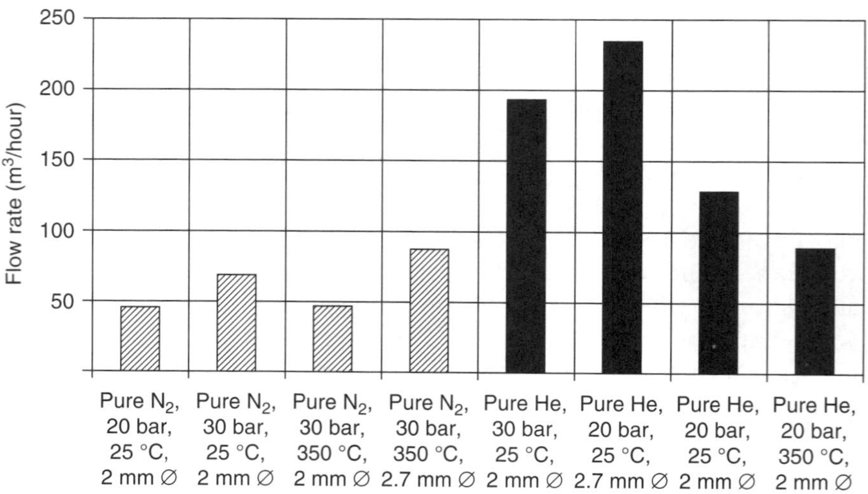

5.5 Gas consumption rates for different cold spray system configurations.

Gas supply/storage systems

In cold spray, the weekly, monthly or annual gas consumption dictates the choice of gas supply, including its storage and delivery. This gas supply needs to be able to deliver flow rates of up to 120 m^3/hour and pressures in excess of 30 bar. The capacity may be relatively low for research and development, medium for batch processing in job shops and high for continuous processing in certain industrial applications. Single cylinders are impractical, as typical industrial cylinders only have a storage capacity of around 10 m^3, which provides only a few minutes of spray time.

Nitrogen can be supplied as a compressed gas in vessels of different sizes, but its main advantage is that it can be stored cheaply and efficiently in liquid form; this is shown in Fig. 5.6, although not all methods are available in all countries. At standard temperature and pressure, the gas to liquid volume ratio for nitrogen is 651 : 1. Thus, there is a substantial cost advantage when using liquid nitrogen storage for bulk gas supplies. This is illustrated in the comparison of unit volume gas costs shown in Fig. 5.7. Note that gas costs vary depending on geographic location, market availability and consumption as previously discussed. Liquid nitrogen needs to be converted to gas at the relevant flow rates and correct pressure, which requires equipment for external vaporisation and compression. These are commercially available but add to the footprint of the cold spray process. In spite of the increased capital costs associated with liquid nitrogen storage, its use

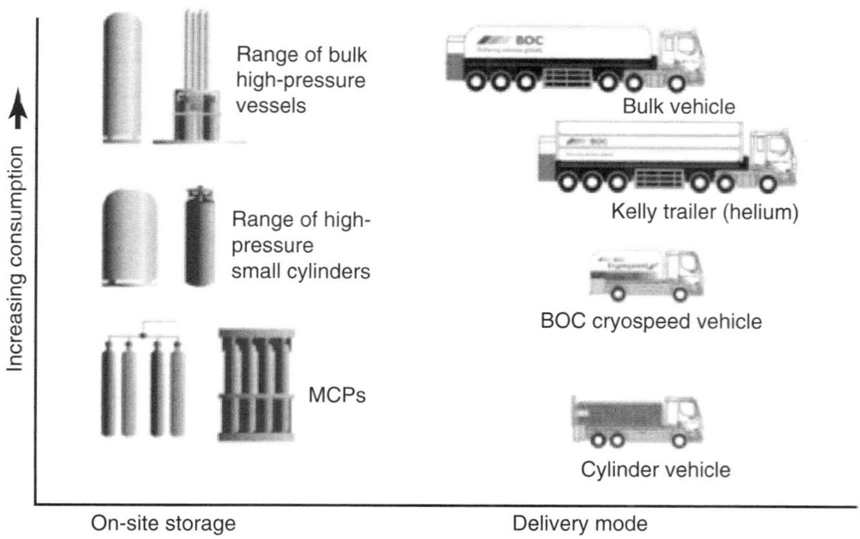

5.6 Gas supply options for nitrogen and helium showing the range of capacities. MCP, manifold cylinder pallet.

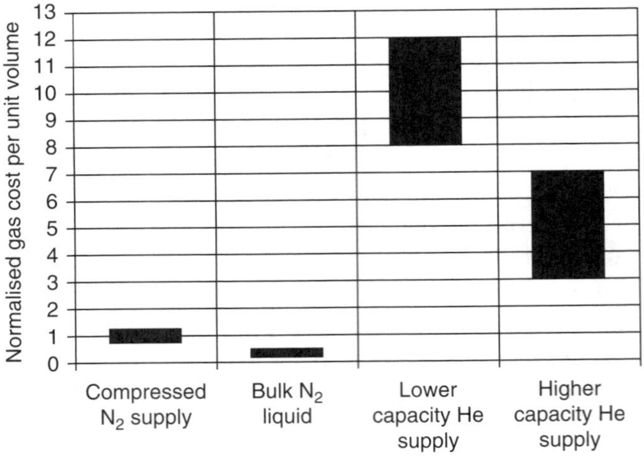

5.7 Comparisons of unit gas costs for nitrogen and helium with different supply options.

becomes economically beneficial when gas consumption exceeds 500 m^3 per week.

Helium is always stored as a gas as it is not economically practicable to store it in liquid form. The compressed gas is stored in vessels of different sizes, with capacities ranging from 200 m^3 for clusters of cylinders to 4000 m^3 for tube trailers. Examples of the various storage equipment options are presented in Fig. 5.6. Manifolded cylinder pallets (MCPs) supply gas at high pressures but the flow rates and capacity are generally low and limited. Trailers deliver both high pressure and high capacity. In both cases, waste occurs due to the residual gas that is below the pressure needed for cold spray equipment. The storage equipment is usually rented or leased, with costs being insignificant compared with the total cost of the gas.

There is a cost saving through economies of scale, but the compressed helium cost per unit volume is still much greater than that for compressed nitrogen gas, as can be seen in Fig. 5.7. When both compressed gases are in the bulk supply mode, the cost of helium is about 5 times that of nitrogen in unit volume terms. However, the relative cost per unit volume of helium becomes 10 times greater when compared with liquid nitrogen storage. The delivery systems needed to transport the gas from storage to the cold spray equipment have to be designed for the appropriate flow rates and pressures; these systems include pipelines, regulators and safety devices.

It should now be apparent that providing direct comparisons between the costs of using different gases is not straightforward. Around the world, gas costs are dependent on external factors including market conditions, global pricing, the overall availability of the product within a region and the distance from key sources. These are largely beyond the control of gas users. In addition, the gas costs for the end-user depend on consumption. However, it is possible for a gas user to minimise costs by optimising their supply system. For a given consumption, total costs will vary significantly depending on the choice of compressed versus liquid supply, storage vessel capacity and delivery frequency. There is also a relatively large fixed cost element for high-pressure supply systems. The costs of storage vessel maintenance, vessel rental, system and pipework installation, as well as management accounting practices, will have a considerable impact on the unit cost of gas at the point-of-use. Therefore, the relative costs illustrated can only ever be indicative and it is always worthwhile optimising supply systems with gas suppliers or experts.

5.3.3 Helium recycling

As we have shown, due to the elevated cost per unit volume and higher flow rates, the cost of spraying with helium is significantly greater than spraying with nitrogen. In fact, helium is such a high-cost gas that gas recovery and recycling become imperative. A helium-recycling system can reduce the gas costs to such an extent that it becomes economically feasible, and even competitive with nitrogen, when high-value materials that are difficult to deposit are sprayed. Helium recycling has been successfully implemented in a number of industries. The type of helium-recycling apparatus most suited for cold spray processing is based on pressure swing adsorption (PSA) technology that allows operation independent of gas input purity levels.[28] This is in contrast to conventional membrane-based systems, which rely on fixed operating conditions.

The capital or lease cost of recycling equipment needs to be weighed against the expense of helium gas losses without it. This depends on the gas consumption over different timescales; however, the size of the recycling equipment can be optimised for this. As shown in Fig. 5.8, recycling efficiencies rapidly change the economics of using helium in cold spray based on processing costs, not gas costs. With a little over 80% recycling efficiency, the unit volume costs for helium become the same as for nitrogen in bulk compressed gas supply, although liquid nitrogen supplies will always be considerably more economical. The BOC helium-recycling system installed in the cold spray facility at the University of Cambridge, achieves 80–85% overall efficiency.[29] The overall recycling efficiency includes the percentage of gas recovered from the cold spray facility and the operational efficiency of the recycler itself. In order to recover the helium without being overwhelmed by atmospheric impurities, spraying needs to be performed within a chamber, which may be a limiting factor for product size and process capability.

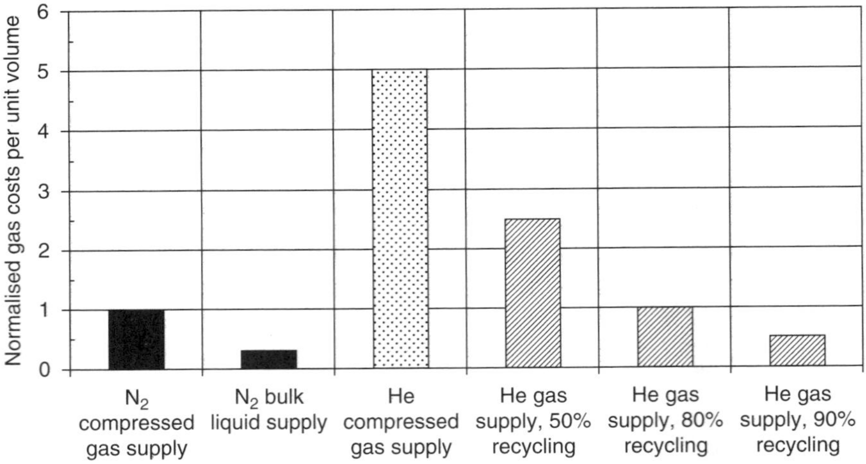

5.8 Relative unit volume (gas) costs of nitrogen and helium with and without recycling. Note, these costs do not include capital recovery.

5.3.4 Deposition efficiencies

When spraying with helium, the greater deposition efficiencies experienced generate savings in powder cost. These savings are dependent on the value of the powder feedstock and need to be weighed up against increased gas costs. While recycling reduces the gas costs significantly, it still cannot be justified for relatively cheap powders. The chart shown in Fig. 5.9 represents the gas and powder costs associated with depositing 1 kg of copper and titanium powder, using helium – with and without recycling – and nitrogen. Here, a cost ratio of 30:1 was applied to helium and nitrogen, whilst a ratio of 3:1 was applied to the cost of titanium and copper. When using high-cost, high critical velocity powders such as titanium, the increase in deposition efficiency can make the use of helium recycling cost-effective.

5.3.5 Other consumables: electricity and nozzles

After gas and powder, the next most significant consumable is electricity (when using gas heating). Without gas heating, the electrical power consumption of a powder feeder and actuator is negligible; although electrical consumption is becoming more significant with the development of higher power commercial gas heaters that can be rated up to 47 kW – such as for the heating units in the CGT 4000 Kinetic system.[8] However, gas heating introduces significant cost reductions through improved deposition efficien-

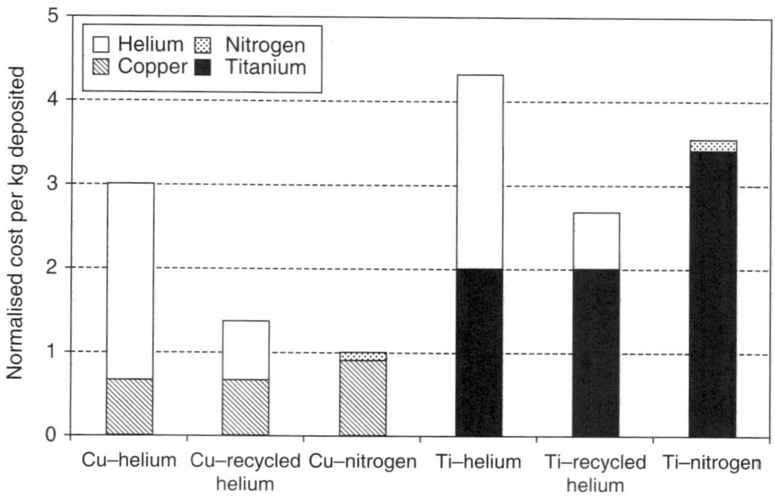

5.9 Chart showing the cost of deposition in terms of gas and powder used. Data apply to a kilogram of copper and titanium deposited with helium (with and without recycling) and nitrogen.

cies and lower gas consumption, and extends the applicability and performance of nitrogen gas.

The only other significant consumable cost after gas, powder and electricity is the spray nozzle. Powder flow through the nozzle inevitably leads to wear, which in turn distorts the internal profile leading to poor gas flow conditions and non-optimised performance. There are three main types of nozzle: single-piece sintered carbide nozzles, single-piece metallic nozzles and two-piece metallic nozzles. The sintered carbide nozzles are hard and have low wear rates and hence long lifetimes; however, they are expensive to produce. Alternatively, nozzles may be made from a softer metallic material and these are changed on a more regular basis. In some cases, plastic nozzles have been used to prevent fouling.

5.4 Comparison with other thermal spray techniques

Thermal spraying is the name given to a group of processes – that is said to include cold spray – where materials are deposited in a molten or semi-molten state.[19] The variety of processes can deposit metallic and non-metallic materials using feedstock in the form of powder, wire or rod (ceramics). The coatings produced are used in a wide range of applications and industries from aerospace and power generation (for turbine blade coatings and consumable turbine tips), to railway lines for wear and corrosion resistance, and to paper manufacture and electronic goods manufacture. This extensive application range is due to the particular properties of the coatings and their relatively low manufacturing costs.

In this brief economic comparison of cold spray and other thermal spray methods, however, it should be noted that the various techniques are often complementary rather than substitutes for one another. This could be in terms of the types of materials used, the spray characteristics or the properties of the resultant coatings. Several of the processes that involve higher temperatures, including flame spray and plasma spray, are more applicable to high-melt-temperature materials such as ceramics and refractory metals. Some processes are limited by the type of material they can deposit; for instance, arc spraying requires the raw material to be in the form of a conductive wire. Both flame and arc spraying techniques are conducive to manual spraying with large area coverage, whereas vacuum/low-pressure plasma systems require automated spraying in acoustic chambers. However, low-cost flame spray and arc spray systems produce coatings with high oxide content (5–15%) and porosity levels (10–20%) and tend to be used for low-value applications.

5.4.1 Capital costs

As with cold spray, the cost of thermal spray systems depends on their complexity, level of control and gas source. The capital equipment costs of various thermal spray techniques are shown in Fig. 5.10. Many of the thermal spray methods require relatively little capital equipment, in particular flame and arc spray. The main contributors to capital costs in these two processes come from the gun and the feedstock apparatus, whether it is a wire feeder or a low-pressure powder feeder. In arc spraying there is the additional cost of the electrical power supply and transformer. Flame and arc spray are typically used for low-cost applications, where the quality of the deposit will not compromise the performance of the component or surface coating. High-velocity oxy-fuel (HVOF) spraying straddles the cheaper and more expensive thermal spray processes. Its capital costs are associated with a more sophisticated gun that operates with gases and combustion reactions at higher pressures.

In Fig. 5.10, plasma spraying is shown to have the greatest cost range because it includes relatively inexpensive systems, which are operated in atmospheric conditions with a simple gun, and the more expensive systems that are operated in large, sealed chambers for containing protective atmospheres or partial vacuums. Plasma spraying also involves more sophisticated guns that are needed to generate the plasma by electrical arc or radio frequency, while the electrical power units also add to the capital cost.

The cold spray capital costs represented in Fig. 5.10, range from the simplest systems, built in-house, to a full turnkey system with the maximum gas heating capacity. As discussed in Section 5.2.1, at the minimum expenditure

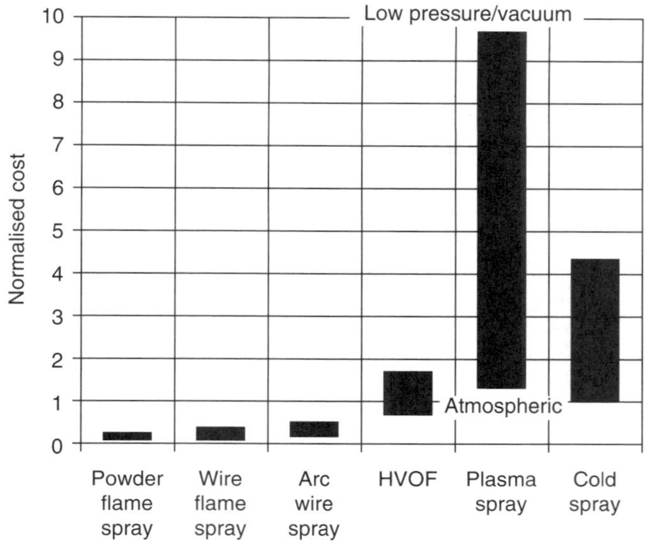

5.10 The relative capital costs of various thermal spray techniques. Note that HVOF is high-velocity oxy-fuel. The costs in the graph are based on those when for spraying copper. The calculations are based on proprietary feedstock and gas cost information.

level the majority of costs come from the powder feeder and gas heater, which both must be rated for higher pressures than other thermal spray processes. Nevertheless, the capital equipment costs of cold spray compare favourably with those of low-pressure or vacuum plasma spray and thus have the potential to displace these processes in the manufacture of high-quality (i.e. low oxide and porosity) metallic coatings.

5.4.2 Deposition rate and efficiency

Before going on to discuss consumable costs we need to consider and distinguish between the thermal spray processes in terms of their different deposition rates and efficiencies (see Fig. 5.11). Greater deposition efficiencies reduce feedstock waste and hence the cost of raw materials. The greater the deposition rates and efficiencies, the shorter the spray time, which leads to lower costs in terms of labour, energy and gases.

Arc spray is recognized to have by far the greatest deposition rates, followed by wire flame spray. Using wire feedstock not only gives a cost advantage in material terms, but also in deposition rate. However, the quality of coatings and deposits are generally inferior to those made by HVOF, plasma spray and cold spray; the higher deposition efficiencies reflect this. Cold spray can deliver very high deposition efficiencies and this

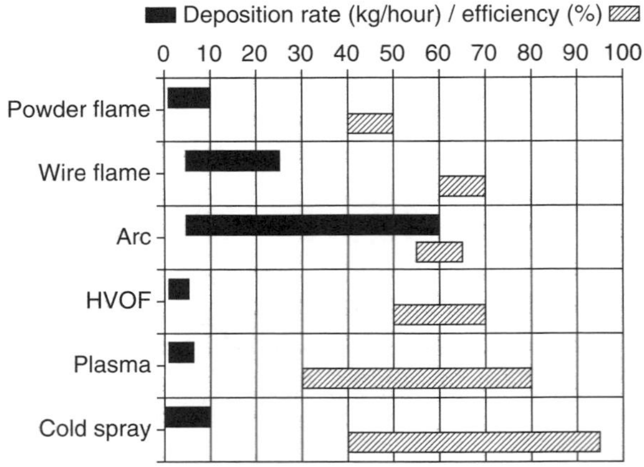

5.11 Deposition rates and efficiencies of different thermal spraying techniques.

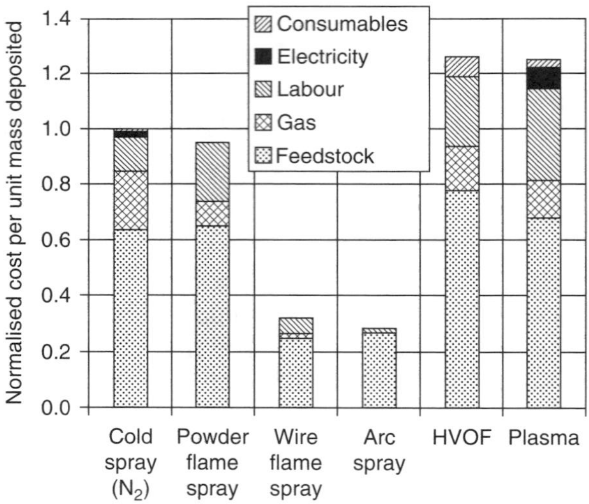

5.12 A normalised comparison of running costs for various thermal spray techniques. The calculations are based on the average values of deposition rates and efficiencies for copper.

has a significant impact on the economics of the process, especially with high-value materials such as tantalum and titanium.

Figure 5.12 shows the contributions to running costs for different thermal spraying techniques. Note that this is in terms of per unit mass deposited and takes into account different deposition rates and efficiencies. As mentioned above, those processes with high deposition rates or high deposition efficiencies or both, produce cost savings in labour, energy consumption and feedstock waste for every unit mass deposited. If the comparison was performed on a per unit time basis, then processes like arc spray become the most expensive because of the high consumption of feedstock. As can be seen in the figure, this is the dominant component of the running costs. This comparison does not take into account the capital equipment costs and other associated factors.

5.4.3 Feedstock costs

In general, wire feedstock is considerably cheaper than powder feedstock for a given material. While most thermal spray techniques use fine powder, some can use wire – such as flame spray and arc spray. Pure metals, alloys and some composites are inherently cheaper in wire form than in powdered form, especially if specific and narrow cuts of particulate sizes are required. In some cases, reel or wire can be 5 to 10 times cheaper per unit mass. Consequently, flame and arc spray have a significant cost advantage over the other thermal spray techniques (see Fig. 5.12).

The size range of powders used in flame spray is typically 5–100 μm, which is large compared with other thermal spray techniques. HVOF is similar to cold spray with a size range of 5–60 μm and vacuum plasma spray is typically narrower at about 5–20 μm. For those techniques that use powders, flowability is the most important requirement. Flowability arises from spherical-shaped powders with narrow size ranges of 20–30 μm, e.g. –25 to +5 μm. Again similar to HVOF, cold spray requires a narrow size range for optimal process performance, in terms of deposition efficiencies and coating quality. To realise cold spray's full potential, in particular its high deposition efficiencies, the metal powder feedstock itself needs to be of high quality, especially with respect to oxide content. This is not necessarily the case with other thermal spray techniques because of the high temperatures leading to material oxidation regardless of their initial quality. Process developments in cold spray are directed towards broadening the powder size range so that the powder feedstock costs are brought down to be comparable with those of HVOF or even less.[30]

5.4.4 Gas and energy costs

Gas costs are significant for all thermal spray techniques. Generally, the cost of gases for the processes that give high-quality coatings (cold spray, HVOF and plasma spray) are similar for equivalent amounts of material deposited (see Fig. 5.12). This is in spite of the disparity in consumption rates and types of gas used. Flame spray and HVOF use oxygen–fuel mixtures that are combusted to generate the energy to heat the material feedstock. In addition, the high-pressure combustion products propel the particulates onto a substrate. The consumption of gas mixtures in flame spray varies greatly for different guns, but generally ranges from 0.1 to 5 m^3/hour with oxygen being the major component. In HVOF, the combustion gas mixtures include the more expensive fuels of hydrogen and propane, where the cost is further compounded by the higher consumption rates of oxygen (up to 50 m^3/hour). Arc spray, flame spray and HVOF can also use significant quantities of propellant or shielding gases, but these are usually the much cheaper gases of air or nitrogen. Plasma spraying requires plasma-forming gases consisting of pure argon or mixtures with hydrogen, helium or nitrogen. The consumption of such gases ranges from 1 to 5 m^3/hour, but they are considerably more expensive than the pure nitrogen used in cold spray.

The energy input in thermal spray techniques is used to melt, or at least soften, the feedstock material, and often to produce the gas stream to accelerate the particulates. The most electrical energy intensive processes are arc and plasma spray, and now cold spray. Arc spray consumes 5–10 kW of electrical power while plasma spray systems typically operate at 30–60 kW (some up to 100 kW). As can be seen in Fig. 5.12, the electrical energy costs are only significant for the moderate-to-high power consumption and moderate-to-low deposition rate processes of cold spray and plasma spray. However, the energy costs are small compared with feedstock and gas costs.

5.4.5 General comparison with thermal spraying

For the manufacture of high-quality coatings of metals, alloys and composites, cold spraying with nitrogen is a competitive alternative to HVOF and plasma spraying. Cold spraying with helium may still be competitive but only for materials that are high value, oxygen sensitive and difficult to deposit (e.g. titanium) because of the higher deposition efficiencies and coating quality it provides.

5.5 Comparison with other additive fabrication techniques

Additive fabrication techniques are commonly used to produce freeform components on a layer-by-layer basis. Each layer is created by the selective addition of material corresponding to a cross-sectional slice of the part to be built. Components are generated directly from computer aided design (CAD) data without any part-specific tooling or significant human intervention.[31] This allows for multiple design iterations to be made without incurring the time and cost penalties usually associated with the manufacture of high-volume tooling. By implementing this type of additive fabrication technology – as opposed to the traditional subtractive and formative fabrication techniques, such as milling and forging – it becomes possible to manufacture functional engineering components in a single step, where the time and cost of manufacture is independent of component complexity.[32]

Additive fabrication techniques dedicated to the production of metallic components can be broadly grouped by their material consolidation process. Whereas selective laser sintering (SLS) employs a sintering process to selectively consolidate the build material, processes such as direct metal deposition (DMD) and laser engineered net shaping (LENS) rely on melting and re-solidification. While many of these techniques are being employed commercially, they can all suffer the detrimental effects of high-temperature processing such as large residual stresses, poor mechanical properties, unwanted phase transformations and component distortion.[33] In contrast, the non-melting nature of cold spray's deposition mechanism ensures that the sprayed material is free from thermally induced tensile stresses while the underlying substrate remains in solid state.

Because cold spray operates with little or no heat, the process has frequently been used to deposit oxygen-sensitive materials such as titanium[2] and tantalum,[3] as well as temperature-sensitive materials such as nanocrystalline[34] or amorphous[35] materials. However, it is the process's ability to rapidly develop thick coatings and free-standing shapes that distinguishes it from its thermally based counterparts (see Fig. 5.13).

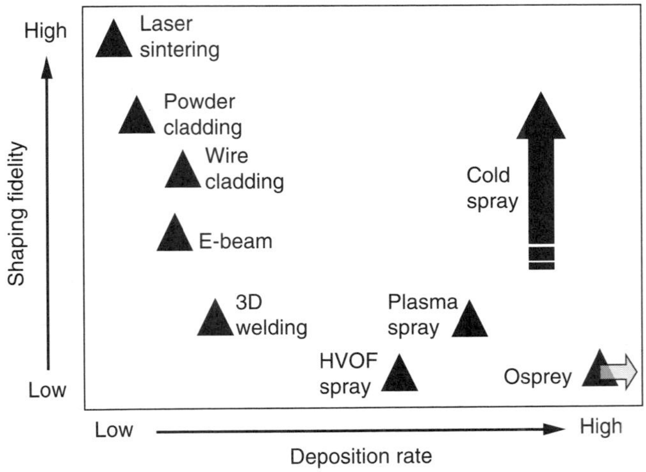

5.13 Comparison of cold spray with other techniques used for additive fabrication.

Table 5.2 Comparison of cold spray with some other techniques that have been applied to additive fabrication

Process	Type	Accuracy	Layer thickness	Deposition/ build rate	Deposition efficiency	Deposition materials
SLS[33]	Sintering	±50 μm	<75 μm	<2.5 cm³/min	N/A	Steels, numerous
LENS[34]	Cladding	±100 μm	<0.4 mm	<8 cm³/min	<14%	SS, alloys, numerous
DMD[35]	Cladding	High	<0.254 mm	<4.1 cm³/min	N/A	H13, Al, numerous
HVOF[12]	Thermal spraying	Low	<1.5 mm	20–80 g/min	<70%	Carbides, numerous
PS[12]	Thermal spraying	Low	<0.5 mm	50–100 g/min	<60%	Ceramics, numerous
CS	Thermal spraying	Medium	0.05–10 mm	10–120 g/min	<100%	Al, Cu, Ti, numerous

PS, plasma spray; CS, cold spray; SS, stainless steel

Unfortunately, high deposition rates generally result in low shaping fidelity; although for cold spray, this may be increased through the application of aerodynamic focusing[36] and/or novel build strategies[29] – this is shown by the elongated arrow in Fig. 5.13. When combined with a computer numerical control (CNC) positioning system, cold spray becomes a viable method for additive fabrication. Table 5.2 compares cold spray with other techniques that have been applied to additive fabrication, in terms of some common process characteristics.

These favourable comparisons have led to interest in the application of cold spray for the production of bulk forms and near-net shapes.[37] Large sizes and shapes have been fabricated and geometrical features incorporated during spray preparation and machine finished. Moreover, by controlling the feedstock composition, one can vary the deposit composition and microstructure to produce functionally graded materials (FGMs) and other unique structures.[29]

When considering cold spray for the production of these bulk forms and near-net shapes, it is important to consider three factors that influence its cost-effectiveness: (a) part size and shape, (b) production volume and (c) part quality.[38]

(a) Part size is important since costs rise in proportion to part volume; the larger the part the greater the cost of consumables. Although greater than either plasma or HVOF spraying, the current shaping fidelity of the cold spray process is limited by the spray 'footprint' or 'spot size' of the particle stream. A typical spot size would be of the order of

5–10 mm – roughly equivalent to the nozzle exit diameter but dependent on standoff distance. Thus without back machining, spray resolution and part accuracy is low. This situation may be improved by scaling down the size of the nozzle; however, this would come at the expense of reduced throughput and hence build rates.

(b) In terms of large production volumes, cold spray – and most other additive fabrication techniques for that matter – will never be able to compete with die-casting, which offsets its high tooling costs with large production runs. Similarly, for smaller production volumes, CNC machining is often more suitable; however, for one-off jobs, or very low production volumes – where initial setup can add cost – cold spray can compete.

(c) Although its deposition rates and efficiencies are high, parts produced by cold spray are often of inferior quality to those produced by conventional manufacturing techniques – in terms of dimensional accuracy and surface finish. Thus, when part quality is essential, cold spray may be unsuitable; however, it is likely to be more successful when the design requirements are less stringent.

Therefore, at present, cold spray seems best suited to the production of bulk forms and near-net shapes that are difficult to construct by any other means. An example of this would be the direct fabrication of tantalum targets for thin-film sputtering processes. Such targets are difficult to manufacture by conventional methods due to tantalum's extremely high melting point (3017 °C); however, cold spray can deposit this material in a fast and efficient manner, without melting.

Clearly, the process has its limitations and will not be able to compete with thermally fused components for certain applications. However, mould tools suffer considerably when repaired by thermal methods due to the detrimental microstructural changes that occur. In addition, recent results show material exhibiting similar strengths to ductile failure in the wrought material (after annealing)[39,40] and make the process potentially useful under tensile as well as compressive loading.

5.5.1 Cost comparisons

In a similar way to Section 5.4, cost comparisons can be made between cold spray and other additive fabrication techniques based on capital costs, consumable costs, energy costs and deposition characteristics. Whereas the costs associated with the thermal spray processes are similar to those of cold spray, the costs related to many of the additive fabrication techniques are not. Furthermore, because additive fabrication is a relatively new field of manufacturing, cost data is that much more difficult to come by.

In general, the capital costs for additive fabrication techniques are closely linked to the energy source used to consolidate the build material, which differs from technique to technique. Whereas an electric arc may be used to melt metal in the most economical way, as in three-dimensional welding,[41] a laser beam can generate high energy densities over a small region, thus providing greater part accuracy and resolution, but at greater cost.[32] When considering a basic cold spray system, the capital costs are dominated by the costs of the powder feeder and gas heater, which have a total cost equivalent to that of a basic laser source. However, many of the additive fabrication techniques require additional systems that substantially increase the overall capital cost (thermal management, powder/wire delivery, beam delivery systems, etc.). Thus the overall cost of an additive fabrication system can far exceed that of a cold spray system.

In terms of consumable costs, many of the additive fabrication techniques use a gas to aid the process in some way; either for feedstock delivery, shrouding or to provide an inert build environment. Although the cost of some of these gases may be higher than the typical gas used in cold spray, the volume of gas consumed is an order of magnitude lower. Hence, whereas the gas costs are disparate, the feedstock costs are not. For additive fabrication processes that employ a powder as the build material, the size distributions used are similar to those used in cold spray and as such the costs are comparable. However, for those processes that use a wire or foil as the feedstock material, the costs are substantially reduced.

While the consumable costs in cold spray are greater than in many additive fabrication techniques, the same cannot be said of the energy costs. The main contributing factor to energy costs in cold spray is the power requirements for the gas heater; a typical system is rated at 25 kW. Although these devices are not 100% efficient, the power rating is that which is taken from the outlet. In contrast, a 3 kW CO_2 laser may have a wall-plug efficiency of less than 5%, meaning the overall power consumption is closer to 60 kW. This is not to say that every laser-based additive fabrication system consumes this much power, but in general, they consume more than a typical cold spray system.

The deposition rates and efficiency of the cold spray process help to reduce the consumable costs and energy costs by minimising the amount of material and time needed to build a certain part. Table 5.2 compares these figures for some processes that have been applied to additive fabrication. Furthermore, by minimising the length of time needed to build a part, the costs associated with operating times and man hours can be reduced.

In an attempt to assess the cost benefits of the process, Karthikeyan[5] used value-stream analysis to compare cold spraying with casting. It was found that cold spray reduced material input, rework and finishing, and increased overall material utilisation. In his model, if raw material input was reduced by 50% and rework and finishing reduced by 75%, the total value-stream was reduced by 70%. (Note that mould and melt-pour costs were eliminated, while the cost of fabricating the part using cold spray was added retrospectively.) By applying these ideas, it was found that cold spray could be successfully commercialised for the production of coatings and innovative freeforms, as well as for repair/refurbishment, in sectors ranging from defence, aerospace and power generation, to electronic/electrical, biotechnology and manufacturing.

As an example of this, cold spray could be used for the production of high-temperature, oxidation-resistant MCrAlY coatings that are currently produced using vacuum plasma spray or physical vapour deposition. Because of the vacuum systems they employ, these processes are extremely expensive to install and operate. In contrast, cold spray can produce such coatings with comparable microstructure and properties, but in atmospheric conditions.[42] Furthermore, using cold spray it is possible to collect the overspray and reprocess the expensive raw materials.[29]

5.6 Conclusions and future trends

Cold spray is a competitive process for producing high-quality metallic coatings – whether in the form of pure metals, alloys or composites – with low oxide content and porosity. The process will continue to find particular application with oxygen-sensitive materials such as tantalum and titanium because of its non-thermal deposition mechanism. Furthermore, it is a cost-effective alternative to other thermal spray techniques that provide deposits of inferior quality. Since the process is a relatively low-energy means of depositing high melting point, expensive materials – with little or no waste – cold spray may find key markets not only in coatings but also in additive fabrication.[5,29] It will also play an important role in replacing coating technologies that use toxic chemical processes such as chrome plating, which will become restricted as new environmental legislation comes into force.[43]

The process is open to further technical development that will change its economics, especially regarding capital equipment costs. There is scope for more efficient gas heaters, low-pressure powder feeders and gas recycling. Hybridisation of cold spray may also introduce new and improved process capabilities and application areas. Coating applications will expand as the technical and economic benefits become recognised and established. This means that the economics of cold spray will also evolve with respect to capital equipment and raw materials.

Gas and powder heating will provide improvements in coating quality, deposition efficiency and gas consumption that will further improve the economics of the process.[44] This will make cold spray more competitive with HVOF and plasma spraying in the high-quality coating market.

While cold spray was conceived using helium, nitrogen is now the predominant process gas with future developments in its supply coming from either liquid storage or on-site generators. Performance improvements in nitrogen-based systems will allow the deposition of difficult-to-deposit materials that were previously the domain of helium process gases.[45] However, some materials or applications will always require the highest performance, thus ensuring helium remains an integral part of cold spray's capabilities. Moreover, because of the increasingly high temperatures employed, spraying with nitrogen can lead to embrittlement due to the changes in chemical composition that occur. Helium recycling then plays a crucial role in the economics of the process. Establishing the performance implications of various impurity levels for helium and nitrogen is another area that will improve the economics further still.

Powders are another area of future development as knowledge increases on how exactly the particulate characteristics such as size and shape affect process performance. Tolerance to broader powder size ranges for instance will decrease the costs of the feedstock material.[45,46] At the same time, powder manufacturers are adapting their processes to be more efficient at producing the required powders for cold spray.

5.7 Sources of further information and advice

General books on thermal spraying, which discuss all the techniques aside from cold spray techniques, are:

- *The Science and Engineering of Thermal Spray Coatings*, L. Pawlowski, Wiley, Chichester, UK, 1995.
- *Handbook of Thermal Spray Technology*, edited by R. Davis, ASM International, Materials Park, Ohio, USA, 2004.

A good description of performing an economic analysis of a new manufacturing process, using lasers in this case, is given in Chapter 10: Economics, by P. J. Oakley, of the *Handbook of the EuroLaser Academy*, volume 2, edited by D. Schuocher, Chapman and Hall, London, 1998.

An overview of additive fabrication and rapid manufacturing processes is given in *Wohlers Report 2006*, by T. Wohlers, published by Wohlers Associates Inc., 2006.

5.8 References

1 Alekseev Yu A, Mogorychnyi V I, Volkov V T, Krysa V K and Mukhametzyanov A G, Use of Gas Dynamic Spraying to Fin Tubes of Heat Exchangers, *Chemical and Petroleum Engineering*, **32** (4), 1996, 393–396.

2 Lima R S, Kucuk A, Berndt C, Karthikeyan J, Kay C M and Lindemann J, Deposition Efficiency, Mechanical Properties and Coating Roughness in Cold-Sprayed Titanium, *Journal of Materials Science Letters*, **21** (21), 2002, 1687–1689.

3 Steenkiste T H and Gorkiewicz D W, Analysis of Tantalum Coatings Produced by the Kinetic Spray Process, *Journal of Thermal Spray Technology*, **13** (2), 2004, 265–273.

4 Blose R E, Walker B H, Walker R M and Froes S H, Depositing Titanium Alloy Features to Forgings and Extrusions using the Cold Spray Process, Proceedings of the 2006 International Thermal Spray Conference, May 15–19, 2006, Seattle, Washington, USA.

5 Karthikeyan J, Cold Spray Technology, *Advanced Materials and Processes*, **163** (3), 2005, 33–35.

6 Supersonic Spray Technologies Auto Body Repair System, CenterLine (Windsor) Ltd, Windsor, Canada. www.supersonicspray.com.

7 Tevtech LLC, Wilmington, Massachusetts, USA. www.tevtechllc.com.

8 Cold Gas Technology GmbH, Ampfing, Germany. www.cgt-gmbh.com.

9 Richter P and Hoell H, Latest Technology for Commercially Available Cold Spray Systems, Proceedings of the 2006 International Thermal Spray Conference, May 15–19, 2006, Seattle, Washington, USA.

10 Novoselova T, Fox P, Morgan R and O'neill W, Experimental Study of Titanium/Aluminium Deposits Produced by Cold Gas Dynamic Spray, *Surface and Coatings Technology*, **200** (8), 2006, 2775–2783.

11 Lee H Y, Jung S H, Lee S Y, You Y H and Ko K H, Correlation Between Al_2O_3 Particles and Interface of Al–Al_2O_3 Coatings by Cold Spray, *Applied Surface Science*, **252** (5), 2005, 1891–1898.

12 Kang H K and Kang S B, Tungsten/Copper Composite Deposits Produced by Cold Spray, *Scripta Materialia*, **49** (12), 2003, 1169–1174.

13 Lee H Y, Jung S H, Lee S Y and Ko K H, Alloying of Cold-Sprayed Al–Ni Composite Coatings by Post-Annealing, *Applied Surface Science*, **253** (7), 2007, 3496–3502.

14 Haynes J, Pandey A, Karthikeyan J and Kay A, Cold Spray Discontinuously Reinforced Aluminium (DRA), Proceedings of the 2006 International Thermal Spray Conference, May 15–19, 2006, Seattle, Washington, USA.

15 Srivatsan T S (ed.), *Composites and Functionally Graded Materials*, American Society of Mechanical Engineers, New Youk, 1997.

16 Zhao Z B, Gillispie B A and Smith J R, Coating Deposition by the Kinetic Spray Process, *Surface and Coatings Technology*, **200** (16–17), 2006, 4746–4754.

17 Moxson V S and Froes F H, Fabricating Sports Equipment Components via Powder Metallurgy, *Journal of Materials*, **53** (4), 2001, 39–41.

18 Pawlowski L, *The Science and Engineering of Thermal Spray Coatings*, Wiley, Chichester, UK, 1995.

19 Davis J R (Ed.), *Handbook of Thermal Spray Technology*, ASM International, Materials Park, Ohio, USA, 2004.

20 Mccune R C, Donlon W T, Popoola O O and Cartwright E L, Characterization of Copper Layers Produced by Cold Gas-Dynamic Spraying, *Journal of Thermal Spray Technology*, **9** (1), 2000, 73–82.

21 Conrad H, Effect of Interstitial Solutes on the Strength and Ductility of Titanium, *Progress in Materials Science*, **26** (2–4), 1981, 123–403.

22 Alkhimov A P, Papyrin A N, Kosarev V F, Nesterovich N I and Shushpanov M M, Method and Device for Coating, European Patent 0 484 533 B1, January 25, 1995.

23 Gilmore D L, Dykhuizen R C, Neiser R A, Roemer T J and Smith M F, Particle Velocity and Deposition Efficiency in the Cold Spray Process, *Journal of Thermal Spray Technology*, **8** (4), 1999, 576–582.

24 Schmidt T, Gärtner F, Assadi H and Kreye H, Development of a Generalized Parameter Window for Cold Spray Deposition, *Acta Materialia*, **54** (3), 2006, 729–742.

25 Fukanuma H, Ohno N, Sun B and Huang R, In-Flight Particle Velocity Measurements with DPV-2000 in Cold Spray, *Surface and Coatings Technology*, **201** (5), 2006, 1935–1941.

26 Balani K, Laha T, Agarwal A, Karthikeyan J and Munroe N, Effect of Carrier Gases on Microstructural and Electrochemical Behaviour of Cold-Sprayed 1100 Aluminium Coating, *Surface and Coatings Technology*, **195** (2–3), 2005, 272–279.

27 United States Geological Survey website. www.minerals.usgs.gov.

28 Johnson A, Helium Recycle – a Viable Industrial Option for Cold Spray, Cold Spray 2004: An Emerging Spray Coating Technology, September 27–28, 2004, Akron, Ohio, USA, ASM International, Materials Park, Ohio, USA, 2004.

29 Pattison J, Celotto S, Morgan R, Bray M and O'Neill W, Cold Gas Dynamic Manufacturing: a Non-Thermal Approach to Freeform Fabrication, *International Journal of Machine Tools and Manufacture*, **47** (3–4), 2007, 627–634.

30 Gärtner F, Stoltenhoff T, Schmidt T and Kreye H, The Cold Spray Process and its Potential for Industrial Applications, Proceedings of the 2005 International Thermal Spray Conference, May 2–4, 2005, Basel, Switzerland.

31 Pham D T and Gault R S, A Comparison of Rapid Prototyping Technologies, *International Journal of Machine Tools and Manufacture*, **38**, 1998, 1257–1287.

32 Santos E C, Shiomi M, Osakada K and Laoui T, Rapid Manufacturing of Metal Components by Laser Forming, *International Journal of Machine Tools and Manufacture*, **46** (12–13), 2006, 1459–1468.

33 Amon C, Beuth J, Kirchner H, Merz R, Prinz F, Schmaltz K and Weiss L, Material Issues in Layered Forming, Proceedings of Solid Freeform Fabrication Symposium, August 9–11, 1993, The University of Texas at Austin, USA.

34 Ajdelsztajn L, Jodoin B and Schoenung J M, Cold Spray Deposition of Nanocrystalline Aluminium Alloys, *Metallurgical and Materials Transactions A*, **36** (3), 2005, 657–666.

35 Li C J, Li W Y and Wang Y Y, Formation of Metastable Phases in Cold-Sprayed Soft Metallic Deposit, *Surface and Coatings Technology*, **198**, 2005, 469–473.

36 Smith M F, Brockmann J E, Dykhuizen R C, Gilmore D L, Neiser R A and Roemer T J, Cold Spray Direct Fabrication – High Rate, Solid State, Material Consolidation, Proceedings of the Materials Research Society, November 30–December 1, 1998, Boston, Massachusetts, USA.

37 Karthikeyan J, Cold Spray Technology, *Advanced Materials and Processes*, **163** (3), 2005, 33–35.

38 Wohlers T, *Wohlers Report 2006*, Wohlers Associates Inc., Fort Collins, Colorado, USA, 2006.

39 Blose R E, Spray Forming Titanium Alloys Using the Cold Spray Process, Proceedings of the 2005 International Thermal Spray Conference, May 2–4, 2005, Basel, Switzerland.

40 Hall A C, Cook D J, Neiser R A and Roemer T J, The Effect of a Simple Annealing Heat Treatment on the Mechanical Properties of Cold Sprayed Aluminium, Proceedings of the 2005 International Thermal Spray Conference, May 2–4, 2005, Basel, Switzerland.
41 Song Y A, Park S, Choi D and Jee H, 3D Welding and Milling: Part I – A Direct Approach for Freeform Fabrication of Metallic Prototypes, *International Journal of Machine Tools and Manufacture*, **45**, 2005, 1057–1062.
42 Kang H K and Kang S B, Tungsten/Copper Composite Deposits Produced by a Cold Spray, *Scripta Materialia*, **49**, 2003, 1169–1174.
43 Legislation for chemicals. http://ec.europa.eu/environment/chemicals.
44 Kim H J, Lee C H and Hwang S Y, Fabrication of WC–Co Coatings by Cold Spray Deposition, *Surface and Coatings Technology*, **191** (2–3), 2005, 335–340.
45 Schmidt T, Gaertner F and Kreye H, New Developments in Cold Spray Based on Higher Gas- and Particle Temperatures, Proceedings of the 2006 International Thermal Spray Conference, May 15–19, 2006, Seattle, Washington, USA.
46 Kreye H, Schmidt T, Gartner F and Stoltenhoff T, The Cold Spray Process and Its Optimization, Proceedings of the 2006 International Thermal Spray Conference, May 15–19, 2006, Seattle, Washington, USA.

Part II

Cold spray process parameters

6 Cold spray process parameters: powders

F. J. BRODMANN, F. J. Brodmann & Co., USA

6.1 Introduction

The global powder manufacturing community is in the process of adapting to increasingly challenging product requirements related to higher purity, narrower particle size fractions, and controlled particle morphologies. The major markets served by engineered powder products include the coating industry, parts manufacturing, functional fillers, and powder coating applications. The cold spray process offers the opportunity to compete in all of the listed product market segments. Functional fillers have been used in combination with ductile matrix materials for cold sprayed coatings and net-shape solid parts manufacturing.

The particle size specifications for cold spray powders range from nano-size to over 400 micron (40 mesh). A powder particle size range of this magnitude benefits the manufacturing yield for most powder production processes. Within this range the 'FloMaster' cold spray powders exhibit a very narrowly sized particle spectrum which aids in the cold spray production of dense coatings and also wrought-like net-shaped parts. The required particle size fractions are specific for each of the various cold spray and kinetic spray equipment designs. The most frequently requested particle size fractions are typically within the ranges of 5–25, 15–45, 63–90, 75–150, and 200–400 micron particle diameter. The challenge for the powder manufacturer is to find customers for all powder fractions within the entire spectrum of the particle sizes. The particle size range for 'FloMaster' cold spray powders extends beyond the size distribution of the original feedstock material produced by either atomization, friction alloying, vaporization, or precipitation. The powder particle sizes are routinely increased by sintering, fusion, and agglomeration and decreased by crushing, and ball or jet milling. Process variations related to the choice of atomization gases, melting temperatures, and processing time control the particle morphology.

Both the price and the output yield for cold spray powders are directly related to the manufacturing yield which is a direct function of the particle size and morphology specifications. Coarse metal powders in the 50–100 micron particle size range air atomized with irregular particle shape are produced at the highest yield and lowest cost.

6.2 Selection criteria for cold spray feedstock

Starting in 1997, the first commercial orders for 'FloMaster' cold spray powders were placed primarily from customers representing the automotive parts industry, research institutions, and initial cold spray equipment manufacturers. The early specifications criteria for cold spray powders were limited to chemistry and particle diameter expressed in microns and/or mesh size. The common test method for measuring the particle size of powder products is ASTM B 214 using USA Standard E11 testing sieves and a tapping sieve shaker. The particle size screening test also yields qualitative measures regarding the flowability of test powders. A quantitative test method for powder flowability is ASTM B213 using a certified Hall flowmeter. The equipment simultaneously measures the bulk density of the test powder per ASTM B212.

Hall flow measurements together with bulk density tests were adopted as standard quality control procedures for 'FloMaster' cold spray powders. Comparing the flowmeter data from various cold spray powders indicates that the flowrate is a function of the particle density, particle size and particle shape. Micron-sized dense silver particles with spherical particle morphology produced Hall flowrates in the range of less than 10 s/50 g. Figure 6.1 illustrates the microstructure of 5–20 micron silver powder atomized with nitrogen gas.

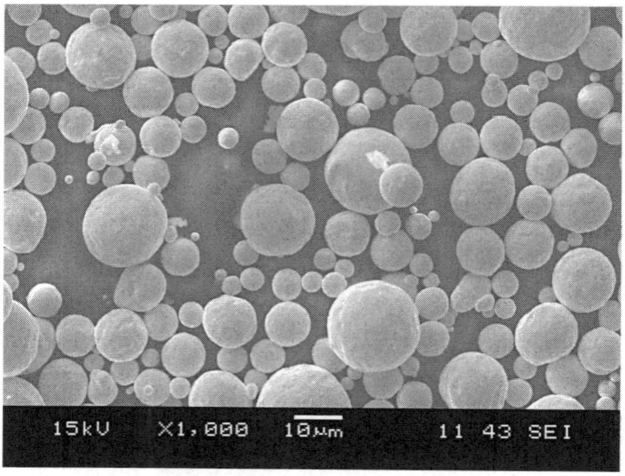

6.1 Scanning electron micrograph of nitrogen gas atomized silver metal powder, 5–20 micron particle size.

The flowrate of a powder is an essential characteristic that determines the feasibility for cold spray processing. Powders with flowrates above 60 s/50 g tend to build up and block the gas flow in the Laval subsonic and supersonic nozzles. The early materials of choice for cold spray feasibility evaluations were zinc, tin, aluminum, aluminum bronze, and copper powders sized to –45/+15 micron with and without surface treatment. The surface treatment was performed by annealing the powder in air using a time–temperature controlled oven environment. Figure 6.2 shows air atomized aluminum powder, 32–45 micron particle size, after annealing in air at 350 °F for 100 hours.

The chemical elements selected for cold spray processing have in common low melting points and low resistance to deformation. The metal groups meeting these properties include pure copper, silver, zinc, tin, bismuth, indium, and aluminum and its alloys. These criteria are also met by high-performance thermoplastic polymer compounds. The potential cold spray polymer feedstock group is represented by high-density aromatic polyimide, aromatic polyester, polyetherketone, and high-temperature fluorocarbon resins. Figure 6.3 illustrates the microstructure of high-temperature aromatic polyester, 45–100 micron particle size and angular particle shape.

The thermoplastic polymer compounds are the common feedstock for powder coatings using electrostatic or fluidized-bed coating methods. During powder coating the parts are exposed to up to 450 °F in a curing oven to melt the polymer powder on the part surface. Temperature-sensitive material must be removed from the substrate before the start of the oven

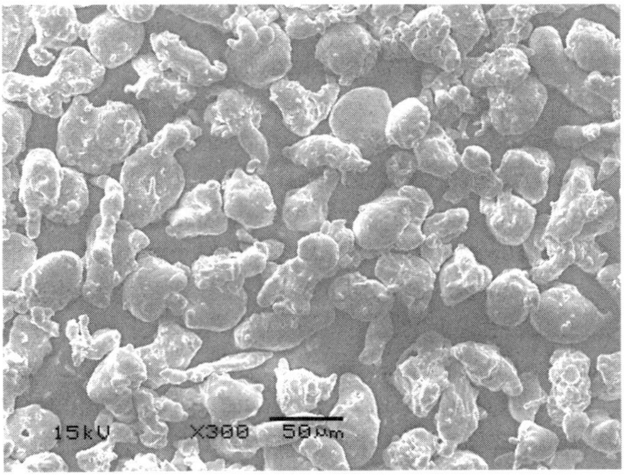

6.2 Scanning electron micrograph of annealed air atomized aluminum, 3600 ppm surface oxygen.

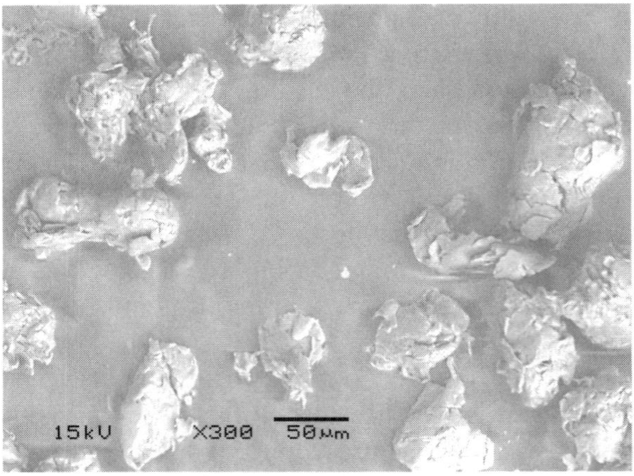

6.3 Scanning electron micrograph of aromatic polyester resin, 45–100 micron particle size.

curing cycle. Cold spraying of polymer powders eliminates the need for the oven cure requirements.

The deformation and softening of solid matter is a function of temperature and the architecture of its structural building blocks such as crystal structure and molecular chain length. Metals crystallize most frequently in the cubic structure in either a face-centered or body-centered lattice. The face-centered cubic-structured metals exhibit the lowest resistance to deformation. The *ASM Metals Reference Book*[1] lists the following metals with face-centered symmetry: Ag, Al, Au, Cu, gamma Fe, beta Co, gamma Mn, Ge, Ir, Ni, Pb, Pd, Pt, Rh, Si, and Sn. A comprehensive study investigating the material characteristics for cold spray materials is outlined in reference 2. The report describes that cold sprayed ceramic feedstock was coated on aluminum indicating that the properties of the substrate material must be integrated in any cold spray coating feasibility study. In the example, a brittle-phase ceramic material is bonded to a ductile metal substrate. A test sample of cold sprayed aluminum metal on window glass prepared by Anatolii Papyrin illustrates that a brittle substrate can be coated with a ductile material.

6.3 Key powder properties

During the initial commercialization phase of the cold spray process involving equipment and process development, it became obvious to the applicators that the feedstock powder properties are the most important variable and require close control limits. Among the property variables, the particle size of the powder is the most controllable. Commercial screeners using vibration or tapping modes for the mesh screen sizes 40–635 mesh (20–425 micron) have an accuracy of ±0.5%. The subsieve (–635 mesh) fractions are separated by water/air elutriators at an accuracy of ±10%. The material properties that are the most difficult to control for cold spray powders are the particle morphology and topography. The particle morphology affects the Hall flowmeter measurements, bulk density, and screening results.

The spectrum of particle shapes includes perfect solid spheres, hollow spheres, and near-spherical, dendritic, sponge, flakes, acicular, and fibrous appearances. A powder composed of single-dimensional particles (spheres) screens faster and at higher accuracy than a material composed of multi-dimensional particles (acicular). The transformation from powder to cold spray coating layer requires a minimum critical flight velocity of the particle at the moment of impact with the substrate surface to facilitate cold weld-type bonding. Assuming that the cold spray parameters are optimized, the particle morphology and density affect the actual impact velocity, and subsequently the coating density and deposition efficiency. In reference 3 the critical velocities for cold spray powders were measured as a function of particle size and particle temperature. It was observed that the particle velocity decreases with an increase in particle diameter. Higher particle temperatures associated with higher preheat gas temperatures resulted in lower critical velocity requirements. An increase in particle temperature ultimately leads to higher surface oxidation levels for non-noble metals. High oxygen contents of the cold spray powder require in turn higher critical velocities for coating formation.

6.4 Powder manufacturing process

The current consumption and production quantities for cold spray powders indicate that the process remains in the experimental phase. Commercial quantities of coarse tin and copper cold spray powders with spherical particle morphology and low oxide content were quoted to automotive customers. The most economical manufacturing process for powders suitable for cold spraying is atomization through a pressure nozzle using air or inert gases. The formation of oxide layers on the surface of the molten metal droplets during air gas atomization prevents the development of spherical particles (Fig. 6.4).

The use of inert gases such as helium is required for the manufacturing of spherical metal particles. The oxygen content of commercially produced, passivated, –45/+15 micron-sized helium atomized aluminum is typically 2500 ppm; air atomized aluminum measures 2720 ppm; after annealing in air for 100 hours the oxygen concentration increases to 3600 ppm. Figure 6.5 shows the microstructure of helium atomized aluminum powder containing 2500 ppm surface oxygen.

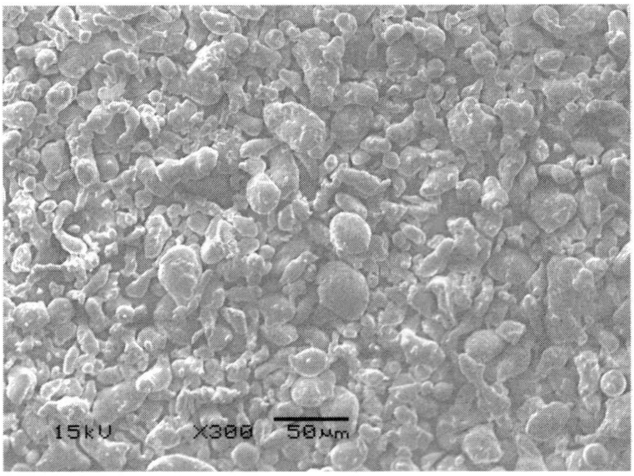

6.4 Scanning electron micrograph of air gas atomized aluminum metal powder, –45/+15 micron particle size.

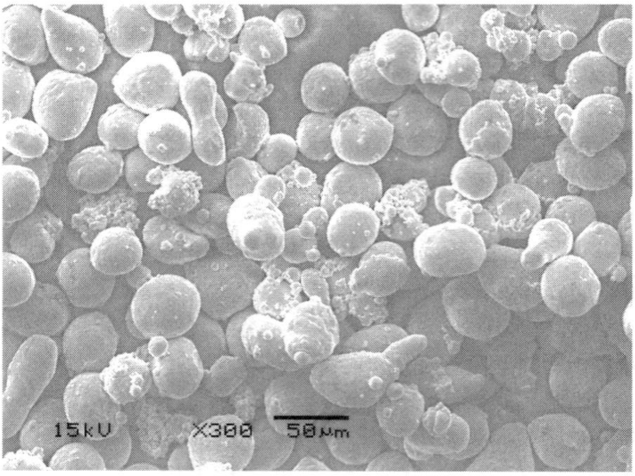

6.5 Scanning electron micrograph of helium atomized aluminum, 32–45 micron particle size.

Pure aluminum, silicon, tin, and zinc powders are classified as hazardous substances by the US Department of Transportation (DOT) and are not regulated after passivation by oxide surface coatings. This procedure explains the small difference in oxygen content between helium gas and air atomized aluminum powder. Copper metal has a high affinity for oxygen and fine powder is safely produced in the submicron particle size ranges via air or water atomization. Water atomized copper particles exhibit very irregular morphologies from spheroidal to spongy. Figure 6.6 shows the microstructure of water atomized high-oxide copper powder.

Low oxide containing copper powder for electronic applications is produced from air atomized feedstock via annealing in a hydrogen atmosphere.[4] The particle size yield for most metal atomization processes is limited to particle sizes below 100 micron. Metal powders tailored for the kinetic spray process require particle sizes up to 100 micron diameter. A special agglomeration process using friction welding was developed to convert a 100 micron powder to a 600–1400 micron powder. Figure 6.7 illustrates the morphology of friction-welded tin metal powder composed of agglomerated particles.

6.6 Scanning electron micrograph of water atomized copper powder, 15–45 micron particle size.

6.7 Scanning electron micrograph of friction welded tin metal powder, 600–1400 micron particle size.

High-temperature flame and/or plasma processes are employed to produce large-sized spherical particles from metal and ceramic feedstock. Figure 6.8 shows the particle morphology of acetylene flame-treated friction-welded, 60–90 micron particle size, steel powder.

At high scanning electron microscope (SEM) magnifications the steel particles were identified as thin-walled hollow spheres. Potential applications for cold spray coatings composed of hollow metallic spheres made from incramute and other alloys include acoustical and/or thermal insulation products. A manufacturing process for aluminum powder composed of hollow spheres is under development. Other particle morphologies such as sponge and flake are commercially available for the fabrication of controlled porosity and high-surface-area coatings or solids. Metal flakes and sponge metals expose a higher surface area than spherical shapes and are therefore classified as hazardous materials. To ensure the safe handling of metal flake and sponge powders the particles are coated with organic liquids. The same substances are also used during the ball milling process for the production of flake powders. Commercially available metal flakes include aluminum, chromium, copper, silver, and some of their alloys. Figure 6.9 shows the microstructure of metal flakes prepared from aluminum alloy 5056 using a high-speed milling process. Recently, stainless steel 316 has been commercially produced as a flake powder for applications requiring corrosion- and abrasion-resistant coatings.

The melting temperatures of refractory metals such as zirconium, hafnium, tungsten, rhenium, tantalum, platinum, iridium, and other platinum group elements exceed the limit for gas atomization. Refractory metal powders

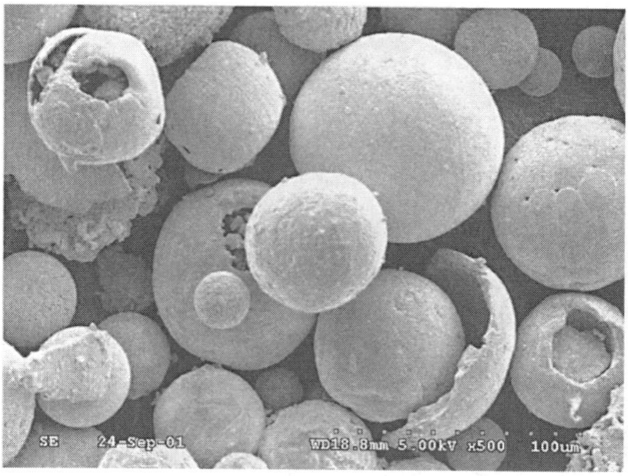

6.8 Scanning electron micrograph of spheridized stainless steel powder, 60–90 micron particle size.

6.9 Scanning electron micrograph of aluminum alloy 5056 flakes, 10–45 micron particle size.

6.10 Scanning electron micrograph of pure rhenium sponge metal powder, 45–63 micron particle size.

are instead produced using hydrogen reduction and/or hydride/dehydride processes. Current research in cold spray is aimed at extending its application to refractory metals. Figure 6.10 exhibits the sponge structure of hydrogen-reduced pure rhenium metal powder

6.5 New material developments

The worldwide materials industry has experienced a revolution related to new materials and their commercial product applications. Three important examples are the development of computer flat screens using indium-based coatings, thin-film batteries using lithium-based electrolytes, and light emitting diodes made from nano-sized titanate particles. The advanced materials are mostly used as coatings to minimize material usage and space requirements. The cold spray process offers advantages to produce, with minimum thermal exposure, coatings from functional materials such as thermo-electric, magneto-caloric, photo-voltaic, piezo-electric, super-magnetic, and high-temperature superconductive formulations.

Crystalline materials that experience phase changes as a function of temperature, quasi-crystalline alloys composed of elements with very different melting points,[5,6] and amorphous structures that crystallize during temperature exposures are prime candidates for cold spray processing. Brittle compounds such as carbides, nitride, silicides, borides, sulfides, and ferrites have potential for cold spraying after the surface is coated with a ductile-phase material. The resultant composite powder structure consists of a core particle surrounded by a ductile coating of controlled thickness. Commercial processes for particle coating are electroplating, electroless plating, vacuum metalization, and cladding. The ductile coating layer for surface bonding is formed from metals and metal alloys such as Al, Cu, Ni, Co, Zn, Sn, Fe, and certain thermoplastic resins. The brittle-core material choices include ceramic, glass, quasi-crystalline and amorphous structures. Commercial applications have been established for composite powders produced from nickel-coated hexagonal boron nitride, synthetic graphite, bentonite, and other core materials.

6.6 Discussion

The emerging cold spray technology developed in two directions – one emphasizing small particle diameters in the 5–40 micron range and the other, described as kinetic spray, using larger particle diameter powders. In both cases the powder particles are injected into a preheated gas stream that propels the particles to supersonic speed before impact onto the substrate surface. The required critical impact speed for larger particles is lower than that for small particles, due to the higher temperatures that the larger particles maintain. Improvements in deposition efficiencies were accomplished at higher gas temperatures, which in turn were responsible for increased particle velocities. At gas temperatures above 375 °F the surface oxygen of the metal particles and the exothermic reaction with non-noble gases such as nitrogen or air further increase the particle temperature. Higher temperatures improve the ductility for both the coating material and the substrate, but might also lead to nitride and oxide formation in the coating. The most reactive metals in contact with non-noble gases are Ti, Al, Si, Mg, Hf, Zr, and Ta; and rare earth metals are classified as flammable solids or spontaneously combustible. Aluminum metal particle surfaces are customarily coated with a metal oxide layer to enable safe handling.

The behavior of the brittle oxide layer during impact with the substrate is illustrated by a so-called 'eggshell' model. The 'eggshell' represents the brittle phase in which the yolk substance is very ductile. During impact of a raw egg with a hard surface the yolk separates from the eggshell. A similar phenomenon is observed during impact of a metal particle that is coated with a brittle oxide or nitride surface layer. At impact the brittle phase is removed from the metal particle and exposes a clean non-oxidized surface before cold welding to the substrate surface. Bond strength and deposition efficiency measurements are required to quantify the function of the brittle-phase coating thickness.

Improvements in bond strength and deposition efficiencies were also reported for low-density metals such as aluminum, after the addition of a high-density less ductile material such as steel to the ductile cold spray feed powder. This so-called 'hammer' material is designed to increase the impact energy for the low-density ductile phase without being deposited in the coating layer. In addition the hammer material at certain concentrations, particle shapes, and grain sizes performs blasting functions and removes impurities from the substrate surface.

Developments related to the feedstock formulations are beneficial for both the fine particle cold spray (5–45 micron) and the coarse particle (60–90 micron) kinetic process. The manufacturing costs and safety concerns for coarse metal powders are generally lower than for fine powders. Micron-size fine powders must be annealed in air or nitrogen to ensure safe handling during processing, shipment, and coating application.

6.7 Conclusions

Since 1997 cold spray coating technology has been emerging from the research and development phase into limited and selective commercial applications. Despite the long commercialization period, the industry involved in Al, Cu, Zn, Sn, Ti, and Fe powder manufacturing remains optimistic about large volume use of cold spray powders in automotive and industrial product applications. The specifications for cold spray powders should include parameters for particle size, particle morphology, and particle surface chemistry. Refinement in the powder manufacturing process and subsequent treatments are required to meet the specification challenges for commercially produced cold spray powders. Rising raw material costs – in particular for Ni, Co, Mo, and Cu – and the more stringent powder processing procedures at low manufacturing volumes are responsible for the high cold spray powder costs. The high prices for metal feedstock equally affect other coating processes such as thermal spray, electrolytic, and sputtering processes. Industrial customers located in the Gulf Coast region are interested in cold spray for coating applications on large structures such as bridges, tunnels, ships, floodgates, and marine platforms. The major concerns related to the cold spray process are equipment availability, affordability, and portability compared with coating methods currently employed.

6.8 Acknowledgments

The author thanks all customers of FloMaster cold spray powders for their feedback especially Dr Anatolii Papyrin, Steve Johnson, John Potter, Rick Blose, Daniel Gorkiewicz, Jerry Moleski, Phil Leyman, Dr Jegan Karthikeyan, Dr Robert McCune, Robert Kasdorf, Wally Birtch, and Carlos Henrique Santos. Many thanks to Dr Sandy Young of ARL, Maryland, USA for the SEM photographs.

6.9 References

1 Gall T L (Ed.), *ASM Metals Reference Book*, Second Edition, 1982, ASM International, Materials Park, Ohio.
2 Vicek J, Gimeno L, Huber H and Lugscheider E, A Systematic Approach to Material Eligibility for Cold-Spray Process, *Journal of Thermal Spray Technology*, **14** (1), 2005, 125–133.
3 Li C-J, Li W-Y and Liao H, Examination of the Critical Velocity for Deposition of Particles in Cold Spraying, *Journal of Thermal Spray Technology*, **15** (2), 2006, 212–222.
4 Yule A J and Dunkley J J, *Atomizations of Melts for Powder Production and Spray Deposition*, Oxford Series on Advanced Manufacturing No. 11, 1994, Clarendon Press, Oxford.
5 Sordelet D J, Besser M F and Anderson I E, Particle Size Effect on Chemistry and Structure of Al-Cu-Fe Quasicrystalline Coatings, *Journal of Thermal Spray Technology*, **5** (2), 1996, 161–174.
6 Goldman A I, Anderegg J W and Besser M F, Quasicrystalline Materials, *American Scientist*, **84**, 1996, 230.

7
The influence of nozzle design in the cold spray process

K. SAKAKI, Shinshu University, Japan

7.1 Introduction

One of the major characteristics of the cold spray process[1] is the high-speed gas jet, which is governed by gas dynamics. In gas dynamics, supersonic flows are obtained with convergent–divergent (CD) nozzles (or de Laval nozzle), which are used for rocket motors. Rocket motors, including their nozzle design, have been studied and analyzed in detail. While the principal purpose of the design of a rocket motor nozzle is to maximize the thrust, in thermal spraying – including cold spray – the main purpose is to obtain better coating quality.[2–13]

Cold spray systems employ various kinds of gun nozzle contours, such as CD (or de Laval nozzle),[1,4–11,13] convergent–divergent-barrel (CDB)[7,13] and convergent-barrel (CB).[7,12]

Previous studies on thermal spraying show that the coating properties are principally determined by the thermal and kinetic energy states of particles upon impact with the substrate. In order to have a balance between these two states, various changes in the design of the high-velocity oxygen fuel (HVOF) gun nozzle have been attempted. However, works concerning the influence of nozzle geometry on the thermal spray process are sparse.[2–4] Previously, the effects of throat diameter and exit divergence of the gun nozzle on the HVOF process have been considered.[4] The combustion gas flow (such as pressure, velocity, temperature and expansion state of gas jet from the nozzle exit), the particle behavior and, therefore, the nature of coatings were found to be significantly influenced by these nozzle parameters. In addition, the effect of the expansion state of the combustion gas jet on the HVOF process was investigated using a diverging nozzle exit. The particle velocity reached a maximum with the correct expansion state of the gas jet due to an increased gas jet velocity. This resulted in an increase in the bonding strength of the NiCrAlY coating.[14]

The nozzle geometry is also important with regard to the cold spray method. In the cold spray method, a coating is formed by exposing a substrate to high-velocity solid-phase particles, which have been accelerated by supersonic gas flow at a temperature much lower than the melting or softening temperature of the feedstock. In this chapter, the influence of nozzle contours, throat diameter and nozzle length on the cold spray process (i.e. the behavior of the carrier gas and spray particles within the nozzle) is investigated by a numerical simulation prior to designing the cold spray equipment and producing coatings.

7.2 Numerical simulation of the cold spray process

There are several techniques that can be used to calculate the gas flow of cold spray systems. Recent analyses have used computational fluid dynamics (CFD) methods to simulate complex phenomenon in two dimensions. In this chapter, the internal nozzle flow was treated as quasi-one-dimensional isentropic flow. This one-dimensional approximation is simple and sufficiently correct for the present purpose of expressing an effect of nozzle geometry on the internal nozzle flow. The applicable gas flow equations can be obtained from any classical book on fluid flow (e.g. references 15–17). In Chapter 11 of this book, gas-particle dynamics are discussed in detail. The modeling can be found in references 14 and 18. A brief description of the simulation is given below.

First, the gas flow model is introduced, and then the particle behavior model is introduced. From the particle behavior solution, optimal conditions are identified that yield maximum particle acceleration. Next, equations are introduced that describe the nozzle shape that yields the optimal condition. Finally, the particle velocity for the optimal nozzle is analytically determined. Figure 7.1 shows a typical cold spray gun nozzle. Compressed gas flows through a CD nozzle to develop supersonic gas velocity. The powder particles are metered into the gas flow immediately upstream

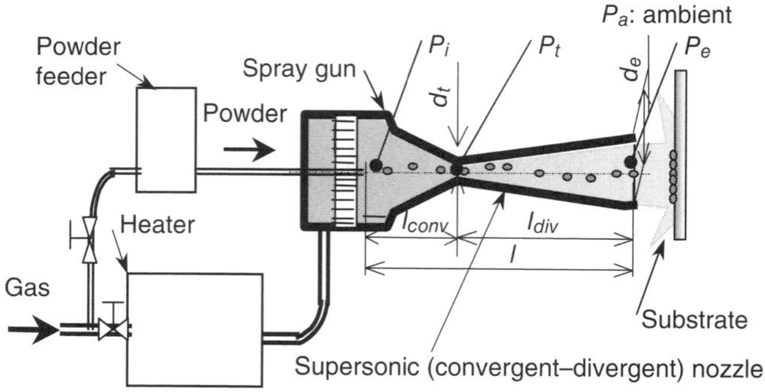

7.1 Schematic diagram of the cold spray equipment.

of the converging section of the nozzle and are accelerated by rapidly expanding gas.

7.2.1 Modeling of process gas flow within the nozzle

The following assumptions were made to model the gas flow in the cold spray gun nozzle.

1 Process gas flow within the nozzle is the quasi-one-dimensional isentropic flow of semi-perfect gas.
2 Process gas flows in the nozzle intake with an initial velocity U_{gi}, temperature T_{gi}, and pressure P_i.

By comparison of gas throat pressure (P_t) and exit pressure (P_e), it is decided whether the optimal nozzle shape becomes a convergent nozzle ($P_e > P_t$) or a CD nozzle ($P_e < P_t$). In cold spray, a CD nozzle is selected because nozzle intake gas pressure P_i ranges up to 2 MPa and a supersonic gas flow is used.

From the nozzle intake pressure (a stagnation pressure) P_i, the throat pressure P_t can be calculated:

$$\frac{P_t}{P_i} = \left(\frac{2}{\kappa+1}\right)^{\frac{\kappa}{\kappa-1}} \qquad [7.1]$$

where κ is the gas specific heat ratio.

If sonic conditions are to be maintained at the nozzle throat, then the throat pressure P_t must be above the ambient. An optimal nozzle throat area A_t, where gas flow is sonic, is calculated as:

$$A_t = m \Big/ \sqrt{\kappa\left(\frac{2}{\kappa+1}\right)^{(\kappa+1)/(\kappa-1)} \frac{P_i}{v_i}} \qquad [7.2]$$

where m is mass flow rate of the process gas and v_i is the specific volume of the gas, and is defined by

$$v_i = \frac{RT_{gi}}{P_i} \qquad [7.3]$$

where R is the gas constant and T_{gi} is the gas temperature at the nozzle intake.

Furthermore, a nozzle exit area A_e in the case of a CD nozzle can be expressed as:

$$\frac{A_t}{A_e} = \left(\frac{\kappa+1}{2}\right)^{1/(\kappa-1)} \left(\frac{P_e}{P_i}\right)^{1/\kappa} \sqrt{\frac{\kappa+1}{\kappa-1}\left[1-\left(\frac{P_e}{P_i}\right)^{(\kappa-1)/\kappa}\right]} \qquad [7.4]$$

Optimal exit area A_e is yielded when the nozzle exit gas pressure P_e is equal to the ambient pressure P_a, and is given by equation [7.4].

The pressure P, temperature T_g and velocity U_g of the gas flow can be calculated from the ratio of the nozzle cross-sectional area A at a given point to the nozzle throat area A_t by:

$$\frac{A}{A_t} = \sqrt{\left(\frac{\kappa-1}{2}\right)\left(\frac{2}{\kappa+1}\right)^{(\kappa+1)/(\kappa-1)} \Big/ \left[\left(\frac{P}{P_i}\right)^{2/\kappa} - \left(\frac{P}{P_i}\right)^{(\kappa+1)/\kappa}\right]} \quad [7.5]$$

$$T_g = T_{gi}\left(\frac{P}{P_i}\right)^{(\kappa-1)/\kappa} \quad [7.6]$$

$$U_g = \sqrt{2\frac{\kappa}{\kappa-1}P_i v_i\left[1-\left(\frac{P}{P_i}\right)^{(\kappa-1)/\kappa}\right]+U_{gi}^2}$$

$$= \sqrt{2\frac{\kappa}{\kappa-1}RT_i\left[1-\left(\frac{P}{P_i}\right)^{(\kappa-1)/\kappa}\right]+U_{gi}^2} \quad [7.7]$$

7.2.2 Modeling of particle behavior within the nozzle

Particle acceleration and heating in a gas flow within the nozzle are given by solving the equations of motion and heat transfer as described below. These equations in this chapter are based on the following four assumptions:

1 The spray particle is spherical with negligible internal temperature gradients.
2 The particle specific heat is independent of its temperature and constant.
3 The gravitational effect and the interaction between particles are ignored.
4 The influences of particles on gas flow are neglected. This is equivalent to stating that the gas energy decrease along the nozzle due to acceleration and heating of the particle is neglected.

Under the above assumptions, the equations of motion of a particle in the cold spray process can be written as:

$$\frac{dU_p}{dt} = \frac{3}{4}\frac{C_d}{D_p}\frac{\rho_g}{\rho_p}(U_g - U_p)|U_g - U_p| \quad [7.8]$$

where U_p is the particle velocity, t is time, C_d is the drag coefficient, D_p is the particle diameter, ρ_g is the process gas density and ρ_p is the particle density. C_d for a sphere is a function of particle Reynolds number.[19]

Heating of a particle in a gas flow can be expressed as follows:

$$\frac{dT_p}{dt} = (T_g - T_p)\frac{6h}{c_p \rho_p D_p} \qquad [7.9]$$

where T_p is the particle temperature, h is the heat transfer coefficient and c_p is the specific heat of the particle. The heat transfer coefficient h in equation [7.9] can be found by means of the semi-empirical Ranz–Marshall equation, in which h is a function of the Reynolds number and the Prandtl number. The influence of radiant heat between the combustion gas and particle was neglected.

For the calculation of the Reynolds number and the Prandtl number, the values of the specific heat of the gas, the gas viscosity and the gas thermal conductivity are used in the film temperature T_f, which is defined by:[20]

$$T_f = (T_g + T_p)/2 \qquad [7.10]$$

7.2.3 Numerical approximation of the cold spray process

An outline of a numerical approximation of the cold spray process is as follows: (1) initial conditions are given; (2) the pressure P, temperature T_g and velocity U_g of the gas flow are calculated from the ratio of the cross-sectional area of the nozzle at the intake point to the nozzle throat area; (3) the above differential equations (equations [7.8] and [7.9]) concerning particle behavior are solved numerically by the Euler method; (4) processes (2) and (3) are repeated from the nozzle intake to the nozzle exit.

The cold spray condition data used and the initial conditions are shown in Table 7.1 (except the powder feed rate and the spray distance). Table 7.2 gives the values of diameter, density and specific heat of the spray power used in the calculation. Figure 7.2 shows schematic diagrams of the cold

Table 7.1 Initial conditions for cold spray process simulation

Gas: nitrogen or helium
Gas initial pressure, P_i: 0.5–5.0 (2.0) MPa (abs.)
Gas initial velocity, U_{gi}: 0–100 (0) m/s
Gas initial temperature, T_{gi}: 300–2000 (750) K
Specific heat ratio of gas, κ: nitrogen, 1.4; helium, 1.66
Gas constant, R: nitrogen, 0.2969; helium, 2.0772 kJ/(kg K)
Particle
Particle velocity, U_{pi}: 0–100 (10) m/s
Particle temperature, T_{pi}: 300 K

Figures in brackets are baseline condition.

Table 7.2 Cold spraying powder properties

Property	Ni-Al bronze	Copper	Titanium
Diameter (μm)	1–50 (20)*	5–45 (10)*	25
Melting point (K)	1340	1356.5	2095
Density (kg/m^3)	7600	8900	4540
Specific heat (J/(kg K))	440	384.8	472

* Mean diameter of powder.

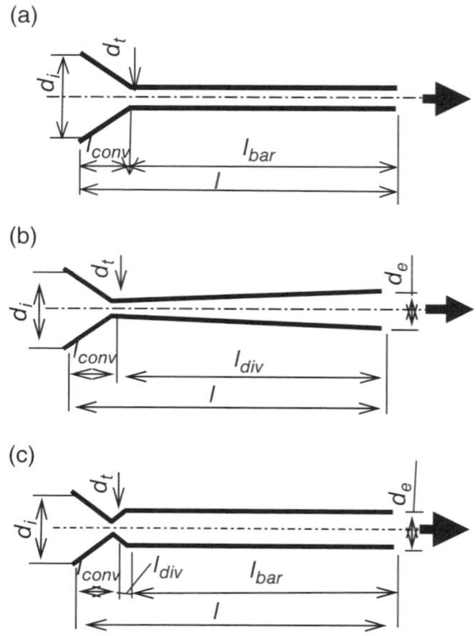

7.2 Schematic cross-sections of the cold spray nozzles: (a) convergent-barrel nozzle; (b) convergent–divergent nozzle; (c) convergent–divergent-barrel nozzle.

spray nozzles used for numerical simulation: (a) CB nozzle; (b) CD (or de Laval) nozzle; (c) CDB nozzle. The cross-sections of these nozzles are circular. Nozzle total length l, entrance converging length l_i, entrance diameter d_i, and throat diameter d_t are shown in Table 7.3.

The results provided by this simulation could be a little larger than the real values because of the above assumptions.

Table 7.3 Cold spray gun nozzle size

Nozzle no.	Nozzle geometry	Powder injection position	d_i (mm)	d_t (mm)	d_e (mm)	Ratio of A_e/A_t	l (mm)	l_{con} (mm)	l_{div} (mm)	l_{bar} (mm)
Convergent–barrel nozzle*		Axial	40	2	4	4	300	50	–	250
Convergent –divergent nozzle*		Axial	40	2	4	4	300	50	250	–
Convergent –divergent–barrel nozzle*		Axial	40	2	4	4	300	50	10	240

* Used in numerical simulations in Fig. 7.3.

7.2.4 Results of numerical simulation of the cold spray process

The numerical simulation results for the effect of nozzle geometry on gas pressure, velocity of gas, and particles, and temperature of gas and particles are given in Fig. 7.3.

In this section, the results of the CD nozzle, which is used by typical cold spray devices (CD nozzle in Fig. 7.3), are explained. The acceleration of the gas takes place predominately in the area of the nozzle throat and in the first third of the diverging section. Here the gas velocity U_g has already reached 85% of its exit velocity. Namely, gas velocity reaches sonic velocity at the nozzle throat and reaches supersonic velocity with the increasing cross-sectional area of the nozzle in the diverging section (see Fig. 7.3(c)). At the same time the gas temperature drops to values far below room temperature as the gas expands in the diverging section of the nozzle. The particle is introduced into the gas flow immediately upstream of the converging section of the nozzle and is accelerated by the rapidly expanding gas. The dwell time of the particle in contact with hot gas is brief, and the temperature of the solid particle at impact remains substantially below the initial gas preheat temperature.

7.3 Nozzle design influences

The numerical simulation results for the effect of nozzle geometry on gas pressure, velocity of gas and particles, and temperature of gas and particles are given in Fig. 7.3.[7] It is clearly seen that for the CD nozzle and the CDB

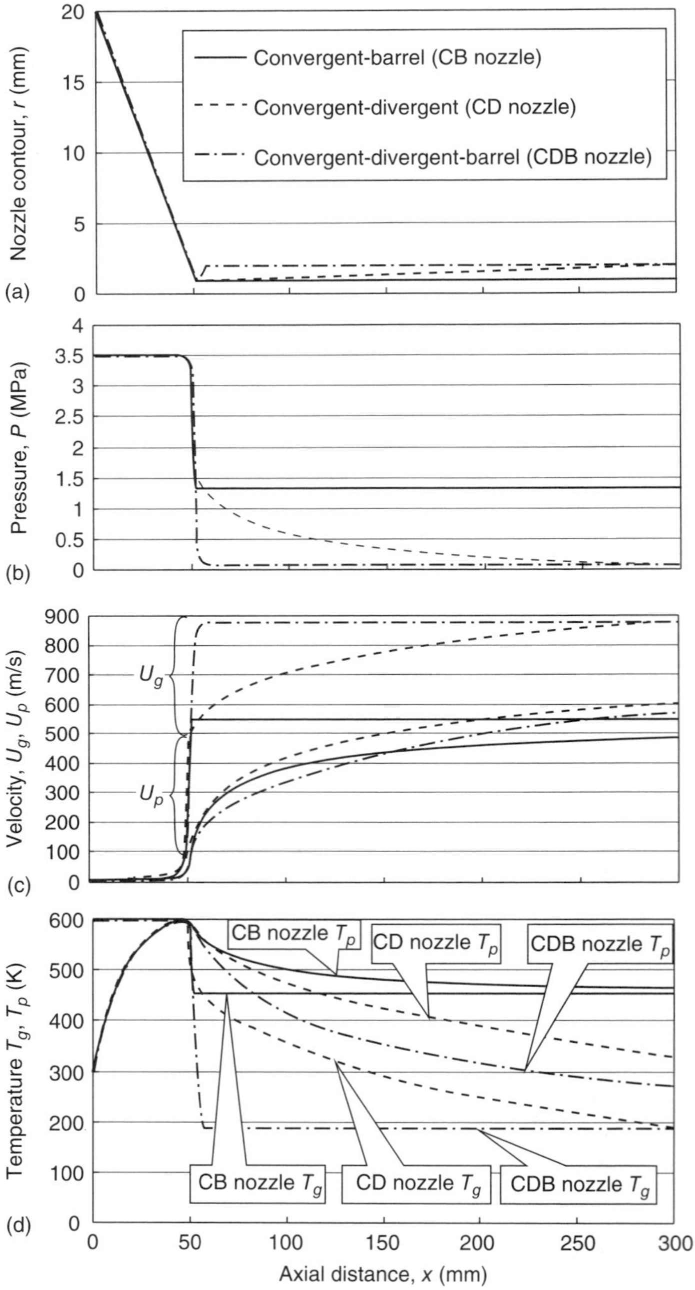

7.3 Effect of nozzle geometry on calculated results of: (a) nozzle contour; (b) gas pressure; (c) velocity of gas and particle; (d) temperature of gas and particles for nitrogen as the process gas, copper particle (15 μm). $P_i = 3.5$ MPa; $T_{gi} = 600$ K, $U_{gi} = 10$ m/s, $T_{pi} = 300$ K; $U_{pi} = 10$ m/s.

nozzle the gas pressure (or density) decreases significantly to a very low value upon leaving the nozzle throat owing to the fast gas expansion (see Fig. 7.3(b)). On the other hand, for the CB nozzle, the gas pressure is not significantly decreased along the nozzle up to the exit. Moreover, the oscillation of the gas variables outside the nozzle exit indicates that expansion waves are generated and these are more obvious for the CB nozzle.[12,17] This is due to the fact that at exit the high-pressure gas expands sharply to ambient for the CB nozzle, while for the CD and CDB nozzles, the gas has expanded gradually from the throat.

Figure 7.3(c) shows that the gas velocity increases remarkably to a high value after leaving the nozzle throat with the CD and CDB nozzle, while the velocity is relatively low (sonic velocity) with the CB nozzle. Although the particle velocity with the CB nozzle is lower than that with CD and CDB nozzles, the ratio of the particle velocity to the gas velocity is higher for the CB nozzle. This shows that particles can achieve more effective acceleration with a CB nozzle, because the particle velocity is not only influenced by the driving gas velocity but also by the gas density, as shown in equation [7.8]. The results show that the particle velocity at the nozzle exit for the CD nozzle is higher than that for the CDB and CB nozzles. On the other hand, the particle temperature at the nozzle exit for the CB nozzle is higher.

The divergent section with the CD nozzle in Fig. 7.3 is conical. The numerical codes of fluid dynamics, in particular the method of characteristics (MOC), were also used to develop new nozzle designs that allow a more uniform particle acceleration.[10,11] According to the MOC, the bell-shaped diverging section nozzle can produce better and more uniform particle acceleration over the diameter of the nozzle cross-section. Using the standard trumpet-shaped nozzle, a copper particle will be accelerated to a velocity of 500 m/s, which is not high enough to allow the bonding of copper. At the same parameter setting using the bell-shaped MOC-designed nozzle, the velocity of a 20 μm copper particle is increased to 580 m/s, which is well above the critical velocity of copper of 550 m/s.[11]

7.4 References

1 Alkimov A P, Papyrin A N, Kosarev V F and Shushpanav M M, '*Gas Dynamic Spraying Method for Applying a Coating*', U.S. Patent 5,302,414, April 12, 1994.

2 Hackette C M and Settles G S, 'The Influence of Nozzle Design on HVOF Spray Particle Velocity and Temperature', in *Thermal Spray Science and Technology*, C C Berndt and S Sampath, Eds, ASM International, Materials Park, OH, 1995, pp. 135–140.

3 Kopiola K, Hirvonen J P, Laas L and Rossi F, 'The Influence of Nozzle Design on HVOF Exit Gas Velocity and Coating Microstructure', *J. Thermal Spray Technology*, 1997, **16**(4), 469–474.

4 Sakaki K and Shimizu Y, 'Effect of Increase in Entrance Length of Gun Nozzle on HVOF and Cold Spray Process', *J. Thermal Spray Technology*, 2001, **10**(3), 487–496.
5 Gilmore D L, Dykhuizen R C, Neiser R A, Romer T J and Smith M F, 'Particle Velocity and Deposition Efficiency in Cold Spray Process', *J. Thermal Spray Technology*, 1999, **8**(4), 576–582.
6 Stoltenhoff T, Kreye H and Richter H J, 'An Analysis of the Cold Spray Process and its Coatings', *J. Thermal Spray Technology*, 2002, **11**(4), 542–550.
7 Sakaki K, Huruhashi N, Tamaki K and Shimizu Y, 'Effect of nozzle geometry on cold spray process', *International Thermal Spray Conference*, E Lugscheider and C C Berndt, Eds, March 4–6, 2002 (Essen, Germany), DVS Deutscher Verband für Schweißen, 2002, 385–389.
8 Han T, Zhabo Z, Gillispie B A and Smith J R, 'Effect of Spray Conditions on Coating Formation by the Kinetic Spray Process', *J. Thermal Spray Technology*, 2005, **14**(3), 373–383.
9 Li W-Y and Li C-J, 'Optimal Design of a Novel Cold Spray Gun Nozzle at a Limited Space', *J. Thermal Spray Technology*, 2005, **14**(3), 391–396.
10 Heinrich P, Kreye H and Stoltenhoff T, '*Laval Nozzle for Thermal and Kinetic Spraying*', U.S. Patent 2005/0001075 A1, January 6, 2005.
11 Gaertner F, Stoltenhoff T, Schmidt T and Kreye H, 'The Cold Spray Process and its Potential for Industrial Applications', *J. Thermal Spray Technology*, 2006, **15**(2), 223–232.
12 Li W-Y, Liao H, Wang H-T, Li C-J, Zhang G and Coddet C, 'Optimal Design of a Convergent-Barrel Cold Spray Nozzle by Numerical Method', *Applied Surface Science*, 2006, **253**(2), 708–713.
13 Sakaki K, Shinkai S, Ebara N and Shimizu Y, 'Effect of Geometry of the Gun Nozzle, the Increase in the Entrance Convergent Section Length and Powder Injection Point on Cold Sprayed Titanium Coatings', *Materials Science Forum* 2006, **534–536**, 413–416.
14 Sakaki K, Shimizu Y, Saitoh N and Gouda Y, 'Effects of Throat Diameter and Exit Diverging of the Gun Nozzle on the Combustion Gas Stream of HVOF Thermal Spraying: Application of the Nozzle Theory to Designing the HVOF Gun Nozzle', *J. Japan Thermal Spraying Society*, 1997, **134**(1), 1–9 (in Japanese).
15 Shapiro A H, '*The Dynamics and Thermodynamics of Compressible Fluid Flow*,' Ronald Press, New York, 1953.
16 Sabersky R H, Acosta A J and Hauptmann E G, *Fluid Flow*, Macmillan, New York, 1971.
17 Matuo K, *Compressible Fluid Dynamics*, Rikougakusha, Tokyo, Japan, 1994 (in Japanese).
18 Sakaki K, Shimizu Y and Gouda Y, 'Effect of Expansion State of Combustion Gas Jet on High Velocity Oxygen-Fuel Thermal Spraying Process', *J. Japan Institute of Metals*, 1999, **63**(2), 269–276 (in Japanese).
19 Clift R, Grace J R and Weber M E, *Bubbles, Drops and Particles*, Academic Press, New York, 1987.
20 Tanazawa I, *JSME Heat Transfer Handbook*, Japan Society of Mechanical Engineers, Tokyo, Japan, 1993, p. 44 (in Japanese).

8
The role of particle temperature and velocity in cold spray coating formation

T. VAN STEENKISTE, Delphi Research Lab, USA

8.1 Introduction

The cold spray process[1–12] is a method for producing a coating or a free-standing structure by utilizing high-velocity particles, which upon impact with a substrate or previously deposited particles convert the particle's kinetic energy into plastic deformation, strain, and ultimately heat. The particles do not melt and consequently produce coatings with a low degree of oxide or residual stress. Previous works have developed the mechanisms by which the interaction of the high-velocity gas streams couple to the particles to produce the high-velocity particles via drag effects.[1–9] The concept of a critical particle velocity for coating formation[1,2] was developed by Alkhimov, Kosarev, and Papryrin, in the early stages of the process optimization and has been expanded on by various authors.[1–12] Several other developments – the kinetic spray process, which uses larger diameter particles,[13–18] and the radial injection gas dynamic spraying (RIGDS)[19] process for example – also produced similar coatings. RIGDS uses a combination of larger hard particles (materials that do not plastically deform such as ceramics or high-strength materials such as tungsten) with smaller ductile particles to produce coatings. All of these processes rely on the conversion of the particle's kinetic energy to produce either a particle–particle or particle–substrate bonding. The kinetic energy of a particle and the momentum of each particle in the coating formation must be converted into other forms of energy via mechanisms such as plastic deformation (both substrate and particle interactions for the initial particles and particle–particle interactions as the coating formation builds), void consolidation, particle–particle rotation, strain, etc., and ultimately heat. Failure to convert all the incident kinetic energy into heat and strain energy results in an elastic collision, i.e. the particle will bounce off. This chapter will discuss the roles that the particle temperature and particle velocity play in this coating formation process.

8.2 Particle velocity and temperature

The literature has many references for the mechanisms that influence how particles accelerate in the high-velocity gas stream;[5,6,8,9,11] however a brief review is necessary in order to understand the role that particle velocity and temperature parameters play in the cold spray process. A typical cold spray system is shown in Fig. 8.1. The cold spray process involves preheating the main gas flow and combining it with the particles–gas mixture from the high-pressure powder feeder in a premixing chamber. The combination of gas and particles is generally injected axially through the de Laval type nozzle and the particles accelerate due to drag effects from the high-velocity gas. The particles can also be injected downstream of the throat region in the diverging section of the nozzle.[19] The conversion of the particle's kinetic energy to mechanical deformation and thermal energy occurs when the particles impact the substrate surface at high velocity. Particle velocity is influenced by several factors: main gas temperature, molecular

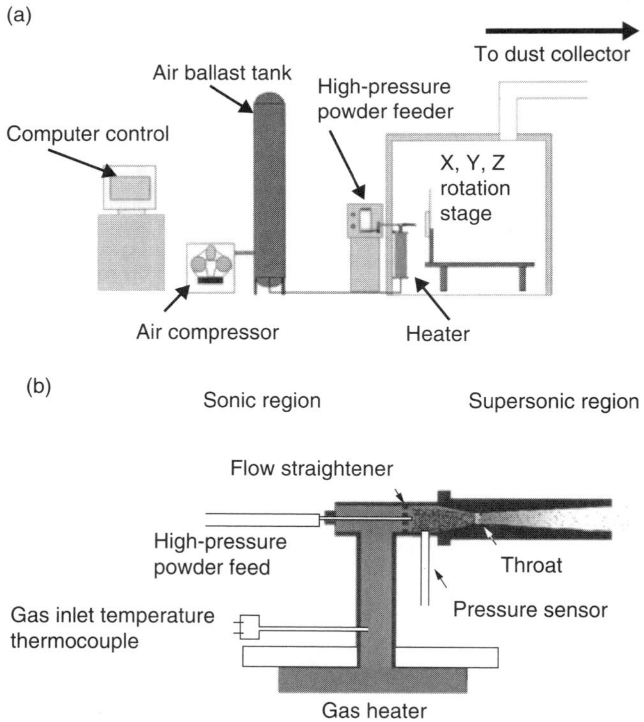

8.1 (a) Schematic diagram of the cold spray equipment. (b) Close-up diagram of the nozzle region.

weight of the main gas, main gas pressure, powder carrier gas effects, particle size, residence time in the high-velocity gas flow, particle shape, particle density, and the interaction with the bow shock wave at the substrate surface to name a few. The relationship that governs the particle velocity inside the nozzle is the coupling of the gas velocity to the particle drag in the gas stream. The cold spray process uses a de Laval converging–diverging nozzle to entrain metal powders in a supersonic airflow. A schematic diagram of the cold spray machine's key components is shown in Fig. 8.1(a) and (b). The nozzle has an entrance cone (converging region), followed by a throat section, and finally down stream of the throat region, the nozzle has a diverging cross-section. Again there are many references[5,20,21] in the literature for optimizing the nozzle design and development. Figure 8.1(b) shows the typical location of the pressure sensor port and the main gas inlet thermocouple in the nozzle assembly. These sensors provide feedback for the computer control of the pre-chamber (before the throat) pressure and the inlet gas temperature. Similarly the powder feeder pressure and feed rate of the particles are monitored and computer controlled. An in-line heater varies the main gas temperature. Generally, the purpose of heating the main carrier gas is not to heat the particles but rather to increase the gas velocity, however increasing the particle temperature does provide significant benefits, as will be discussed later.

Higher gas velocities generally result in higher particle velocities. The speed of sound for a gas is dependent on both the gas temperature as well as the molecular weight of the gas. The equation for the speed of sound v is

$$v = (\gamma RT/M_w)^{1/2} \qquad [8.1]$$

where γ is the ratio of specific heats (1.4 for air, 1.66 for He), R is the gas constant (8314 J/kmol K), T is gas temperature, and M_w is the molecular weight of the gas. A convenient way to increase particle velocity (i.e. increase the gas velocity) is to increase the main gas temperature or use a smaller molecular weight gas such as helium. Increasing the pressure does not result in an increase in gas velocity for the following reason. Once the pressure downstream is 52.8% of the pressure upstream (for air) the flow through the nozzle throat becomes sonic. When the air velocity becomes sonic, further increases in the upstream pressure do not cause a further increase in the air velocity through the nozzle throat (flow is choked). Increasing the upstream pressure, however, increases the density of the gas, since the mass flow rate is also a function of density; the mass flow rate increases linearly with pressure while the gas velocity is constant. This density increase can influence the coupling between the gas velocity and the particle velocity.

Figure 8.2 is a one-dimensional computation of the velocities for the main gas and the particles assuming air as the main gas, inlet gas temperature of 527 °C (980 °F), inlet pressure to be 2.0 MPa (300 psi), and copper as the coating powder. The nozzle has a 100 mm diverging section measured from the throat; with a rectangular exit of 2 mm by 10 mm. Analytic equations were used to compute the gas velocity and temperature in the nozzle from the gas inlet conditions and the nozzle area versus length.[22] Particle velocities in the nozzle were calculated from the drag forces using correlations in the literature.[23] Particle temperatures were calculated using heat transfer correlations[24] (using the same conditions as in Fig. 8.2) and are shown in Fig. 8.3. These simple models, while ignoring boundary layer effects and heat transfer to the nozzle do provide insight into the controlling factors of the particle velocity and particle temperature.

Figure 8.2 shows that the velocity of the gas and particles scale roughly as the square root of the absolute main gas temperature. Figure 8.3 shows the heating of the particles as a function of size and distance from the throat. From Fig. 8.2 and 8.3, one observes that for particles less than 5 microns the response is very similar to that of the gas, a rapid increase in velocity and particle cooling during the trip through the diverging section of the nozzle, while for larger particles (greater than 5 microns), one notes a time-

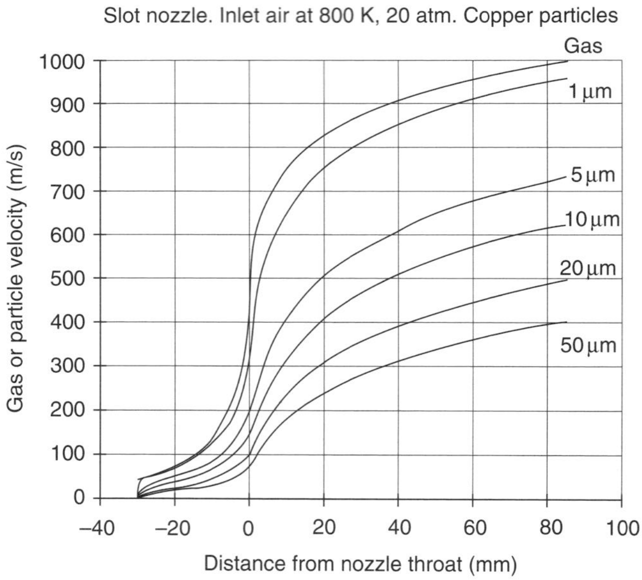

8.2 Computed air and particle velocities as function of distance along the nozzle. Particle diameters denoted in microns for the particle velocity curves.

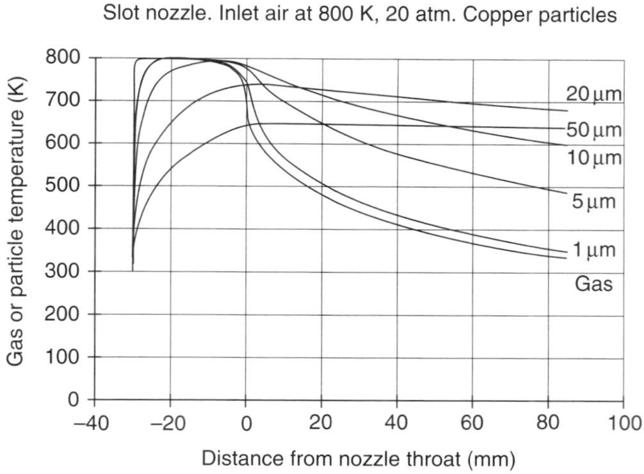

8.3 Computed air and particle temperatures as function of distance along the nozzle. Particle diameters denoted in microns for the particle velocity curves.

dependent rise and fall of the temperature and velocity related to the size of the particles. It is this interaction between the particle and the high-velocity gas that allows one to tailor the different spray processes for coating formation. For example, the kinetic spray process[13–18] uses large particles (greater than 50 microns) to produce coatings. These particles are traveling at lower velocities, but have higher particle temperatures, compared with the cold spray process, which uses higher particle velocities but lower particle temperatures. Again it is this relationship between particle temperature and particle velocity that is important for coating formation and will be discussed further.

As these high-velocity particles approach the substrate they encounter the bow shock wave generated by the high-velocity gas stream impinging on the substrate. Again there are many references in the literature[5,8,21] that describe this phenomenon. The smaller particles (less than 5 microns) follow the gas flow streams and are generally deflected by the bow shock wave, while increasingly larger size particles experience varying degrees of deceleration as they pass through the bow shock wave.[14]

Another consideration is the powder shape. Since one is relying on the coupling of the high-velocity gas stream to accelerate the particles, the particle shape will determine how effective this interaction is. Long aspect shapes (such as needles or flakes) do not present a large cross-sectional area to promote drag effects. This type of particle geometry will tend to orientate into a direction that will present a reduced cross-section perpendicular to the gas flow reducing the drag coupling and consequently the final particle velocity. The particle density is another factor to consider.[12] Particles with a high density, such as tungsten, will accelerate more slowly than particles with a lower density such as aluminum. The very-high-density materials may also require higher velocity powder carrier gas flows to maintain particle suspension in the gas flow for the journey to the nozzle. This higher powder carrier gas flow can lead to a reduced mix gas temperature in the nozzle from the increased dilution of the room temperature powder carrier gas flow.

8.3 The impact of particle velocity and temperature on coating formation

For the particles to adhere, all of the particle's kinetic energy must be transformed into heat and strain energy to the coating and/or substrate. This requires an inelastic collision process such as plastic deformation of the particle and/or the substrate, void consolidation, particle rotation, grain deformation, work hardening, strain, etc. It will be seen that such plastic deformation is in fact found experimentally, with roughly spherical incident particles flattening into a pancake-like structure with an aspect ratio ranging from 3:1 to 5:1 depending on the particle's velocity and temperature, and material composition. Figure 8.4 shows a gas atomized aluminum starting

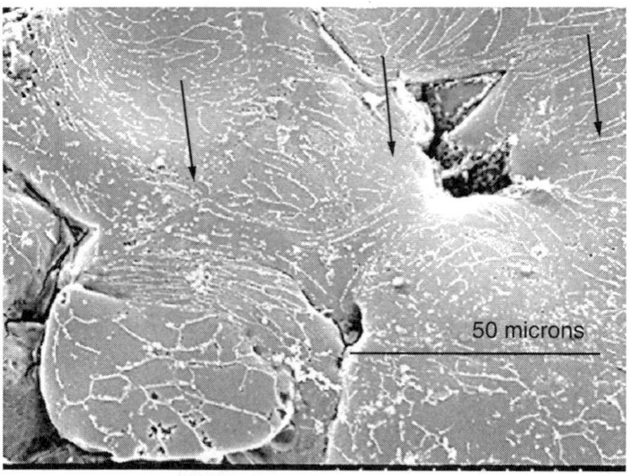

8.4 Electro-polished sample of a coating produced at an inlet temperature of 204 °C (500 °F). Arrows show areas of bonding. Note the degree of plastic deformation that has occurred between particles and the resulting plastic flow of the internal grain structure. The particle in the lower left side of the picture shows internal plastic flow of the grain boundaries at the impact site; however, the original grain structure was preserved in locations removed from the impact area.

powder after impacting on previously deposited particles. In Fig. 8.4 (etched photo) one can clearly see that at the point of impact a high degree of localized plastic deformation has occurred. One can also observe that the particle grain structure for the majority of that particle that is not in the impact zone is relatively unchanged. Evidence for void consolidation and multiple impacts is also shown in Fig. 8.4 as subsequent particle collisions with the existing particles on the substrate surface force material into the voids between the previous deposited particles (see arrows). Kinetic energy and momentum from the incoming high-velocity particles on impact with the deposited particles are used to build the coating. The coating build-up is not a simple one-particle impact, but rather a series of multiple impacts that transfer the incoming particle's kinetic energy to the substrate initially then to the coating. A multistep process has been suggested[15] consisting of substrate cratering and first-layer build-up, followed by particle deformation and realignment, metallurgical bonding and void reduction, etc.; this is shown schematically in Fig. 8.5 as a possible coating formation mechanism. The initial particles impact the substrate and bonding is presumed to be the result of adiabatic shear.[4,25] Except for low melting point metals, one does

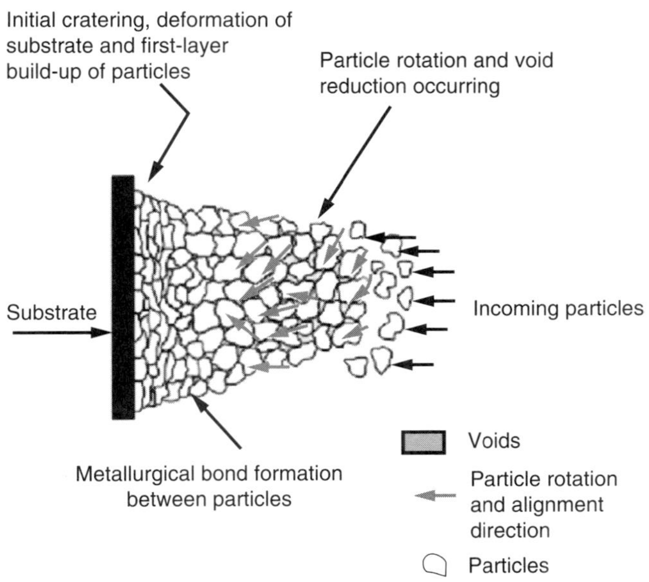

8.5 Multi-stage coating formation using the cold spray process. Substrate cratering and first-layer build-up of particles, particle rotation, void reduction, particle deformation and realignment, and metallurgical bond formation between particles occurs.

not observe any evidence of bulk melting of the particle and the original grain structure is preserved, although highly deformed. The amounts of oxygen in the powders and the coatings are determined to be approximately the same,[6,15,16] supporting the presumption that there is little or no melting of the particles. Evidence has been presented in the literature to support this impact behavior and numerous micrographs in the literature[1–21,25,26] demonstrate the plastic deformation of the particles similar to those observed in Fig. 8.6, 8.7, and 8.8 for example. One also observes that the

(a)

(b)

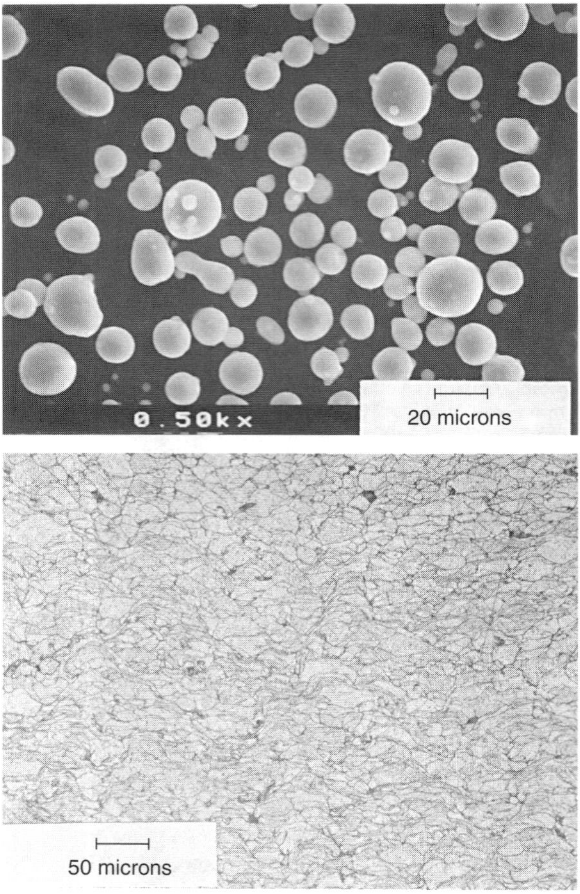

8.6 (a) Scanning electron microscope picture of powders cold sprayed to make the coatings; Al powder, 20 μm average diameter, obtained from Valimet, California, USA. (b) Microstructure of kinetic sprayed coatings using a nozzle inlet pressure of 1.7 MPa, inlet temperature of 340 °C, powder feed rate of 65 g/min, and stand-off 1.9 cm.

(a)

(b)

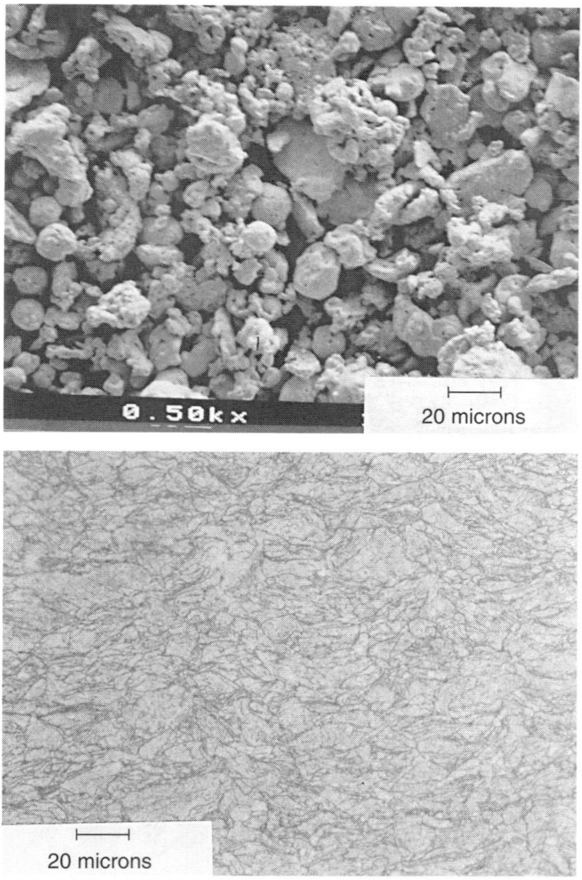

8.7 (a) Scanning electron microscope picture of powder cold sprayed to make the coatings. Cu powders, average particle size <45 μm (–325 mesh), obtained from Miller, Texas, USA. (b) Etched scanning electron microscope photo of the microstructure of cold sprayed coatings: Cu coating on Al substrate using a nozzle inlet pressure of 1.7 MPa, inlet temperature of 590 °C, powder feed rate of 113 g/min, and stand-off 1.9 cm.

very last particles that form a coating are generally not as compacted as the rest of the coating, since these are the last particles to arrive and will not be impacted by any other particles.

Early experiments were conducted at the General Motors Research Laboratories under the NCMS (National Center for Manufacturing Science) consortium to determine what spray parameters are important to coating formation. Figures 8.6, 8.7, and 8.8 demonstrate the plastic deformation that produces the pancake-like structure for a typical cold spray coating.

(a)

(b)

8.8 (a) Scanning electron microscope picture of powder cold sprayed to make the coatings. Fe powders, average particle size <45 μm (–325 mesh), obtained from Praxair, Indiana, USA. (b) Scanning electron microscope photo of microstructure of cold sprayed coatings: Fe coating on Al substrate using a nozzle inlet pressure of 2.0 MPa, inlet temperature of 590 °C, powder feed rate of 55 g/min, and stand-off 1.9 cm.

Figure 8.6(a) is a scanning electron microscope (SEM) photo of the original aluminum gas atomized starting powder. Figure 8.6(b) is the SEM photo of the etched aluminum coating produced on an aluminum substrate using the following spray parameters: nozzle inlet pressure, 1.7 MPa; inlet temperature, 340 °C; powder feed rate, 65 g/min; stand-off, 1.9 cm. Figure 8.7(a) is an SEM photo of the original copper starting powder and Fig. 8.7(b) is an etched SEM photo of the cold sprayed copper coating on aluminum substrate sprayed with a nozzle inlet pressure of 1.7 MPa, inlet temperature of

590 °C, powder feed rate of 113 g/min, and stand-off 1.9 cm. Figure 8.8(a) is the SEM photo of the iron starting powder and Fig. 8.8(b) is the SEM photo of an iron coating on aluminum substrate sprayed using a nozzle inlet pressure of 2.0 MPa, inlet temperature of 590 °C, powder feed rate of 55 g/min, and stand-off 1.9 cm. In all cases one observes plastic deformation of the original particle structure and no bulk melting.

There exists a critical velocity region, below which the particles do not form coatings and above which coatings rapidly build to millimeter or thicker dimensions.[1–19,25,26] This critical velocity is both material and size dependent and is also a function of the physical and mechanical properties of the powder material. Experiments have shown that materials with higher moduli are more difficult to form into coatings,[1–19,25,26] requiring higher main gas temperatures (i.e. increased particle velocity and increased kinetic energy), or incorporation of the materials into a ductile matrix (ceramic–metal composites for example).[18]

Figures 8.9, 8.10, and 8.11 are coating thickness measurements as a function of increasing main gas temperatures (i.e. increasing particle velocity) for aluminum (20 µm size range), copper (<45 µm size range), and iron (<45 µm size range) particles. For the aluminum coatings in Fig. 8.9 the deposition started at around 200 °C to 250 °C, for copper it started at around

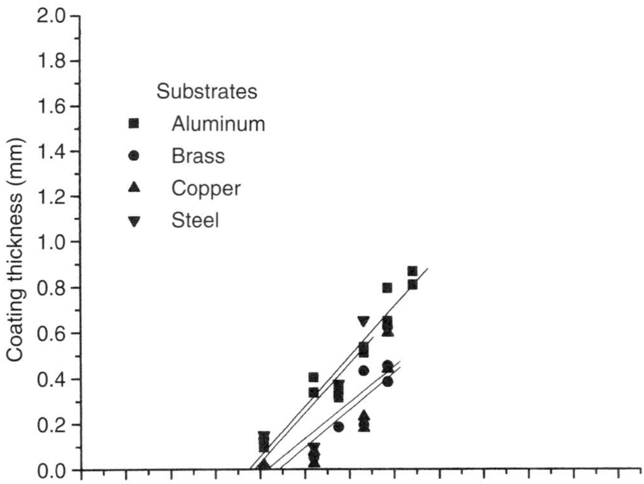

8.9 Coating thickness as a function of inlet nozzle gas temperature for aluminum powder sprayed onto aluminum, brass, copper, and steel substrates.

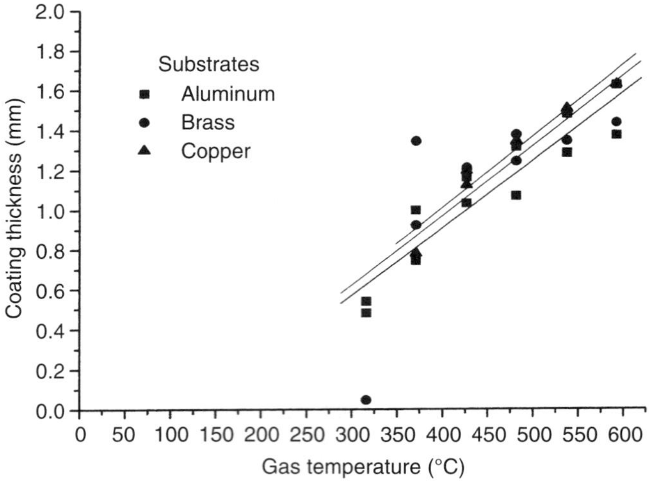

8.10 Coating thickness as a function of nozzle gas inlet temperature for copper powder onto aluminum, brass, and copper substrates.

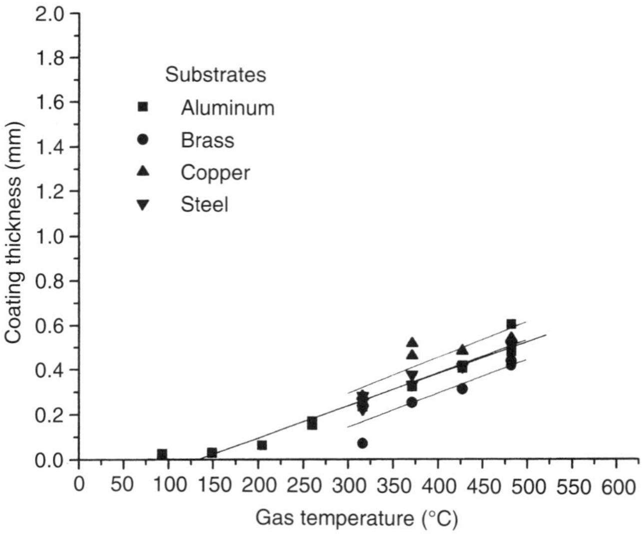

8.11 Coating thickness as a function of nozzle gas inlet temperature for iron powder sprayed onto aluminum, brass, copper, and steel substrates.

325 °C to 350 °C and for iron at around 310 °C to 350 °C, depending on the substrate. As discussed previously and calculated in Fig. 8.2, one can expect that particle velocities will increase roughly as the square root of the inlet gas temperature. Using the approximate square root dependence, one finds from Fig. 8.2 that a 10 μm Cu particle would have a threshold velocity of about 470 m/s at these main gas temperatures. This compares well with the value of around 500 m/s measured by Alkhimov *et al.*,[1] and others,[6,11,25] for a 10 μm Cu particle.

Generally the critical velocity will be a range of velocities, since the starting powder feed stock is usually a distribution of particles sizes.[6,10,11,25] Individual particles with different diameters will achieve different velocities due to the drag interactions with the high-velocity gas flow as previously stated and shown in Fig. 8.2. However, the dependence on a critical velocity for coating formation is only one of the critical aspects used to determine the coating formation. Equally important is the particle temperature. These two parameters mainly determine whether a particle will produce a coating or not. The particle velocity will determine the amount of kinetic energy available for plastic deformation. The particle temperature will determine the mechanical properties of that particle at the moment of impact.

Substrate properties also contribute to coating formation; these are discussed in the literature[25–27] and will not be discussed in this chapter. Generally, a roughened-surface substrate (i.e. sandblasted) will promote higher deposition efficiency, since the substrate surface does not have to be cleaned and roughened using the first incoming particles.

The yield stress of the particle and/or the substrate must be exceeded during the collision, and is a part of the necessary energy transformation. For this reason, powders and substrates typically employed in the cold spray process are metals with relatively high plasticity. Ceramic and other high-strength materials are very difficult to spray and to date the only coatings that have been produced are essentially a few particle layers thick with particles embedding into the substrate and little if any particle–particle bonding. Weak mechanical interlocking, severe plastic deformation of the substrate, and fracturing of the brittle materials are the usual features of such coatings. An example is shown in Fig. 8.12, for a coating produced using a very brittle Al–Cu–Fe quasi-crystal powder. Little evidence for particle–particle bonding is observed and the particle–substrate bonding is a mixture of mechanical interlocking and embedment into the substrate surface. Fracturing of the Al–Cu–Fe quasi-crystal particle material is readily observable as shown in Fig. 8.12, and deposition efficiency was very low. Similar results have also been observed in the literature for other brittle materials.[28]

It has been observed in our laboratory and in the literature[18,28] that mixtures of metals and ceramics can be sprayed successfully. Figure 8.13 shows

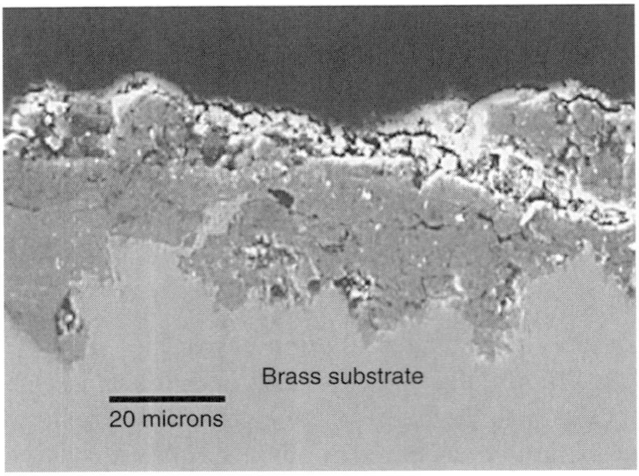

8.12 Quasi-crystal (Al–Fe–Cu) coating on a brass substrate. Note the severe deformation of the brass substrate by the particles. The particles are highly fractured and mechanically interlocked with each other. Deposition efficiency is very low.

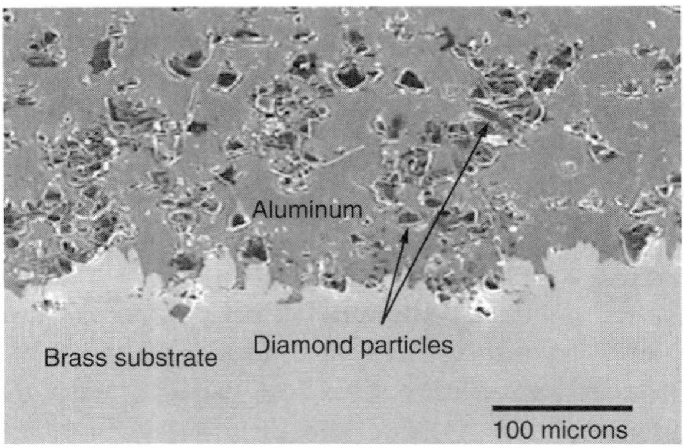

8.13 Cross-sectional scanning electron micrograph of the Al–diamond composite on a brass substrate (bottom). No evidence of plastic deformation of the diamond particles is observed.

an SEM of an aluminum–diamond coating. The diamond particles do not plastically deform and become incorporated into the ductile aluminum matrix during the coating formation. The ductile matrix incorporates the material that does not plastically deform. Deposition efficiencies are generally lower since the amount of incorporation of the non-ductile components into the coatings is usually less than the original starting percentage. The RIGDS method[19] uses mixtures of metals and ceramics to produce coatings:

here the energy of the non-deforming material helps the ductile materials to form coatings. An analogy would be a hammer and anvil method used to form steels, where some of the incoming particles are sandwiched between the substrate and the non-deforming particles. Cold sprayed coatings can also be successfully formed on ceramic and other hard substrates. In this case, presumably the ductility of the powders is sufficient to allow the coatings to form since all the plastic deformation must occur in the particles.[28]

Detailed numerical computations of particle deformations and crater formations in the substrate for smaller particles have been presented in the literature.[4] One's understanding of the phenomena involved is aided by the following simple approximate computations for particle collision stresses and times. The collision time can be approximated as d/v, where d is the particle diameter and v is the incident velocity of the particle. For an example case of a 50 micron diameter particle incident at 500 m/s, the time Δt for the collision process to take place is $\Delta t \cong d/v = 10^{-7}$ s. The impact stress S averaged over the time Δt may be estimated as $\Delta P/A\,\Delta t$, where A is the cross-sectional area of the (flattened) particle and ΔP is the change in particle momentum as a result of the collision. Here $\Delta P = mv$, the incident particle momentum, and $A \cong \pi d^2$, where an average aspect ratio of 4 has been taken. Then the time-averaged impact stress S is given by $S \cong \Delta P/A\Delta t = \rho v^2/6$, where ρ is the density of the material of the incident particle. For aluminum particles, $\rho = 2.4 \times 10^3$ kg/m^3, so $S \cong 100$ MPa. This is significantly larger than a typical yield point of pure aluminum of 34.5 MPa.[29] This is consistent with the observed flattening of the particles in the cold spray process, i.e. plastic deformation is observed. Much higher stresses may also occur over small localized impact regions of the particle due to effects of shock waves and the percentage contact area at the moment of impact. The peak impact stress for copper particles impinging on stainless steel was found in reference 4 to be 5 GPa. Again, typical yield strength values for copper are approximately 250 MPa.[29] Sufficient energy exists to plastically deform these materials by the conversion of the particle's kinetic energy at these velocities, temperatures, and sizes.

One also observes in the literature that comparable coatings can be produced using much larger particles that are traveling slower (300–500 m/s) and are at higher particle temperatures (several hundred Kelvin), as in the kinetic spray process,[13–18] compared with the cold spray process.[1–12] These larger particles also exhibit a critical velocity range that is much slower than the critical velocities of the smaller cold spray particles.

Clearly if the incident velocity were too low, the yield strength of the particle and/or substrate material would not be exceeded and the particle would not stick. Oxide coatings on the incident particle and on the substrate may also play a role in this critical velocity, as oxides tend to be more brittle than metals. Fracturing these oxide shells becomes a necessary part of coating formation, and clearly the incident velocity must be sufficiently high for this to also occur.

So how does one account for the fact that larger, slower particles, with velocities below the critical velocity of the faster, smaller size particles can also form coatings with comparable properties? One needs to understand the particle temperature and the role it plays in impact coating formation. The particle temperature affects the physical and mechanical properties of the particle at the point of impact. It is well known that the yield strength of materials, hardness, shear modulus, ultimate tensile strength, etc. are a function of temperature. Reference literatures for various materials and alloys have measured values for these properties as a function of temperature. Generally increasing temperatures result in a decreasing mechanical performance. Since particles at higher temperatures need less kinetic energy to plastically deform, increasing the particle temperature should allow one to decrease the particle velocity necessary to achieve the same deposition efficiency for the coating.

The measurement of particle velocity is usually done by laser illumination of the particles as they travel between two detectors. Knowing the laser pulse time and the distance traveled (between the laser pulses) by the particles allows one to calculate the particle velocity. Control Vision (Boise, Idaho, USA) and Spray Watch (OSEIR Ltd, Finland) are two manufacturers of such equipment. The particle temperature is much more difficult to measure. Since the particles are 'cold' and do not radiate visible radiation like the incandescent thermal spray particles, the measurement of the particle temperature is usually calculated from the known spray parameters and the equations used to calculate the curves in Fig. 8.3. Many other references[11,14] (larger diameter particles) using more detailed calculations are also found in the literature.

8.4 Altering particle temperature

Particle temperature can be increased by several methods. The simplest way is to increase the main gas temperature. This gives the added bonus of increased particle velocity; however, there are physical limitations on the heater elements if one wants to maintain a heater having small weight, size, and electrical consumption. Generally, for electrical resistance heaters a practical upper bound would be around 700 °C. Another method involves changing the ratio of the powder feed gas, which is injected into a pre-chamber (see Fig. 8.1) with the main gas (at elevated temperature). The powder feed gas is generally at room temperature and carries the powder into the heated main gas flow. This reduces the temperature of the combined gas flows into the de Laval nozzle. Particle temperature increases of the order of 50 K have been reported by changing this ratio.[14] Larger particle feedstock powders also provide for higher particle temperatures since the residence time in the main gas flow is longer (slower velocity due to mass effects and negligible cooling in the diverging section of the de Laval nozzle, see Fig. 8.2 and 8.3); these powders have been used successfully by the kinetic spray process[13–18,26] to form coatings.

Increasing the residence time by inserting an extension before the nozzle has also been successfully used to increase the particle temperature[30] and is an optional part of the new Kinetiks 4000 Cold Spray System.[31] These devices are designed to minimize both the disturbance to gas flow and the wall contact prior to the particles' entry into the nozzle, while increasing the residence time and particle temperature. Increasing the length of the convergent section of the nozzle has also been used to increase the particle temperature.[32]

The use of heaters on the powder feeder lines is another option that heats the powder carrier gas and subsequently heats the particles in this higher temperature gas flow. Again this would give an additional benefit of not reducing the temperature of the main gas flow when injected into the gas stream by virtue of the powder carrier gas being hotter. Depending on the material properties of the powder the delivery can become more complicated and it can be harder to maintain a precise mass flow of powder, since hot powders generally become 'sticky', clumping together and limiting their ability to flow well. Nozzle clogging and sticking to interior walls of the nozzle assembly and/or powder feed lines can also start to occur without properly designed components and attention to detail when using heated particles.[14] Longer nozzles can also increase the particle temperature for larger size particles (see Fig. 8.3), but may reduce the particle temperature for smaller size particles, again due to the expansion of the gas in the de Laval nozzle and increased particle contact with the cooler gas stream. Oxidation of the particle's surface is another factor of concern at higher particle temperatures when the carrier gas is air. Experiments[26] have measured a doubling of the oxide levels with the use of increased powder gas heating, when using air. Nitrogen, helium, or other inert processing gases would be required when elevating particle temperatures above the point of oxidation.

Experiments using nickel as the coating material were undertaken to observe what effect particle temperature would have on coating formation as all other spray parameters are held constant. Powder line heating and larger diameter particles were used to increase the particle temperatures. Figure 8.14 shows the experimental setup used in the experiments Heating tape and insulation were wrapped around the powder feeding line and the temperature of the mixed gas–particle measured at the entrance just before injection into the nozzle. Adjusting the pressure differential between the powder feeder and the main gas pressure in the mixing chamber and/or the

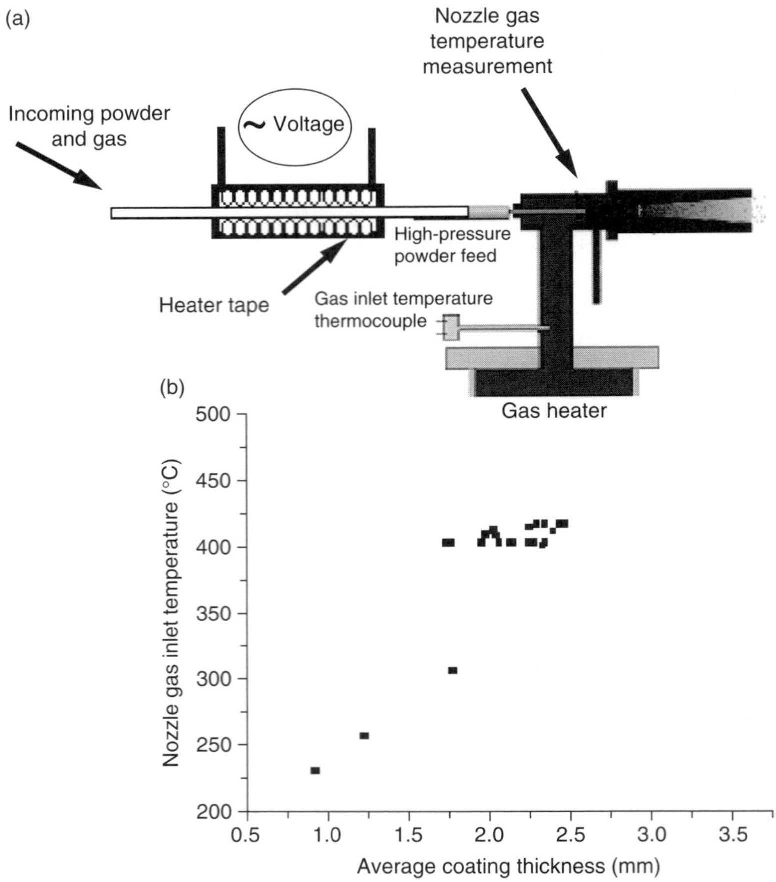

8.14 (a) Experimental setup. (b) Nickel coating thickness as a function of increasing particle temperature. An external electrical powder line heater heats the particle carrier gas and particles, which then mixes with the main gas flow. As nozzle inlet temperature and particle temperature increase, the coating thickness and deposition efficiency increase for the fixed operating spray parameters.

temperature of the heating tape by adjusting the voltage applied to the heating tape controlled the temperature of the carrier gas–particle mixture. Figure 8.14 also shows a graph of gas–particle mixture temperature (all other spray parameters fixed) as a function of coating thickness. By adjusting only the particle carrier gas temperature (i.e. particle temperature) the coating thickness increased from 1 mm to over 2.5 mm for an approximate 200 °C gas temperature increase. The increased particle temperatures (with some increased particle velocity addition from the increased mixed gas temperature) changed the material properties of the particles in such a way that the ensemble was able to bond more efficiently, resulting in thicker coatings and increased deposition.

8.5 Conclusions

Many factors contribute to the formation of cold spray coatings. Optimizing both the particle velocity and the particle temperature are among the most critical. These two parameters closely interact with each other to produce coatings. The particle velocity determines the amount of kinetic energy available for plastic deformation and bonding. The particle temperature determines the physical properties of the particle at the moment of impact. Increasing the particle temperature generally can result in a lowering of the critical velocity, while decreasing the particle temperature will generally result in a need for higher particle velocity. Cold spray coating formation of high-strength, low-ductility, higher service temperature materials will demand that processes operate with higher main gas heater temperatures (to achieve increased critical velocities) in conjunction with higher particle temperature heating methods. Ultimately, it is the balance between the kinetic energy and momentum of each individual particle that must be 100% transformed by plastic deformation, strain, void consolidation, heat, etc. that will allow one to build coatings using the cold spray process. Explosive compaction experiments have successfully compacted ceramic powders.[33] Could the cold spray process be used for ceramic coatings some day? Theoretically if one could further increase particle velocities and increase particle temperatures there should be an operating window of spray parameters that would produce ceramic coatings. The benefits of the cold spray process that one realizes from not melting particles – low residual stress, no oxide formation, no thickness limitations, etc. – have demonstrated that the cold spray process is a viable alternative to some of the more traditional coating processes, such as thermal spray.

8.6 Acknowledgments

The author would like to acknowledge numerous useful conversations with Drs John Smith, Taeyoung Han, Alaa Elmoursi, and Zhibo Zhao, and Mr Bryan Gillispie of the Delphi Research Laboratories, as well as the assistance of Mr Daniel Gorkiewicz in the spraying of the coatings and Dr Richard Teets for the particle temperature and velocity computations.

8.7 References

1 Alkhimov A P, Papyrin A N, Kosarev V F, Nestorovich N I and Shushpanov M M, A Method of Cold Gas-Dynamic Deposition, *Doklady Akademii Nauk SSSR*, **315**, 1990, 1062–1065.

2 Alkhimov A P, Papyrin A N, Kosarev V F, Nesterovich N I and Shushpanov M M, U. S. Patent 5302414, *Gas Dynamic Spraying Method for Applying a Coating*, April 12, 1994.
3 Mccune R C, Papyrin A N, Hall J N, Riggs II W L and Zajchowski P H, An Exploration of the Cold Gas-Dynamic Spray Method for Several Materials Systems, Proceedings of the 8th National Thermal Spray Conference, C.C. Berndt and S. Sampath (Eds), ASM International, Materials Park, OH, USA, 1995.
4 Dykhuizen R C, Smith M F, Gilmore D L, Neiser R A, Jiang X and Sampath S, Impact of High Velocity Cold Spray Particles, *Journal of Thermal Spray Technology*, **8**(4), 1998, 559–564.
5 Dykhuizen R C and Smith M F, Gas Dynamic Principles of Cold Spray, *Journal of Thermal Spray Technology*, **7**(2), 1998, 205–212.
6 Gilmore D L, Dykhuizen R C, Neiser R A, Roemer T J and Smith M F, Particle Velocity and Deposition Efficiency in the Cold Spray Process, *Journal of Thermal Spray Technology*, **8**(4), 1999, 576–582.
7 Smith M F, Brockmann J E, Dykhuizen R C, Gilmore D L, Neiser R A and Roemer T J, Cold Spray Direct Fabrication – High Rate, Solid State, Material Consolidation, *MRS Symposia Proceedings*, Vol. 542, pp. 65–67, Materials Research Society, Pittsburgh, PA, USA, 1999.
8 Alkhimov A P, Kosarev V F and Papyrin A N, Gas-Dynamic Spraying, Study of Plane Supersonic Two Phase Jet, *Journal of Applied Mechanics and Technical Physics*, **38**(2), 1997, 176–183.
9 Alkhimov A P, Kosarev V F and Papyrin A N, Gas-Dynamic Spraying. An Experimental Study of the Spray Process, *Journal of Applied Mechanics and Technical Physics*, **39**(2), 1998, 183–188.
10 Alkhimov A P, Gulidov A I, Kosarev V F and Nesterovich N I, Specific Features of Microparticle Deformation Upon Impact on a Rigid Barrier, *Journal of Applied Mechanics and Technical Physics*, **41**(1), 2000, 204–209.
11 Kreye H and Stoltenhoff T, Cold Spray-Study of Process and Coating Characteristics, *Thermal Spray: Surface Engineering via Applied Research*, C.C. Berndt (Ed.), ASM International, Materials Park, OH, USA, p. 419, 2000.
12 Vicek J, Huber H, Voggenreiter H, Fischer A, Lugscheider E, Hallen H and Pache G, Kinetic Powder Compaction Applying the Cold Spray Process. A Study On Parameters, *Thermal Spray 2001: New Surfaces for a New Millennium*, C.C. Berndt, K. A. Khor, and E. F.Lugscheider (Eds), ASM International, Materials Park, Ohio, USA, 2001.
13 Van Steenkiste T H, Smith J R, Teets R E, Moleski J J and Gorkiewicz D W, U. S. Patent 6139913, *Kinetic Spray Coating Method and Apparatus*, October 31, 2000.
14 Han T, Zhao Z B, Gillispie B and Smith J R, A Fundamental Study of Kinetic Spray Process, *Proceedings of the International Thermal Spray Conference, 2004 Thermal Spray Solutions: Advances in Technology and Application*, Osaka, Japan 2004.
15 Van Steenkiste T H, Smith J R and Teets R E, Aluminum Coatings via Kinetic Spray with Relatively Large Powder Particles, *Surface and Coatings Technology*, **154**, 2002, 237–252.
16 Van Steenkiste T H, Smith J R, Teets R E, Moleski J J, Gorkiewicz D W, Tison R P, Marantz D R, Kowalsky K A, Riggs II W L, Zajchowski P H, Pilsner B,

Mccune R C and Barnett K J, Kinetic Spray Coatings, *Surface and Coatings Technology*, **111**, 1999, 62–71.

17 Zhao Z B, Gillispie B A and Smith J R, Coating Deposition by the Kinetic Spray Process, *Surface and Coatings Technology*, **200**, 2006, 4746–4754.

18 Elmoursi A, Van Steenkiste T, Gorkiewicz D and Gillispie B, Fracture Study of Aluminum Composite Coatings Produced by the Kinetic Spray Method, *Surface and Coatings Technology,* **194**, 2005, 103–110.

19 Gr Maev R and Leshchynsky V, Air Gas Dynamic Spraying of Powder Mixtures: Theory and Application, *Journal of Thermal Spray Technology*, **15**(2), 2006, 1–8.

20 Alkhimov A P, Kosarev V F and Klinkov S V, The Features of Cold Spray Nozzle Design, *Journal of Thermal Spray Technology*, **10**(2), 2001, 375–381.

21 Champagne V, Helfritch D, Leyman P, Lempicki R and Grendahl S, Nozzle Design Influence on the Supersonic Particle Deposition Process, *Cold Spray 2004: An Emerging Spray Coating Technology*, September 2004, Akron, OH, USA.

22 Anderson J D JR., *Modern Compressible Flow*, McGraw-Hill, New York, 1982.

23 Henderson C B, Drag Coefficients of Spheres in Continuum and Rarefied Flows, *AIAA Journal*, **14**, 1976, 707–721.

24 Carlson D J and Hoglund R F, Particle Drag and Heat Transfer in Rocket Nozzles, *AIAA Journal*, **2**, 1964, 1980–1984.

25 Garetner F, Borchers C, Stoltenhoff T and Kreye H, Numerical and Microstructural Investigations of the Bonding Mechanisms in Cold Spraying, *Proceedings of the International Thermal Spray Conference*, 2003, Orlando, FL, USA, 2003.

26 Van Steenkiste T and Smith J R, Evaluation of Coatings Produced via Kinetic and Cold Spray Processes, *Journal of Thermal Spray Technology*, **13**(2), 2004, 274–282.

27 Sakaki K, Tajima T, Li H, Shinkai S, Shimizu Y and Nagano J, Influence of Substrate Conditions and Traverse Speed on the Cold Sprayed Coatings, *Proceedings of the International Thermal Spray Conference*, 2004, Osaka, Japan.

28 Yong Lee H, Jung S H, Lee S Y, You Y H and Ko K H, Correlation Between Al_2O_3 Particles and the Interface of Al-Al_2O_3 Coatings by Cold Spray, *Applied Surface Science*, **252**, 2005, 1891–1898.

29 Buch A, *Pure Metal Properties: A Scientific–Technical Handbook*, ASM International, Materials Park, OH, USA, 1999.

30 Zhao Z B, Gillispie B A, Han T, Smith J R and Fuller B K, U. S. Patent Application 20050040260, *Coaxial Low Pressure Injection Method and a Gas Collimator for a Kinetic Spray Nozzle,* February, 2005.

31 Kinetiks 4000 Cold Spray System, Cold Gas Technology GmbH, Ampfing, Germany.

32 Sakaki K and Shimizu Y, Effect of the Increase in the Entrance Convergent Section Length of the Gun Nozzle on the High Velocity Oxygen Fuel and Cold Spray Process, *Journal of Thermal Spray Technology*, **10**(3), 2001, 487–496.

33 Murr E, Staudhammer K P and Meyers M A (Eds), *Metallurgical Applications of Shock-Wave and High-Strain-Rate Phenomena*, Marcel Dekker, New York, 1986.

9

Particle/substrate interaction in the cold-spray bonding process

M. GRUJICIC, Clemson University, USA

9.1 Introduction

Since a general background for the cold-gas dynamic spray process can be found in references 1 to 5 and in the preceding chapters of this book, we will turn our attention in this chapter to the problem of particle/substrate interactions and bonding during cold spray. The actual mechanism by which the solid particles deform and bond during cold spray is still not well understood. It is well established, however, that the feed-powder particles and the substrate/deposited material undergo an extensive localized deformation during impact. This causes disruption of the thin surface (oxide) films and enables an intimate conformal contact between the particles and the substrate/deposited material. The intimate conformal contact of clean surfaces combined with high contact pressures are believed to be necessary conditions for particles/substrate and particles/deposited material bonding. This hypothesis is supported by a number of experimental findings such as:[3]

1 A wide range of ductile (metallic and polymeric) materials can be successfully cold sprayed while non-ductile materials such as ceramics can be deposited only if they are co-cold sprayed with a ductile (matrix) material.
2 The mean deposition particle velocity should exceed a minimum (material-dependent) critical velocity to achieve deposition, which suggests that sufficient kinetic energy must be available to plastically deform the solid material and/or disrupt the surface film.

Due to their highly transient non-linear dynamic and contact nature, the interactions between particles and the substrate (including the nature of the particle/substrate bonding) during the cold gas process are very difficult to analyze using experimental means. Post-mortem-type microstructural and micro-analytical examinations, while very beneficial, have difficulties in helping to identify the relative contribution of various bonding mechanisms.

As a result, computational simulations of the particle/substrate interactions and the attendant phenomena have been an important technique for elucidating the bonding mechanism(s) accompanying cold spray. Consequently, a brief overview is given in Section 9.2 of the fundamental concepts and the governing equations used during transient non-linear dynamics simulations of the cold-gas particle/substrate interactions.

A number of phenomena that are frequently responsible for inter-phase bonding – such as interfacial melting, atomic inter-diffusion, plastic deformation, etc. – are now believed not to play a significant role in the particle/substrate bonding during cold spray. This conclusion is based on the following arguments:[6]

1 The average kinetic energy of depositing particles at impact is typically significantly lower than the energy required to melt the particle/substrate interfacial region, suggesting that particle/substrate bonding is primarily, or perhaps entirely, a solid-state process. The lack of melting was directly confirmed through micrographic examination of the cold-sprayed materials.[2]
2 Due to very short particle/substrate contact times, atomic inter-diffusion is not expected to play a significant role in particle/substrate bonding. This can be readily proven as follows: the metal–metal inter-diffusion coefficient at temperatures near the melting point is of the order of 10^{-15} to $10^{-13}\,m^2/s$, and for a typical particle/substrate contact time of 40 ns, the atomic inter-diffusion distance is between 0.004 and 0.1 nm. Since this distance is only a fraction of the inter-atomic distance, atomic diffusion at the particle/substrate interface should be excluded as a dominant particle/substrate bonding mechanism under the dynamic cold-spray deposition conditions.[6]
3 Plastic deformation is generally considered to be a mechanism for energy dissipation, i.e. a source of heat in the interfacial region. However, as mentioned earlier, typically the temperature increase in the interfacial region is not sufficient to produce interfacial melting or to significantly promote atomic inter-diffusion, leaving the latter two phenomena (points 1 and 2) to play a relatively minor role in cold-spray particle/substrate interfacial bonding.

Adhesion is an atomic length-scale phenomenon and its occurrence is controlled by the presence of clean surfaces and high contact pressures to make the two surfaces mutually conforming. The strength of adhesion then depends on the (attractive or repulsive) character of atomic interactions (governed by atomic electron orbital localization/delocalization tendencies) and by the crystallographic and orientation details of the contact surfaces. While adhesion can be modeled using quantum mechanical and atomic-scale simulations,[7] such techniques have not yet been applied to the case of cold-spray particle/substrate bonding. Nevertheless, one can generally infer the character of atomic interactions by examining the corresponding binary phase diagram. Specifically, the existence of a solubility limit is an indication of repulsive atomic interactions, while the formation of an intermediate/intermetallic compound is an indication of attractive atomic interactions. Complete mutual solubility of two metals in the solid state indicates weak repulsive or attractive atomic interactions. Again, it should be noted that for efficient adhesion to take place at the particle/substrate contact surfaces, such surfaces must be clean and subjected to high contact pressure. As will be discussed in greater detail in Section 9.3, adiabatic shear instability and the associated localization of the plastic flow into interfacial jets plays a critical role in ensuring cleanliness of the particle and substrate contact surfaces during cold spray. Furthermore, high particle velocities ensure the necessary high levels of particle/substrate contact pressure.

In addition to the factors discussed above, the overall strength of adhesion-based particle/substrate bonding depends on the size of the particle/substrate contact surface area. It is generally observed that stronger bonding occurs when the interfaces are wavy and when they contain vortex-like features.[6] This observation can be, at least partly, attributed to the associated increase in the interfacial area. However, as discussed in the context of explosive welding,[8] interfacial perturbations (roll-ups and vortices) can give rise to nano-/micro-scale material mixing and mechanical interlocking, both of which enhance the interfacial bonding. A more detailed discussion of these interfacial instability-based phenomena is presented in Section 9.4.

9.2 Transient non-linear dynamics analysis of the cold-gas particle/substrate interactions

The majority of the computational analyses of the cold-gas single-particle/substrate interaction process reported in the literature were carried out using one of the following four finite differences/finite elements software packages:

1. The CTH (Eulerian) computer code which was developed at the Sandia National Laboratory for modeling a wide range of solid dynamics problems involving shock wave propagation and material motion.[9]
2. AUTODYN, a general purpose (Lagrangian/Eulerian) transient non-linear dynamics modeling and simulation software particularly suited for modeling explosion, blast, impact and penetration events, developed by Century Dynamics Inc.[10]

3 LS-DYNA, a (Lagrangian) general-purpose transient non-linear implicit/explicit finite element program, developed by Livermore Software Technology Corporation.[11]
4 ABAQUS/Explicit, a (Lagrangian) general-purpose non-linear finite element analysis program developed by Abaqus Inc.[12]

Within a Lagrangian code, the computational mesh is attached to and deforms with the material. In the case of an Eulerian code, on the other hand, the computational grid is fixed in space, while the materials pass through it during a computational analysis. Regardless of the computer program used, a typical computational analysis of the particle/substrate interactions involves the solution of the governing (conservation) equations coupled with the materials modeling equations and subjected to the appropriate initial and boundary conditions. In the remainder of this section a more detailed discussion is presented regarding the basic governing equations implemented in the CTH code, the numerical method used for obtaining the solution as well as the use of the CTH in modeling particle/substrate impact. Most of the details presented are equally applicable to the other computer codes.

9.2.1 Particle/substrate interaction problem definition

While the angle at which cold-spray particles impact the substrate is known to affect the deposition efficiency,[3] most of the simulations reported in the literature deal with normal-angle impacts. There are two main reasons for this: (a) normal-angle impacts enable the use of an axi-symmetric geometry avoiding the need for computationally costly three-dimensional simulations; and (b) typically, the main objective of the computational analyses is the elucidation of the particle/substrate bonding mechanism(s), which is affected by the particles and substrate materials, particle velocity and temperature, but not by the impact angle.

In a typical computational analysis, the substrate is modeled as a thick semi-infinite cylindrical plate while a micron-size sphere is used to represent the particle. The initial temperature within the particle and the substrate is assumed to be uniform (typically equal to the room temperature) while particle velocity is varied in a range (typically 400–1000 m/s). The computational domain containing a single particle and the substrate is discretized using a mesh consisting of quadrilateral cells. To ensure a fine cell size in the particle/substrate contact region, an adaptive mesh refinement strategy implemented in the CTH code is generally used. This strategy is block-based, with the blocks (each containing the same number of cells) connected in a hierarchal manner with the adjacent ones and with a maximum cells-size ratio limited to 2:1. An example of the hierarchal block structure is displayed in Fig. 9.1. The interfacial high-resolution region of the mesh is created by splitting blocks midway along each of the coordinate directions. This produces four child blocks per parent block for a two-dimensional problem. Likewise, four blocks are recombined to form a single block when such blocks are located far away from the particle/substrate interface. A symmetry boundary condition is used along the vertical axis of the particle/substrate assembly (Fig. 9.1). Sound speed-based absorbing (transmitting) boundary conditions are used along the bottom and the side surfaces of the substrate. These boundary conditions allow a mass flow into and out of the computational domain and are generally used to approximate an infinite or a semi-infinite medium. In addition, pressure is set to zero in all so-called 'ghost' cells (the cells in which the volume fraction of the solid material is zero). The use of this boundary condition prevents the mass from entering the computational domain while a mass flow out of the computational domain is allowed.

During numerical modeling, the particle/substrate interaction is typically assumed to be an adiabatic process, that is, the heat transfer is not considered. The validity of this assumption can be assessed by comparing the thermal diffusion distance, $\sqrt{D_{th}t_c}$ (where D_{th} is the thermal diffusivity, t_c is the particle/substrate 'contact' time during which a non-zero pressure acts

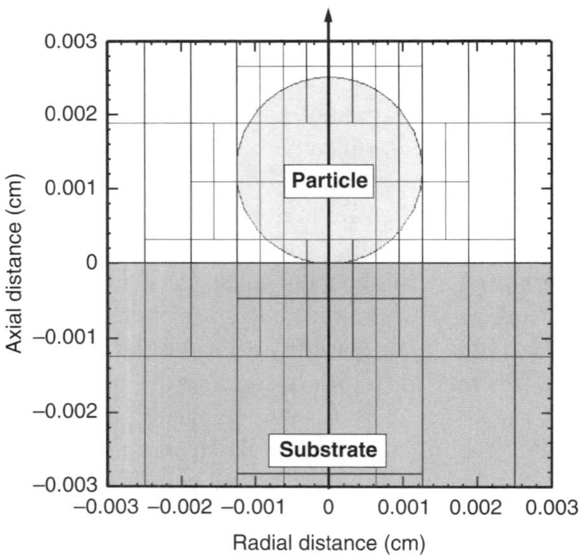

9.1 The computational domain and the hierarchical block structure used in the numerical simulation of the particle/substrate impact.

on the particle/substrate interface), with, x, the characteristic system dimension (the average edge length of a computational grid cell). For the typical values $D_{th} = 10^{-6}\,m^2/s$ and $t_c = 10^{-8}\,s$, the thermal diffusion distance takes on a value $\sqrt{D_{th}t_c} = 10^{-7}$ m. Since this value is relatively small in comparison with the average computational grid-cell edge length, $x \approx 10^{-6}$ m, the neglect of heat conduction during the particle/substrate collision appears justified. Assadi *et al.*[13] further showed that the adiabatic assumption made is justified even if, due to small system dimensions, heat conduction is assumed to be in the so-called 'ballistic' regime and, hence, is controlled by the lattice waves (the phonons).

9.2.2 Conservation equations

The boundary-value problem defined in Section 9.2.1 is solved within the CTH code for the (space and time) dependent variables (the material density, ρ, the velocities, v, and the internal energy density, E) by solving numerically the following conservation equations.

A mass conservation equation:

$$\frac{\partial \rho}{\partial t} = -\rho \nabla \cdot \dot{v} \qquad [9.1]$$

three momentum conservation equations:

$$\rho \frac{\partial v}{\partial t} + \nabla \cdot (\sigma + IP) = -\rho B \qquad [9.2]$$

and an energy conservation equation:

$$\frac{\partial (\rho E)}{\partial t} = (\sigma + IP) \cdot \nabla v + \dot{Q} \qquad [9.3]$$

These equations are solved in conjunction with an appropriate material model and kinematics equations. In equations [9.1] to [9.3], t is used to denote the time, σ represents stress, P is pressure, I is a second-order identity tensor, B is a body force per unit mass and $\dot{Q}$ a mass energy density source term, while ∇ is used to denote the gradient, $\nabla.$ indicates the divergence operators and a dot over a quantity indicates a time derivative.

9.2.3 Material models

Within the CTH code, the total stress tensor is decomposed into a hydrostatic (pressure), P, and a deviatoric part, S. The evolution of the pressure, P, within each material, during an impact event is governed by an equation of state (e.g. $P = f(\rho, E)$, where f is used to denote a function). The deviation portion of the stress, S, within each material, is defined by a kinematic equation that combines a yield criterion, a flow rule and a material constitutive (strength) model. When the boundary-value problem defined in Section 9.2.1 is solved, the equation of state, the kinematic equation and the material constitutive equations are used to update the pressure and the deviation portion of the stress.

Equation of state

The relationships between P, E, ρ and temperature, T, most frequently used within the CTH code are those defined by the following Mie–Grüneisen equations of state:[11]

$$P(\rho, E) = P_R(\rho) + \rho\Gamma\,[E - E_R(\rho)] \qquad [9.4]$$

$$E(\rho, T) = E_R(\rho) + C_V\,[T - T_R(\rho)] \qquad [9.5]$$

where Γ is the Grüneisen function defined as $\Gamma = \rho^{-1}(\partial P/\partial E)_\rho$; subscript R is used to denote the quantities along a 'reference curve', usually the Hugoniot, an isentrope, or the zero-Kelvin isotherm; and $C_V = (\partial E/\partial T)_V$ is the constant-volume specific heat. These equations are most suitable at lower temperatures, near the Hugoniot, and do not account for phase transitions such as melting, vaporization or chemical reactions and electronic excitation. The Hugoniot is represented by a quadratic relationship between the shock velocity υ_S and the particle velocity, υ_P, as:

$$v_s = C_S + S_1 v_p + S_2 \frac{v_p^2}{C_S} \qquad [9.6]$$

where C_S is the speed of sound while S_1 and S_2 are material constants.

Within the CTH code, the pressure is typically updated at the end of each time increment by substituting the updated dependent variables into equations [5.4] to [5.6].

Materials constitutive equations

Three linear-elastic, rate-dependent plastic material constitutive models – the Steinberg–Guinan–Lund model,[14,15] the Zerilli–Armstrong model[16] and the Johnson–Cook plasticity model[17] – describe material changes. All three models are based on a consideration of thermally activated and drag-controlled dislocation dynamics and are typically used to model particle and substrate materials. All three models take account of the effects of isotropic strain hardening, thermal softening, strain-rate dependency and pressure-dependency of the yield strength, Y. In addition, the Steinberg–Guinan–Lund model accounts for the pressure and temperature dependency of the shear modulus, G.

Within the Steinberg–Guinan–Lund model,[14,15] the yield strength is defined as:

$$Y = [Y_T(\dot{\varepsilon}^p, T) + Y_A f(\varepsilon^p)] \frac{G(P,T)}{G_0} \qquad [9.7]$$

where the thermal component of the yield strength, Y_T, is defined by:

$$\dot{\varepsilon}^p = \left(\frac{1}{C_1} \exp\left[\frac{2U_K}{T} \left(1 - \frac{Y_T}{Y_P}\right)^2 \right] + \frac{C_2}{Y_T} \right)^{-1} \qquad [9.8]$$

the function $f(\varepsilon^P)$ as:

$$f(\varepsilon^P) = [1 + \beta(\varepsilon^P + \varepsilon_i)]^n \qquad [9.9]$$

and the shear modulus as:

$$G(P,T) = G_0 \left[1 + A \frac{P}{\eta^{1/3}} - B(T - 0.02585\,\text{eV}) \right] \qquad [9.10]$$

where Y_A, C_0, C_1, C_2, U_K, Y_P, β, ε_i, n, A and B are material-dependent parameters; ε^P is the equivalent plastic strain; $\dot{\varepsilon}^P$ is the equivalent plastic strain rate; $\eta \equiv \rho/\rho_o$ is the compression and ρ_o the initial density. The temperature T is expressed in units of eV and is defined as kT^*, where k is the Boltzmann constant and T^* is the temperature in Kelvins.

Within the Zerilli–Armstrong model,[16] the yield strength is defined as

$$Y = A' + \left(c_1 + c_2\sqrt{\varepsilon^p}\right) e^{-c_3 T} (\dot{\varepsilon}^p)^{c_4 T} + c_5(\varepsilon^p)^N \qquad [9.11]$$

where A', C_1, C_2, C_3, C_4, C_5 and N are material-dependent parameters.

Within the Johnson–Cook plasticity model,[17] a material is treated as being strain hardening, strain-rate sensitive and thermally softening, and its equivalent normal plastic-deformation resistance, $\sigma^e = Y$, is given by:

$$\sigma^e = [A + B(\varepsilon^p)^n][1 + C \ln(\dot{\varepsilon}^p / \dot{\varepsilon}_0^p)] \left[1 - \left(\frac{T - T_{init}}{T_{melt} - T_{init}} \right)^m \right] \qquad [9.12]$$

where ε^P is the equivalent normal plastic strain, $\dot{\varepsilon}^P$ is the equivalent normal plastic strain rate, $\dot{\varepsilon}_0^P$ is a reference equivalent normal plastic strain rate, T is the temperature, and the subscripts *init* and *melt* are used to denote the initial and the melting temperatures, respectively. A, B, n, C, m and T_{melt} are the Johnson–Cook plasticity model parameters.[9]

It should be noted that within the Eulerian framework, when a cell contains two or more materials, the yield strength is defined as the volume fraction weighted yield strength of the individual materials. Consequently, single material cells containing voids have a reduced yield strength since the material volume fraction in such cells is less than unity.

Yield criterion and flow rule

The von Mises yield criterion is implemented into the CTH code, according to which (plastic) yielding occurs when the magnitude of the deviatoric stress, $|S| = \sqrt{S \cdot S}$ (a raised dot is used to denote the tensor inner product) reaches (or exceeds) a material-state, temperature and loading-rate dependent value of:

$$|S| = \sqrt{2/3}Y \tag{9.13}$$

where the yield strength Y is given by equations [9.7], [9.11] or [9.12]. Upon yielding, an *associated* flow rule is used to describe the subsequent plastic deformation. According to the associated flow rule, plastic flow occurs in such a way to most rapidly accommodate the deviatoric part of the stress which causes the plastic flow and, hence, the (deviatoric) plastic strain rate, $\dot{\varepsilon}^P$, is defined as:

$$\dot{\varepsilon}^p = |\dot{\varepsilon}^p| \frac{S}{|S|} \tag{9.14}$$

where $|\ |$ is used to denote the magnitude of a tensor. The total deviatoric strain rate, $\dot{\varepsilon}$, is defined as:

$$\dot{\varepsilon} = D - \frac{1}{3} tr(D) I \tag{9.15}$$

where D, given as:

$$D = \frac{1}{2}(\nabla v + \nabla v^T) \tag{9.16}$$

is the rate-of-deformation tensor, I is the identity tensor, superscript T denotes the tensor transpose and tr the tensor trace. The total deviatoric strain rate, $\dot{\varepsilon}$, can be additively decomposed into its elastic, $\dot{\varepsilon}^e$, and plastic, $\dot{\varepsilon}^p$, components as:

$$\dot{\varepsilon} = \dot{\varepsilon}^e + \dot{\varepsilon}^p \tag{9.17}$$

The elastic deviatoric strain rate is assumed to be governed by Hooke's law and can be defined as:

$$\dot{\varepsilon}^e = \frac{1}{2G}(\dot{S} - WS + SW) \tag{9.18}$$

where the quantity within parentheses is known as the Jaumann co-rotational stress rate and

$$W = \frac{1}{2}(\nabla v - \nabla v^T) \tag{9.19}$$

is the spin tensor.

The evolution of the deviatoric stress, S, between the times t and $t + \Delta t$, where Δt is a time increment, can be generally expressed using a backward Euler scheme as:

$$S_{t+\Delta t} = S_t + \dot{S}_{t+\Delta t}\,\Delta t \qquad [9.20]$$

where $\dot{S}$ is defined by inverting equation [9.18] as:

$$\dot{S} = 2G\dot{\varepsilon}^e + WS - SW \qquad [9.21]$$

The deviatoric stress at the end of a time step $t = \Delta t$, $S_{t+\Delta t}$, is updated within the CTH code using the radial return procedure developed by Kreig and Kreig,[18] which uses the total deviatoric strain rate in place of the elastic deviatoric strain rate to obtain a trial deviatoric stress, $S'_{t+\Delta t}$, as:

$$S'_{t+\Delta t} = S'_t + 2G\dot{\varepsilon}\Delta t \qquad [9.22]$$

The magnitude of the trial deviatoric stress is next relaxed to account for the effect of plasticity in accordance with the yield criterion, equation [9.13], as:

$$|S_{t+\Delta t}| = |S'_{t+\Delta t}| - 2G|\dot{\varepsilon}^p|\Delta t = \sqrt{2/3}Y_{t+\Delta t} \qquad [9.23]$$

Since the shear modulus is a function of T and P, equation [9.10], the total strain rate a function of υ, equations [9.15] and [9.16], and the yield strength is a function of P, T, $|\varepsilon^p|$ and $|\dot{\varepsilon}^p|$, equations [9.7], [9.11] or [9.12], and equation [9.22] can be written in a general form as:

$$S_{t+\Delta t}(v, T, P, |\dot{\varepsilon}^p|) = \sqrt{2/3}Y_{t+\Delta t}(P, T, |\bar{\varepsilon}^p|, |\dot{\varepsilon}^p|) \qquad [9.24]$$

and can be considered as the governing equation for the magnitude of the (deviatoric) plastic strain rate, $|\dot{\varepsilon}^p|$. Once, $|\dot{\varepsilon}^p|$ is computed, the deviatoric plastic strain rate is determined using equation [9.14], then the deviatoric elastic strain rate is computed using equation [9.17], and finally the deviatoric stress is updated as:

$$S_{t+\Delta t} = S'_t + 2G\dot{\varepsilon}^e\Delta t \qquad [9.25]$$

Thus equation [9.25] in conjunction with equations [9.13] to [9.24] can be considered as the governing equations for the evolution of the deviatoric stress, S.

9.2.4 Numerical solution

Equations [9.1] to [9.5], [9.7], [9.11] or [9.12], [9.10], [9.23] and [9.25] define a system of nine coupled partial differential and algebraic equations with the time t and the spatial coordinates x, y and z as independent variables and ρ, υ, E, P, T, Y, G, S, and ε^p as nine dependent variables. After assigning the initial and boundary conditions for the dependent variables, the coupled system of partial and algebraic equations is numerically solved within the CTH code using a finite difference based two-step Eulerian computational scheme. In the first step, the Lagrangian step, the computational mesh is attached to the material and, hence, the cells distort in order to follow the material. In the second step, the Eulerian step, remeshing is carried out in order to map the distorted cells back to the original Eulerian mesh.

An example of the typical results pertaining to the evolutions of particle and substrate morphologies obtained using the computational analysis presented in this section is shown in Fig. 9.2 and 9.3.[6] The simulation results of the collision between a single 20 μm diameter aluminum particle with an incident velocity of 650 m/s and a flat semi-infinite copper substrate at four times (5 ns, 20 ns, 35 ns and 50 ns) following the initial particle/substrate contact are shown in Fig. 9.2(a) to (d), respectively. According to the results presented in Fig. 2 in reference 6, 650 m/s is slightly above the minimum critical incident particle velocity for aluminum cold-spray deposition onto

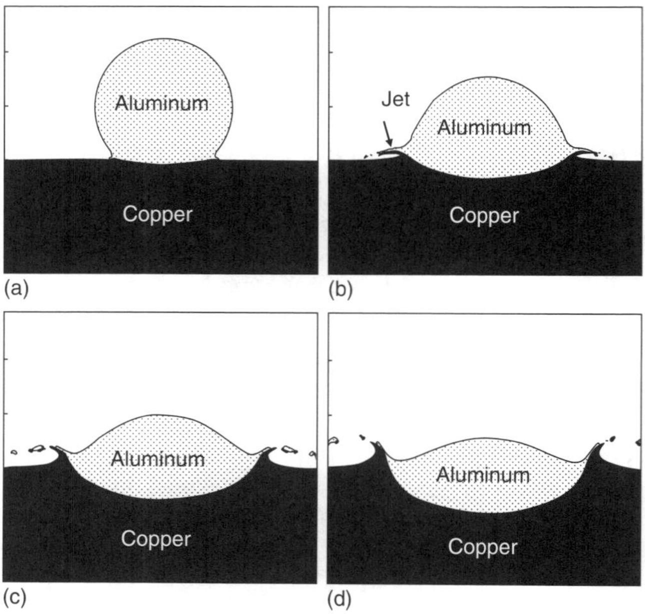

9.2 The evolution of shapes of a 20 μm diameter aluminum feed-powder particle and a copper substrate crater for the incident particle velocity of 650 m/s at the following times: (a) 5 ns; (b) 20 ns; (c) 35 ns; (d) 50 ns.

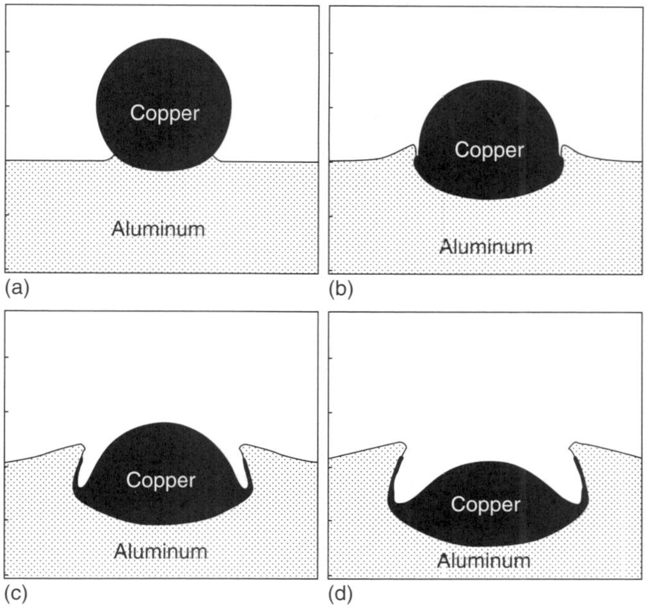

9.3 The evolution of shapes of a 20 μm diameter copper feed-powder particle and an aluminum substrate crater for the incident particle velocity of 650 m/s at the following times: (a) 5 ns; (b) 20 ns; (c) 35 ns; (d) 50 ns.

a copper substrate. The results displayed in Fig. 9.2(a) to (d) show that as the particle/substrate contact time increases, the particle (height-to-width) aspect ratio decreases while the substrate crater depth and width increase. At the same time, a jet composed of both the particle material and the substrate material is formed at the particle/substrate contact surface. When this jet reaches the free surface, it forms a lip which points away from the flattened particle. Fracture of the material at the jet front, as seen in Fig. 9.2(b) to (d), is controlled by a hydrostatic tensile stress condition and, since it is not the focus of the present work, will not be discussed any further.

The evolutions of shapes of a single 20 μm diameter copper particle with an incident velocity of 650 m/s and a flat semi-infinite aluminum substrate at the same four times as in Fig. 9.2(a) to (d) are shown in Fig. 9.3(a) to (d), respectively. According to the results presented in Fig. 2 in reference 6, the deposition efficiency of copper on aluminum substrate at an incident particle velocity of 650 m/s is ~ 50–60%. The results displayed in Fig. 9.3(a) to (d) show, as with the results displayed in Fig. 9.2(a) to (d), that as the particle/substrate contact time increases, the crater depth and width increase while the flattened particle aspect ratio decreases. However, in this case, the crater is much deeper and, due to the fact that the particle is flattened to a lesser extent, the crater width and the final particle/substrate contact surface area (excluding the one associated with the interfacial jet) are somewhat smaller. These differences can be attributed to the larger kinetic energy associated with the heavier copper particle and to the larger strength of copper in comparison to that of aluminum. One of the most striking differences between the corresponding results shown in Fig. 9.2(a) to (d) and Fig. 9.3(a) to (d) is in the length of the interfacial jets. The interfacial jet is much longer in the case of the copper deposition on aluminum than in the case of the aluminum deposition on copper.

The shapes of the cold-spray particles and the substrate craters 50 ns after the initial contact for the aluminum-particle velocity of 400 m/s and the copper-particle velocity of 500 m/s are shown in Fig. 9.4(a) and (b), respectively. These two velocities are lower than the corresponding minimum critical particle velocity for cold-spray deposition. By comparing the results displayed in Fig. 9.4(a) and (b) with those displayed in Fig. 9.2(a) to (d) and 9.3(a) to (d), one can observe that, in the former case, the interfacial jet is

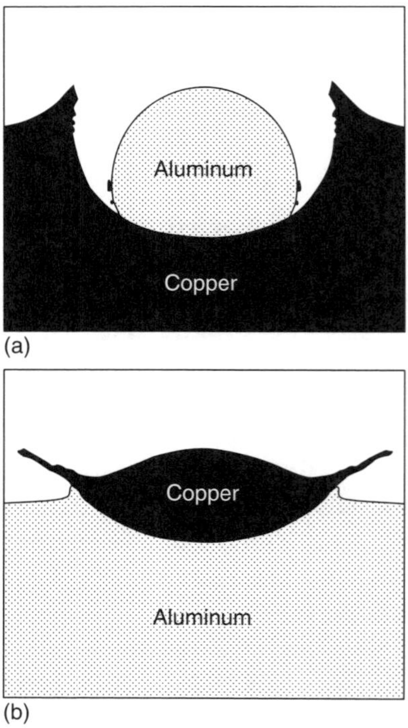

9.4 The shapes of the feed-powder particle and the substrate crater 50 ns after the initial contact for: (a) the aluminum particle velocity of 400 m/s and (b) the copper particle velocity of 500 m/s.

composed of only one material at the sub-critical particle velocities. This finding seems to confirm that the role of the interfacial jets in removing the oxides and other surface films from both the particle surface and the substrate surface is a critical factor for attaining a good particle/substrate bonding (i.e. a high level of deposition efficiency).

The time evolutions of the maximum temperature, the maximum plastic strain rate, the maximum pressure and the maximum von Mises equivalent stress, all at the aluminum-particle/copper-substrate interface at incident particle velocities of 400, 600, 800 and 1000 m/s are shown in Fig. 9.5(a) to (d), respectively. The corresponding results for the copper cold-spray deposition onto an aluminum substrate are shown in Fig. 9.6(a) to (d).

The results depicted in Fig. 9.5(a) and 9.6(a) show that the maximum temperature at the particle/substrate interface briefly (for ~11 ns in the

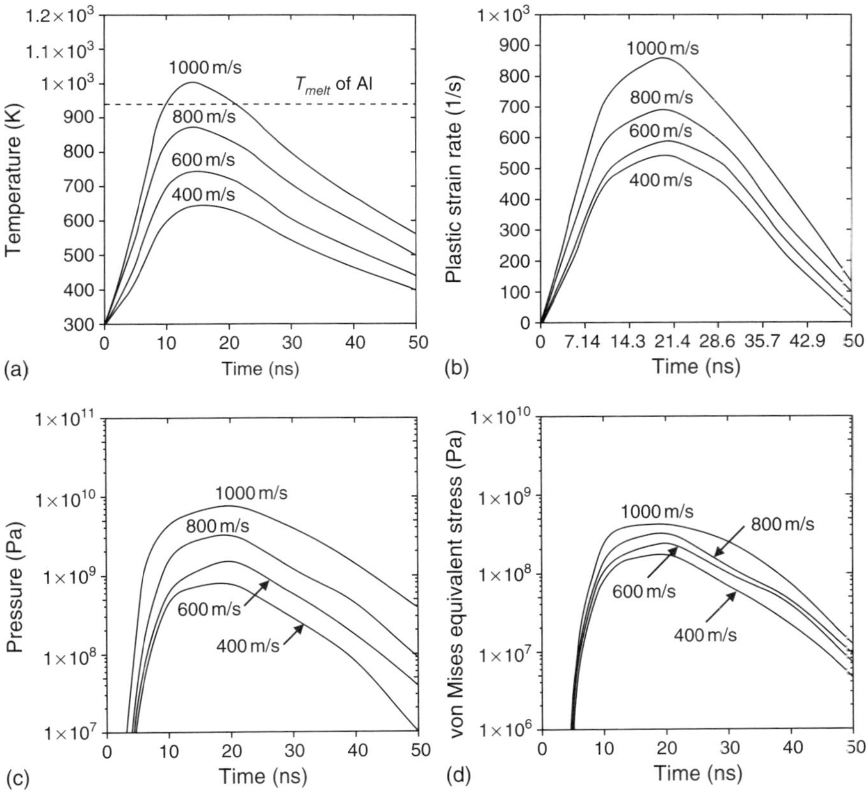

9.5 The evolution of the maximum values of: (a) temperature; (b) plastic strain rate; (c) pressure; and (d) von Mises equivalent stress at the particle/substrate interface during the deposition of aluminum on copper at four particle velocities.

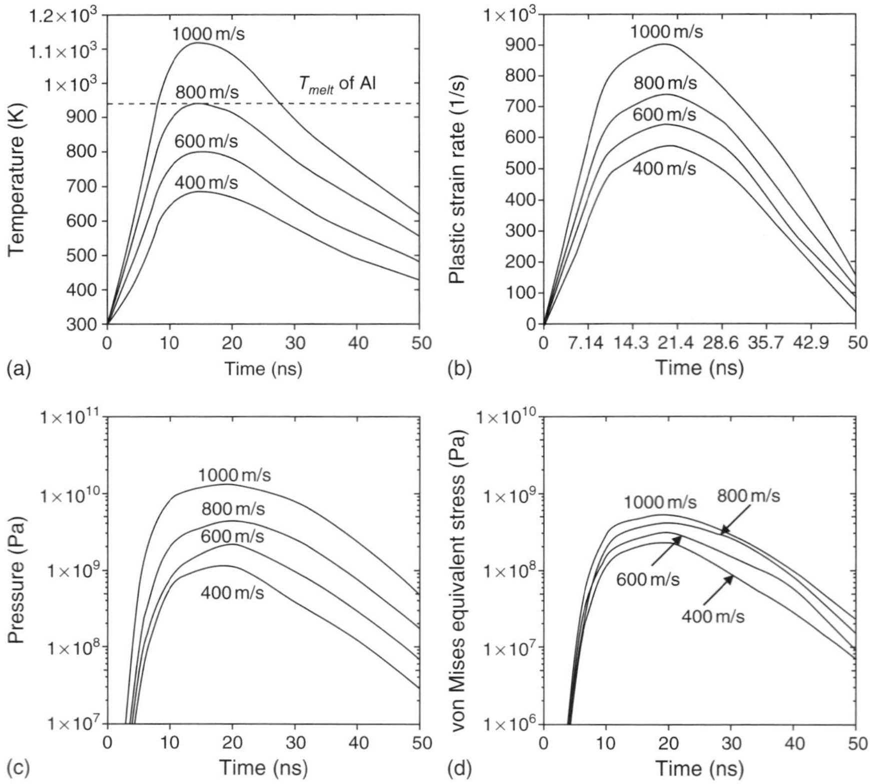

9.6 The evolution of the maximum values of: (a) temperature; (b) plastic strain rate; (c) pressure; and (d) von Mises equivalent stress at the particle/substrate interface during the deposition of copper on aluminum at four particle velocities.

copper-on-aluminum case and for ~20 ns in the copper-on-aluminum case) exceeds the melting point of aluminum (~940 K) and only at the highest incident particle velocities (1000 m/s) used. The melting point of copper (~1356 K) is never exceeded. It should be noted that the computed maximum temperature values at the particle/substrate interface are overestimated due to the fact that the CTH code does not enable modeling of the transfer of heat from the particle/substrate interface region where the heat is generated. A close examination of the temperature distribution over the particle/substrate interface shows that the conditions for melting of the aluminum particle or substrate are met over a very small portion of the interface (approximately 14% and 20% of the particle/substrate interfacial area in the aluminum-on-copper and the copper-on-aluminum cases, respectively). In addition, the heat diffusion distance in the copper particle or the substrate during a typical collision time period of 50 ns is about 3–4 μm, suggesting that heat conduction can play a significant role in reducing the maximum interfacial temperature during the particle/substrate collision process. These findings suggest that melting, even if it takes place, most probably does not play a key role in particle/substrate bonding. This is fully consistent with numerous metallographic examination results that generally show a cold-worked grain structure in the region surrounding the particle/substrate interface in the systems with good particle/substrate bonding.[19]

The results displayed in Fig. 9.5(b) and 9.6(b), as well as those depicted in Fig. 9.2(a) to (d) and 9.3(a) to (d), show that, due to a large impact velocity of the cold-spray particle with the substrate, the plastic deformation is highly localized into the region surrounding the particle/substrate interface. Consequently, a jet is formed that removes the oxidized and/or otherwise contaminated material from the prior particle and the substrate surfaces. As a result, clean particle and substrate materials are brought into contact (under high pressures) at the particle/substrate interface during the collision process. The formation of such jets and their role in removing the contaminated material from the particle and substrate surfaces is believed to be one of the major (necessary) conditions for attaining high particle/substrate bonding strength during cold spray.

The results shown in Fig. 9.5(c) and (d), and 9.6(c) and (d), show that, for both aluminum deposition on copper and copper deposition on aluminum, the contact pressure and the von Mises equivalent stress values (the latter is numerically equal to the material yield strength) are very high in comparison to the yield strength in the un-deformed ($\varepsilon^p = 0$) constituent materials at the maximum interfacial particle/substrate interface temperatures and plastic strain rates. For example, for the case of aluminum deposition on copper at the particle velocity of 800 m/s, at the maximum interfacial temperature (~870 K) and at the maximum plastic strain rate (~660/s), zero plastic strain yield strengths of aluminum and copper are 0.3×10^8 Pa and 0.8×10^8 Pa, respectively. The corresponding maximum interfacial pressures and the maximum interfacial von Mises equivalent stress (~4.1×10^9 Pa and ~3.8×10^8 Pa, respectively) are, thus, higher than the zero plastic strain yield strengths of the two constituent materials by a factor of between 5 and 130. This finding suggests that, due to these extremely high pressure and stress levels, it may be perhaps more appropriate to treat the material adjacent to the particle/substrate interface as a viscous 'fluid-like' material rather than a rate-dependent elastic–plastic solid material. Such a suggestion was previously made in the context of explosive welding,[8] using the same argument as the one presented above. As will be shown in Section 9.4, the presence of the 'fluid-like' material adjacent to the particle/substrate interface can lead to interfacial instabilities which, in turn, can play an important role in particle/substrate bonding.

9.3 Adiabatic shear instability and plastic-flow localization

As mentioned earlier, adiabatic shear instability and the resultant plastic-flow localization are the phenomena that are believed to play a major role in the particle/substrate bonding during the cold-spray process. Consequently, these phenomena are discussed in greater detail in the present section.

The phenomenon of adiabatic shear instability and the associated formation of shear bands was first considered in sufficient detail by Wright.[20,21] To understand this phenomenon on a simple physical level, typical dynamic stress–strain curves (obtained during experiments such as the thin-walled tube-torsion Kolsky bar experiment) are shown in Fig. 9.7(a). For a typical work-hardening material under non-adiabatic conditions, the stress–strain curve (denoted as 'Isothermal' in Fig. 9.7(a)) shows a monotonic increase of the flow stress with plastic strain. However, under adiabatic conditions, the plastic strain energy dissipates as heat increases the temperature, causing material softening. Consequently, the rate of strain hardening decreases and the flow stress reaches a maximum value, past which a monotonic decrease in the flow stress with plastic strain takes place (the curve labeled 'Adiabatic' in Fig. 9.7(a)). In an ideal material with uniform distributions of stress, strain, temperature and materials microstructure, softening can continue indefinitely. In real materials, however, fluctuations in stress, strain, temperature or microstructure, and the inherent instability of strain softening, can give rise to plastic-flow (shear) localization. Under such circumstances, shearing and heating (and consequently softening) become highly localized, while the straining and heating in the surrounding material regions practically stops. This, in turn, causes the flow stress to quickly drop to zero (the curve denoted 'Localization' in Fig. 9.7(a)). Simple schematics are used in Fig. 9.7(b) and (c) to demonstrate the basic difference between the uniform (simple) shear and the localized (simple) shear.

9.3.1 A simple one-dimensional adiabatic shear localization model

In order to analyze the onset of strain softening and adiabatic shear localization, a simple one-dimensional model is presented below. The model is used to reveal the thermo-mechanical behavior of a small material element at the particle/substrate interface during the particle/substrate collision. The model consists of the following governing equations and initial conditions.

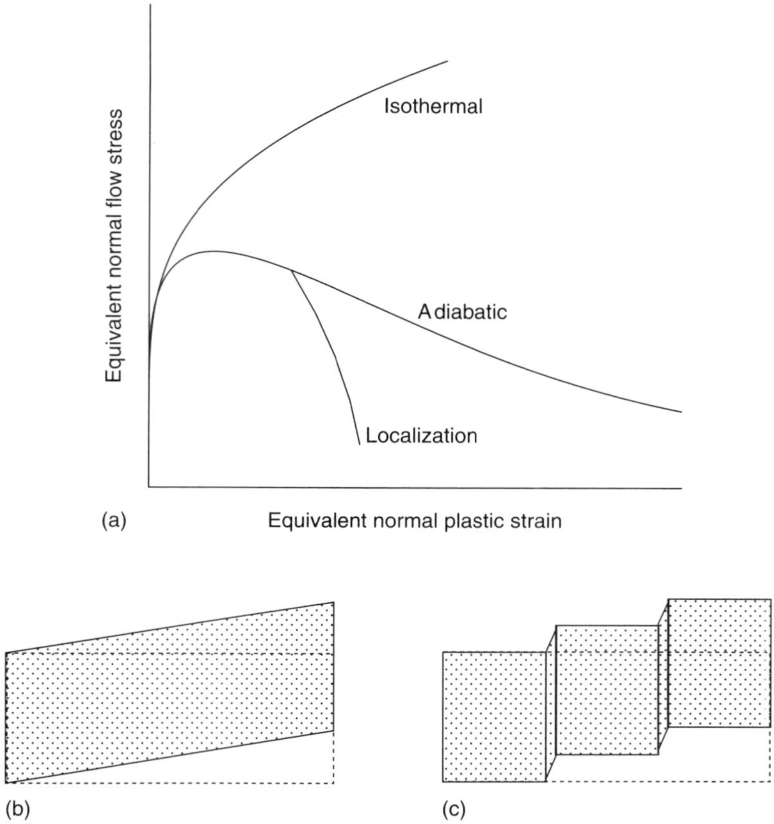

9.7 (a) Schematic of the stress–strain curves in a normal strain-hardening material ('Isothermal'), an adiabatically softened material ('Adiabatic') and in a material undergoing adiabatic shear localization ('Localization'). (b) and (c) Schematics of the uniform and the localized simple shears, respectively. Initial material elements are denoted using dotted lines while sheared elements are denoted using solid lines in (b) and (c).

The equivalent plastic strain rate, $\dot{\varepsilon}^p$:

$$\dot{\varepsilon}^p = \dot{\varepsilon}_c^p \frac{\rho v_p^2}{2} \frac{1}{\sigma^e}, \qquad \dot{\varepsilon}^p (t = 0) = \dot{\varepsilon}_c^p \frac{\rho (v_p^0)^2}{2} \frac{1}{A} \qquad [9.26]$$

The equivalent plastic strain, ε^p:

$$\varepsilon^p = \int_0^t \dot{\varepsilon}^p \mathrm{d}t, \qquad \varepsilon^p (t = 0) = 0 \qquad [9.27]$$

The heading rate, $\dot{T}$:

$$\dot{T} = \frac{\sigma^e \dot{\varepsilon}^p}{\rho C_p}, \qquad \dot{T}(t=0) = \frac{A\dot{\varepsilon}^p(t=0)}{\rho C_p} \tag{9.28}$$

The temperature, T:

$$T = \int_0^t \dot{T}\,\mathrm{d}t, \qquad T(t=0) = T_{init} \tag{9.29}$$

The particle velocity, v_p:

$$v_p = v_p^0\left(1 - \frac{t}{t_c}\right), \qquad v_p(t=0) = v_p^0 \tag{9.30}$$

The equivalent plastic flow strength, σ^e:

$$\sigma^e = [A + B(\varepsilon^p)^n][1 + C\ln(\dot{\varepsilon}^p/\dot{\varepsilon}_0^p)]\left[1 - \left(\frac{T - T_{init}}{T_{melt} - T_{init}}\right)^m\right], \quad \sigma^e(t=0) = A \tag{9.31}$$

where $\dot{\varepsilon}_0^P$ is a strain rate proportionality constant, ρ is the (constant) material mass density, C_p is the (constant) specific heat and a dot over a symbol is used to denote the time derivative of a quantity.

In the model presented above the following assumptions are made.

1 The particle velocity is assumed to decrease linearly with time from its initial value v_p^0 to a zero value at $t = t_c$.
2 The contact pressure at the particle/substrate or particle/deposited material interface in assumed to be proportional to the kinetic energy of the particle per unit volume, $\rho v_p^2/2$.
3 The particle deformation energy is taken to be completely dissipated in the form of heat, i.e. the energy of deformation stored in the form of various deformation-induced microstructural defects is assumed to be negligibly small.
4 Variations of the materials properties such as ρ and C_p with plastic strain, stress or temperature are ignored.

The model described above enables determination of the time evolution of the plastic strain rate, plastic strain, heating rate, temperature and the equivalent stress in a typical material element at the particle/substrate interface during a particle/substrate collision. The model can be solved using a simple forward difference method. As will be shown later in this section, this procedure clearly demonstrates a transition of the stress–strain curve from a strain-hardening type to a strain-softening type at high impact particle velocities.

To quantify the tendency of a strain-softening material to undergo strain (shear) localization, the approach of Schoenfeld and Wright[22] is utilized. According to Schoenfeld and Wright,[22] the tendency for strain localization (measured by the inverse of the amount of uniform plastic strain, taking place past the strain at which the flow stress experiences a maximum, needed to obtain strain localization) scales with the SL parameter defined as:

$$SL = \left(- \frac{\partial^2 \sigma^e / (\partial \varepsilon^p \partial \varepsilon^p)}{(\partial \sigma^e / \partial \dot{\varepsilon}^p) \sigma^e} \right)_{\sigma^e = \sigma^e_{max}} \tag{9.32}$$

where σ^e_{max} denotes the maximum value of the (adiabatic) flow stress at a given impact particle velocity. A numerical solution of the model presented above enables evaluation of the SL parameter and, hence, quantification of the relative tendency of different materials for shear localization. Detailed computational analyses of the particle/substrate interactions carried out by Assadi *et al.*[13] and Grujicic *et al.*[4] established that plastic-flow localization in a material element at the particle/substrate interface occurs when the shear localization parameter, SL, takes on a value larger than $1.6 \times 10^{-4} \pm 0.2 \times 10^{-4}$ s.

In the remainder of this section, the one-dimensional model for adiabatic shear localization presented above is used to predict the minimal impact particle velocity needed to give rise to adiabatic shear localization in the particle/substrate interfacial region. Assuming that $SL = 1.6 \times 10^{-4}$ s can be considered as a critical condition for the onset of adiabatic shear localization, the one-dimensional model is used to determine the minimum value of the particle velocity required to achieve this critical value of SL. This was done by using a simple computational procedure within which the initial particle velocity is varied in increments of 10 m/s, and the value of SL evaluated at the peak level of the equivalent normal stress using equation [9.32]. The results obtained are listed in the last column of Table 9.1. A comparison of these results with their finite element counterparts reported by Assadi *et al.*[13] and Grujicic *et al.*,[4] also displayed in Table 9.1, shows that the agreement between the three sets of results is quite reasonable. This finding suggests that the simple one-dimensional model for the onset of adiabatic shear localization can be used to assess the critical (minimal) impact particle velocity required for successful cold-spray deposition. Once such a velocity is determined, an iso-entropic fluid dynamics model such as those developed by Dykhuizen and Smith[23] and by Grujicic *et al.*[24] can be used to identify the corresponding cold-gas dynamic spray parameters (the propellant gas, the gas temperature and pressure, etc.) needed to obtain the desired average particle velocity.

Table 9.1 A comparison of the threshold impact particle velocities obtained by Grujicic *et al.*[4] with the corresponding velocities reported by Assadi *et al.*[13]

Particle material	Substrate material	Threshold particle velocity (m/s), Assadi *et al.*[13]	Threshold particle velocity (m/s), finite element analysis, Grujicic *et al.*[4]	Threshold particle velocity (m/s), shear localization analysis, Grujicic *et al.*[4]
Copper	Copper	570–580	575–585	571
Aluminum	Aluminum	760–770	760–770	766
Nickel	Nickel	600–610	620–630	634
316L	316L	600–610	620–630	617
Titanium	Titanium	670–680	650–670	657
Copper	Aluminum	N/A	510–530	507
Aluminum	Copper	N/A	600–630	634
Copper	Nickel	N/A	570–580	571
Nickel	Copper	N/A	570–580	576
Copper	316L	N/A	570–580	574
316L	Copper	N/A	570–580	573
Copper	Titanium	N/A	520–550	514
Titanium	Copper	N/A	570–590	582

9.3.2 Detailed computational analyses of adiabatic shear localization

Typical results obtained in detailed transient non-linear dynamics analyses of the cold-spray deposition process, such as those displayed in Fig. 9.2(a) to (d) and 9.3(a) to (d), are discussed in the present section. The main finding arising from detailed transient non-linear dynamics analyses of the cold-spray particle/substrate interactions can be summarized as follows.

1 As the particle penetrates the substrate, a crater develops in the substrate.
2 The diameter and the depth of the crater increase as the particle/substrate contact time increases, while the height-to-width aspect ratio of the particle decreases.
3 The plastic deformation in the particle and in the substrate is concentrated in a narrow region surrounding the particle/substrate interface and, consequently, an interfacial jet composed of the highly deformed material is formed.
4 Due to localization of the plastic deformation to a region surrounding the particle/substrate interface, a significant temperature increase is observed only in this region.
5 As the impact particle velocity increases, for a given combination of the particle and substrate materials, the thickness of the interfacial jet

decreases, indicating an increased level of plastic strain localization in the interfacial region.

6 At a given impact particle velocity and for a given combination of the particle and the substrate materials, the effect of the particle diameter (in the 5–25 μm range) on the evolution of the particle and the substrate shapes during impact is not significant.

7 For a given impact particle velocity, and when the particle and the substrate are composed of different materials, inversion of the particle and the substrate materials generally has a significant effect on the crater diameter and the crater depth, as well as on the thickness of the interfacial jet. In general, when the particle material possesses a larger density, the crater diameter and the depth are larger. In addition, the thickness of the interfacial jet is larger when the particle material is stiffer and less dense. The last observation suggests that as the sound velocity (scales with the square root of a ratio of the elastic modulus and the density) in the particle material increases, the extent of the plastic strain localization decreases.

To reveal the tendency for the development of adiabatic shear localization in the particle/substrate interfacial region, the temporal evolution of the equivalent plastic strain rate, the equivalent plastic strain and the temperature and the equivalent normal stress are monitored in several elements at the lower surface of the particle and at the upper surface of the substrate. An example of the typical results obtained is given in Fig. 9.8(a) to (d). The results displayed in Fig. 9.8(a) to (d) pertain to the case of a copper particle with a 25 μm diameter and a copper substrate. The results displayed in Fig. 9.8(a) to (d) can be summarized as follows.

1 At lower impact particle velocities (v_p = 400–550 m/s), temporal evolutions of the equivalent normal plastic strain rate, the equivalent normal plastic strain, the temperature and the equivalent normal stress show a monotonic change with the particle/substrate contact time.

2 At the highest particle velocities used (v_p = 600 m/s, in the case of Fig. 9.8(a) to (d)), temporal evolutions of the equivalent plastic strain rate, the equivalent plastic strain and, the temperature and the equivalent normal stress are monotonic to a certain particle/substrate contact time. Past this contact time, the equivalent plastic strain rate, the equivalent plastic strain and the temperature undergo an abrupt increase, while the equivalent normal stress undergoes a precipitous decrease to a value near zero.

These findings are fully consistent with the temporal evolution of a material element that undergoes adiabatic softening culminating in adiabatic shear localization (curve denoted 'Localization' in Fig. 9.7(a)).

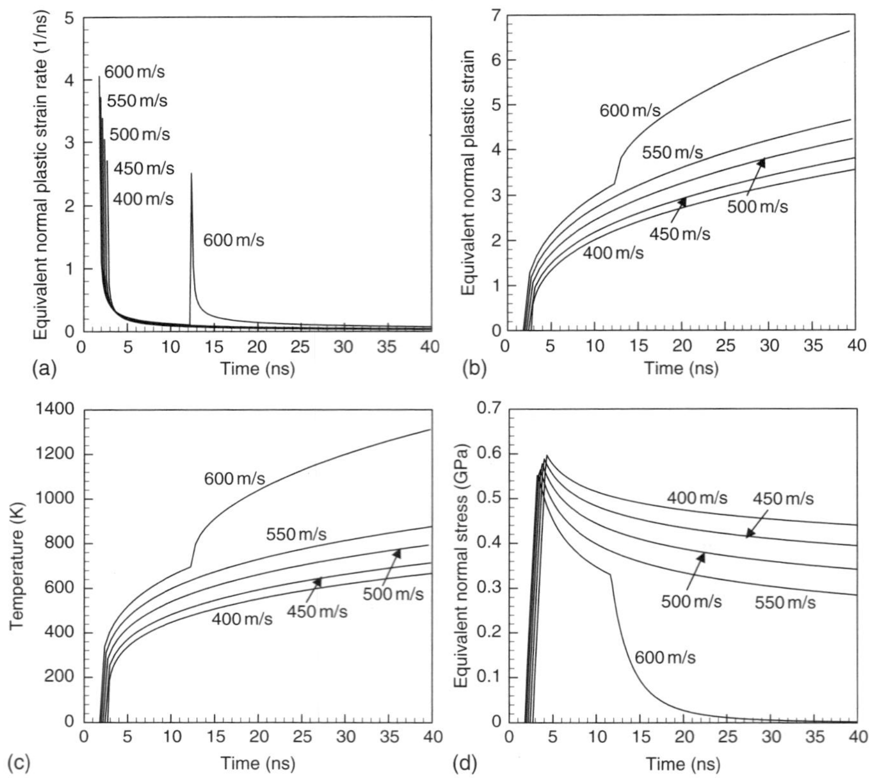

9.8 Temporal evolutions of: (a) the equivalent plastic strain rate; (b) the equivalent plastic strain; (c) the temperature; and (d) the equivalent normal stress in an element at the copper particle surface during the particle collision with a copper substrate for various initial impact particle velocities.

A comparison between the minimal impact particle velocity needed to produce shear localization in the particle/substrate interfacial region obtained by Grujicic *et al.*[4] and the threshold velocity for cold-spray deposition reported by Assadi *et al.*[13] is given in Table 9.1. It is seen that in general there is a good correlation between the two sets of values for a number of materials, suggesting that shear localization indeed plays a critical role in the cold-spray deposition process.

The results presented above suggest that adiabatic shear instability and plastic-flow localization play an important role in particle/substrate bonding during the cold-spray deposition process. Clearly, adiabatic shear localization and the associated formation of the 'surface-scrubbing' jets produce clean contacting surfaces whose presence, in addition to high contact pressures, can be considered as a prerequisite for adhesion-based particle/substrate bonding. Furthermore, adiabatic softening of the material in the particle/substrate interfacial region combined with high contact pressures promote the formation of mutually conforming contacting surfaces via plastic deformation. Hence, once the conditions for the onset of adiabatic shear localization (and adiabatic softening) are attained at sufficiently high impact particle velocities, the conditions for extensive adhesion of the particle and substrate surfaces are reached resulting in particle/substrate bonding. In other words, high contact pressure-, adiabatic softening- and adiabatic plastic flow localization-enhanced adhesion appear to be one of the major, if not the dominant, cold-spray particle/substrate bonding mechanism.

9.4 The role of interfacial instability

The discussion presented in the previous sections suggests that, considering the very short times during which the particle and the substrate are subjected to high temperatures and high contact pressures, interfacial melting and inter-atomic diffusion probably do not play an important role in the cold-spray particle/substrate bonding. High contact pressures, adiabatic shear instability and plastic localization, which give rise to the formation of interfacial jets and clean contact surfaces, have been established (as discussed in Section 9.3) as providing the necessary conditions for particle/substrate bonding. Once such conditions are present, bonding can be established via adhesion. However, the experimentally observed high interfacial bonding strengths suggests that, in addition to adhesion, some type of nano/micro length-scale mechanical material mixing/interlocking mechanism may also contribute to particle/substrate bonding. This section discusses two such interfacial-instability-based mechanisms by which interfacial mixing/interlocking may occur.[6]

Interfacial instability and the resulting formation of interfacial roll-ups and vortices can be, perhaps, best understood within the context of the Kelvin–Helmholtz instability phenomenon. This phenomenon can arise when two fluids, in contact, are moving at different velocities in a direction parallel to their interface. Instability can occur even if the two fluids have identical densities. When the interface is subjected to a (non-zero curvature) perturbation, then as one fluid flows around the other a centrifugal force is generated. This, in turn, gives rise to a change in pressure which may promote amplification of the interfacial perturbations. As shown schematically in Fig. 9.9, these instabilities may subsequently lead to the formation of interfacial roll-ups and vortices. Such interfacial roll-ups and vortices may enhance the overall strength of interfacial bonding in at least three ways:

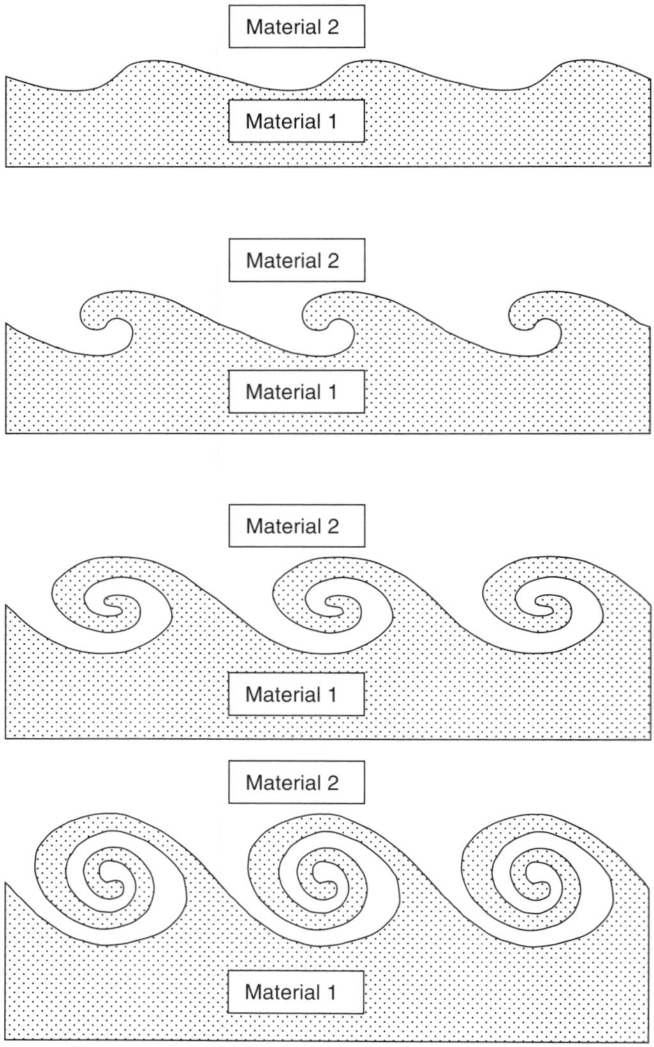

9.9 A schematic of the instability-based evolution of the particle/substrate (Material 1/Material 2) interface and the accompanying formation of interfacial roll-ups and vortices.

(a) by significantly increasing the interfacial area available for adhesion;
(b) by producing a fine length-scale mixing of the two materials;
(c) by creating mechanical interlocking between the two materials.

Interfacial instabilities are typically understood within the context of inertial instabilities, where flow inertia acts to promote the instability and viscosity acts to dampen it. The ratio of these two effects determines if the instability actually grows or decays and is quantified by the Reynolds number as:

$$\mathrm{Re} = \frac{UL}{\nu} \qquad [9.33]$$

where U is the characteristic velocity, L is the characteristic length scale, and ν (= μ/ρ) is the kinematic viscosity. For the situation considered here, a characteristic velocity would be the particle impact velocity which is of the order of 1000 m/s, and a characteristic length scale would be the particle diameter which is of the order of 10 μm. To assess the viscosity of the particle and substrate materials, portions of these materials that are adjacent to the particle/substrate interface and subjected to extremely high pressures are treated as viscous 'fluid-like' materials rather than rate-dependent elastic–plastic solid materials. Such an assumption is commonly made in the context of explosive welding,[8] using the same argument as the one presented above. By defining the viscosity, μ, as the proportionality constant between the equivalent shear stress and the equivalent shear strain rate, the viscosity of copper and aluminum under the deposition conditions modeled in reference 6 is estimated, using equations [9.7], [9.11] or [9.12], to be between 10^5 and 10^6 poise. These values of μ translate into kinematic viscosities of ~100 m^2/s assuming a metal density of the order of 1000 kg/m^3. Using these numbers, a value of Re ~10^{-4} is obtained. Since instabilities arise when the Reynolds number is substantially larger than one, inertial instability is not expected to occur in the present situation, the rather rough estimates of length, velocity and kinematic viscosity notwithstanding.

On the other hand, the work of Yih[25] has demonstrated that an interfacial instability based on viscosity differences between the two flowing streams may occur, even for vanishing Reynolds number. The governing equations for this viscous instability are quite complex and their detailed analysis is beyond the scope of the present work. Instead, a brief overview of the key equations is presented to illustrate the feasibility of this instability and its potential role as a particle/substrate bonding mechanism. Following the analysis of Yih,[25] the amplitude of an interfacial perturbation η is described by the equation:

$$\eta \sim e^{i\alpha(x'+ct)} \qquad [9.34]$$

where x' is a position along the interface, $\alpha = 2\pi/\lambda$ is the wave vector and λ and c are the wavelength and the growth velocity of the perturbation, respectively. Accordingly, the interface becomes unstable when the imaginary part of the perturbation velocity c, c_i, is positive and grows at a rate that is proportional to the magnitude of c_i. Yih[25] developed the following equation for the imaginary part of the perturbation velocity:

$$c_i = \alpha \mathrm{Re} J \qquad [9.35]$$

where J is a function of several variables and is given by equation (5.42) in reference 20. The magnitude of J is generally of the order of 10^4–$10^5\,\mathrm{m^2/s}$. According to equation [9.34], the amplitude of an unstable perturbation increases by a factor of e over a time period of $\tau \sim 1/{*}ac_i$. After substituting c_i from equation [9.35], this time period can be expressed as:

$$\tau = \frac{1}{\alpha^2 \mathrm{Re} J} \qquad [9.36]$$

The analysis presented above enables one to ascertain whether a perturbation with the wavelength comparable with that of experimentally observed waves (ripples) along the particle/substrate interface will grow at a rate high enough to form roll-ups and vortices during the time scale of the particle/substrate collision event. For example, taking a wavelength of 1 µm (smaller than the particle size by more than an order of magnitude), and using the value of the Reynolds number of 10^{-4} obtained earlier and a value of $J = 10^4$–$10^5\,\mathrm{m^2/s}$ yields a time period over which an unstable perturbation undergoes a significant growth of $\tau \sim 10^{-14}$–10^{-15} s. Typical particle/substrate collision times are of the order of tens of nanoseconds, indicating that a 1 µm wavelength perturbation is unstable and would grow during a collision event. Furthermore, equation [9.28] reveals that perturbations with smaller wavelengths (larger wave vectors) will grow even more rapidly, suggesting that this type of instability can be expected to result in the formation of interfacial roll-ups and vortices. Since heavier copper particles are associated with larger kinetic energies, their collisions with the substrate are expected to give rise to the formation of interfacial perturbations with smaller wavelengths in the case of copper deposition on aluminum relative to those encountered in the aluminum-on-copper deposition case. This may be one of the reasons for the experimentally observed higher deposition efficiencies in the case of copper deposition on aluminum.

Yih's interfacial instability analysis[25] presented above can further be used to rationalize the experimentally observed stronger bonding of copper-on-aluminum than that for aluminum-on-copper. Figure 2(a) in reference 25 shows that the magnitude of the function J is dependent on the viscosity ratio $m = \mu_1/\mu_2$ and on the thickness ratio of the two fluid streams, $n = h_1/h_2$, where subscripts 1 and 2 are used to denote the streams with a lower and a higher thickness, respectively. For the case of aluminum deposition on copper, Fig. 9.2(b) suggests that a thinner jet is associated with the aluminum particle, and n has been estimated as 0.8–1.0. Using the procedure described above, the aluminum viscosity to copper viscosity ratio m has been evaluated as 0.5–1.0. Based on Fig. 2(a) in reference 25, the corresponding magnitude of the J function is assessed as 0.5–$1.0 \times 10^4\,\mathrm{m^2/s}$. For the case of copper deposition on aluminum, Fig. 9.3(d) suggests that a thinner jet is associated with the copper particle, and n has been estimated as 0.1–0.3. The copper viscosity to aluminum viscosity ratio m is 1.0–2.0. Based on Fig. 2(a) in reference 25 the corresponding magnitude of the J function is assessed as $1.0–3.0 \times 10^5\,m^2/s$. According to equation [9.36] and these estimates for J, the growth velocity of interfacial perturbations is expected to be larger, by at least an order of magnitude, for the copper-on-aluminum deposition case relative to that for the aluminum-on-copper deposition case. Thus, the role of interfacial instability in promoting particle/substrate bonding is expected to be more pronounced in the case of copper-on-aluminum deposition, as observed experimentally.[19]

Finally, in addition to, or instead of the interfacial instability mechanism described above, a simpler topological bonding mechanism is also suggested. Figure 9.3(a) to (d) shows the evolution of the particle and the substrate crater shapes during an impact event when the particle is copper and the substrate is composed of aluminum. A particularly important feature to note in this figure is that the cavity formed by the impact is reentrant, and that the copper particle completely conforms to the boundaries of this reentrant cavity. This is important because such a cavity, once formed, prevents the copper material from being separated from the aluminum substrate. For example, even if the two materials do not adhere firmly to each other, such a topological geometry would prevent separation of the materials due to the rivet-like nature of the interface. If such a shape were observed only when deposition efficiencies were greater than zero, and was absent when deposition efficiencies were equal to zero, then the presence of this rivet-like shape might be the feature that controls the bonding process. To further explore this idea, the topologies of the two material systems (copper-on-aluminum and aluminum-on-copper) under the conditions where their deposition efficiencies are zero (but a further increase in the particle velocity would yield a non-zero deposition) are examined. These situations are illustrated in Fig. 9.4(a) and (b) for the aluminum-on-copper system and for the copper-on-aluminum system, respectively. In spite of a difference in the impact velocities, the two craters appear to be on the verge of forming a reentrant shape. That is, the shapes are not reentrant, but it seems clear that a slightly larger impact velocity would result in a reentrant cavity. Hence, it is possible that the particle velocities at which the cold-spray process begins to show a non-zero deposition efficiency coincide with the impact velocities at which reentrant cavities begin to form, creating a rivet-like particle/cavity assembly.

To summarize, the discussion presented in this section suggests that:

(a) An interfacial instability due to differing viscosities and the resulting interfacial roll-ups and vortices may promote interfacial bonding by increasing the interfacial area, giving rise to material mixing at the interface and by providing mechanical interlocking between the two materials.

(b) A particle length-scale, rivet-like mechanism may also be operative and its onset may be linked with the minimum critical particle velocity.

9.5 Conclusions

In this chapter, a comprehensive overview is provided of the current understanding of the key phenomena accompanying particle/substrate interactions and bonding during cold spray. In addition, a detailed description is given of the continuum-type transient non-linear dynamics modeling techniques that are widely used to study the particle/substrate interactions and the attendant modeling mechanisms. The overview presented in this chapter shows that it is generally accepted that interfacial melting and inter-diffusion play minor roles in particle/substrate bonding. The prevailing opinion is that the phenomena that must be responsible for the particle/substrate bonding are: (a) adiabatic shear instability and flow localization which give rise to the formation of 'surface scrubbing' sets and (b) high contact pressure-promoted adhesion of the 'cleaned' particle and substrate contact surface. In addition, the formation of sub-micron-scale particle/substrate interfacial roll-ups and vortices, and particle-scale rivet-like interlocks, plays a significant role in particle/substrate bonding.

9.6 References

1 Alkhimov A P, Papyrin A N, Dosarev V F K, Nestorovich N I and Shushpanov M M, *Gas Dynamic Spraying Method for Applying a Coating*, U.S. Patent 5,302,414, 12 April 1994.

2 Tokarev A O, 'Structure of Aluminum Powder Coatings Prepared by Cold Gas Dynamic Spraying', *Met. Sci., Heat. Treat*, **35** (1996) 136–139.

3 Mccune R C, Papyrin A N, Hall J N, Riggs W L and Zajchowski P H, 'An Exploration of the Cold Gas-Dynamic Spray Method for Several Material Systems', in *Thermal Spray Science and Technology*, C C Berndt and S Sampath, Eds., ASM International, Materials Park, Ohio, USA, 1995, pp. 1–5.

4 Grujicic M, Zhao C L, Derosset W S and Helfritch D, 'Adiabatic Shear Instability based Mechanism for Particle/Substrate Bonding in the Cold-gas Dynamic-spray Process', *Mater. Des.*, **25** (2004) 681–688.

5 Bishop C V and Loar G W, 'Practical Pollution Abatement Method for Metal Finishing', *Plat. Surf. Finish*, **80** (1993) 37–39.

6 Grujicic M, Saylor J R, Beasley D E, Derosset W S and Helfritch D, 'Computational Analysis of the Interfacial Bonding Between Feed Powder Particles and the Substrate in the Cold-gas Dynamic-spray Process', *Appl. Surf. Sci.*, **219** (2003) 211–227.

7 Grujicic M, Cao G and Singh R, 'The Effect of Topological Defects and Oxygen Adsorbates on the Electronic Transport Properties of Single-Walled Carbon Nanotubes', *Appl. Surf. Sci.*, **211** (2003) 166–183.
8 El-Sobky H, 'Mechanics of Explosive Welding', Chapter 6 in *Explosive Welding, Forming and Compaction*, T Z Blazynski, Ed., Applied Science Publishers, London, 1983.
9 Hertel E S, Bell R L, Elrick M G, Farnsworth A V, Kerley G I, Mcglaun J M, Petney S V, Silling S A and Yarrington L, 'Lance CTH: A Software Family for Multi-dimensional Shock Physics Analysis', Presentation at the 19th International Symposium on Shock Waves, Marseille, France, 26–30 July 1993.
10 AUTODYN-2D and 3D, Version 6.1, User Documentation, Century Dynamics Inc., Concord, California, USA, 2006.
11 LS-DYNA User's Manual, Livermore Software Technology Corporation, Livermore, California, USA, 2006.
12 ABAQU/Explicit ABAQUS/Explicit 6.6 User Manual, Hibbitt, Karlsson & Soerensen, Pawtucket, Rhode Island, USA, 2006.
13 Assadi H, Gärtner F, Stoltenhoff T and Kreye H, 'Bonding Mechanism in Cold Gas Spraying', *Acta Mater.*, **51** (2003) 4379–4394.
14 Steinberg D J, Cochran S G and Guinan M W, 'A Constitutive Model for Metals Applicable at High Strain Rate', *J. Appl. Phys.*, **51** (1980) 1498–1522.
15 Steinberg D J and Lund C M, 'A Constitutive Model for Strain Rates from 10^{-4} to $10^6 s^{-1}$', *J. Appl. Phys.*, **65** (1989) 1528–1539.
16 Zerilli F J and Armstrong R W, 'Dislocation-Mechanics-Based Constitutive Relations for Material Dynamics Calculations', *J. Appl. Phys.*, **61** (1987) 1816–1829.
17 Johnson G R and Cook W H, 'A Constitutive Model and Data for Metals Subjected To Large Strains, High Strain Rates, and High Temperatures', in Proceedings of the Seventh International Symposium on Ballistics, The Hague, The Netherlands, 1983, pp. 541–547.
18 Kreig R D and Kreig D B, 'Accuracies in Numerical Solution Methods for the Elastic Perfectly-plastic Model', *Trans. ASME: J. Pressure Vessel Technol.*, **99** (1977) 510–522.
19 Van Steenkiste T H, Smith J R and Teets R E, 'Aluminum Coatings Via Kinetic Spray With Relatively Large Powder Particles', *Surf. Coat. Technol.*, **154** (2002) 237–251.
20 Wright T W, 'Shear Band Susceptibility: Work Hardening Materials', *Int. J. Plasticity*, **8** (1992) 583–602.
21 Wright T W, 'Toward A Defect Invariant Basis for Susceptibility to Adiabatic Shear Bands', *Mech. Mater.*, **17** (1994) 215–222.
22 Schoenfeld S E and Wright T W, 'A Failure Criterion Based on Material Instability', *Int. J. Solids Struct.*, **40** (2003) 3021–2037.
23 Dykhuizen R C and Smith M F, 'Gas Dynamic Principles of Cold Spray', *J. Therm. Spray Technol.*, **7** (1998) 205–212.
24 Grujicic M, Zhao C L, Tong C, Derosset W S and Helfritch D, 'Analysis of the Impact Velocity of Powder Particles in the Cold-gas Dynamic-spray Process', *Mater. Sci. Engng A*, **368** (2004) 222–230.
25 Yih C S, 'Instability due to Viscosity Stratification', *J. Fluid Mech.*, **27** (1967) 337–349.

10

Supersonic jet/substrate interaction in the cold spray process

V. F. KOSAREV, S. V. KLINKOV and A. N. PAPYRIN,
Cold Spray Technology (CST), USA

10.1 Introduction

Despite extensive investigation of jet gas dynamics, some issues have not been adequately studied. There are many aspects of this problem, and, with respect to cold spray, one of the most important tasks is to increase the particle velocity as high as possible. The development of the boundary layer on the nozzle walls, the structure and stability of the jet in different exhaust modes, interaction of a supersonic jet with the substrate and heat transfer between the jet and the substrate, should be studied to obtain the optimal velocity and temperature of a particle at the moment of its impact onto the substrate surface.

At present, a wide spectrum of research in the field of gas dynamics of the cold spray process has been conducted. Various aspects of this problem have been considered including the nozzle shape optimization. Some results of modeling and experiments are presented in the following references: Dykhuizen and Smith 1998; Shukla and Elliott 2000; Alkhimov *et al.* 2001; Sakaki *et al.* 2002; Stoltenhoff *et al.* 2002; Grujicic *et al.* 2004; Kwon *et al.* 2005; Pattison *et al.* 2005; Jodoin *et al.* 2006.

In order to accelerate particles we suggested the use of two types of nozzles: nozzles with circular and rectangular sections. Historically, the gas dynamics of jets exhausting from conical nozzles with circular cross-sections, i.e. symmetrical axis flows, has been examined in great detail. An analysis of the features of such jets, as applied to the cold spray method, revealed that the use of nozzles with rectangular cross-sections is also promising. With the same ratio of the nozzle-exit and throat cross-sections, nozzles with a rectangular section can provide, on the one hand, a wider spray beam in the direction of the smaller size of the section and, on the other hand, a narrower beam (to 1–2 mm) in the direction of the larger size of the section. Such nozzles can also decrease the effect of particle deceleration in the compressed layer in front of the substrate by decreasing the thickness of the layer itself. The unusual nozzle shape leads to the formation of a jet that differs considerably from the well-known symmetrical axis jet with a uniform distribution of gas parameters at the nozzle exit.

The purpose of the present chapter is to discuss some of the above-stated gas dynamic and thermal effects associated with a supersonic jet exhausting from the nozzle and its interaction with the substrate in the cold spray method. This chapter presents an overview of some results of studies conducted at the Institute of Theoretical and Applied Mechanics of the Siberian Branch of the Russian Academy of Science where the cold spray process was developed.

10.2 Investigation of supersonic air jets exhausting from a nozzle

Because of the unusual shape of the nozzles studied, the jets formed differ from the well-examined symmetrical axis jets or plane jets with a uniform distribution of parameters at the nozzle exit. In practice, one often has to choose the optimal nozzle shape for successful spraying. One of the criteria used for comparing nozzles is the set of characteristics of jets exhausting from a given nozzle. Therefore, we performed experiments with jets whose characteristics were similar to the parameters of jets used for cold spraying (Alkhimov *et al.* 1997). In some cases, it is necessary to vary the spraying distance, which invokes questions regarding the limits of varying this distance without significant violations of the spraying process and changes in coating properties. To answer this question, we studied the supersonic part of the jet. It should be noted that many results presented relating to jets from rectangular nozzles can also be used to examine nozzles with a conical supersonic part, which are more commonly used.

10.2.1 Experimental setup and research techniques

It is known that all of the parameters of gas flow can be determined from the initial values of three quantities. Let us choose the Mach number as the first quantity, stagnation temperature as the second quantity and static pressure as the third quantity. Thus, in studying isenthalpic (T_0 = constant), isobaric (p = constant) jets, one needs to find only one quantity, namely, Mach number. It is convenient to calculate the Mach number by the Rayleigh formula from the measured stagnation pressures behind the normal shock wave formed on the tip of a thin tube (Pitot tube). In the first approximation, we can assume that p_0' is proportional to M^2 and, hence, to the dynamic pressure ρv^2. Jets exhausting from three difference nozzles were considered (Table 10.1). The angle of the nozzle in the direction of one transverse coordinate y was roughly identical for all three nozzles, namely, 3×10^{-3} radian, and the angle in the direction of the other transverse coordinate

Table 10.1 Parameters of gas flow at the nozzle exit

No.	M*	p_{0cal} (MPa)	h (mm)	H/h
1	1.75	0.53	1	8
6	2.3	1.25	3	3.3
8	2.75	2.51	4.5	2.7

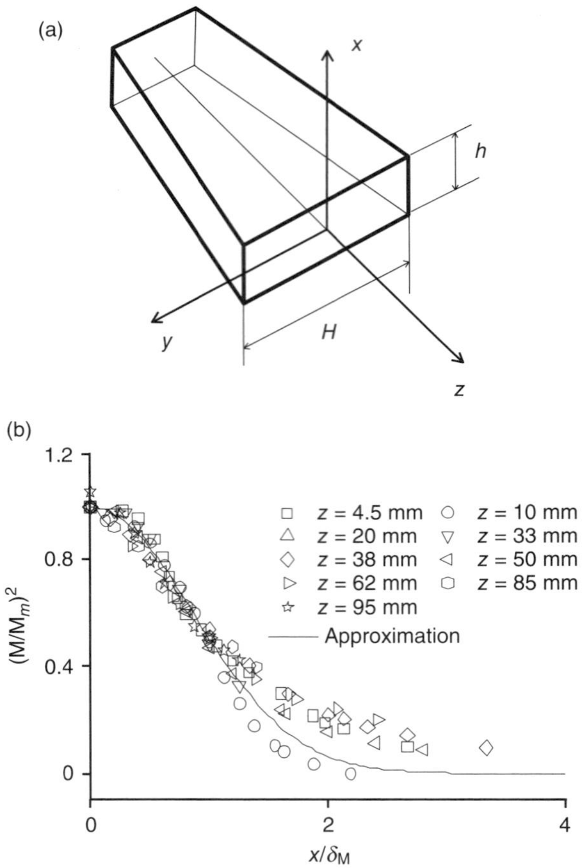

10.1 (a) Nozzle shape and coordinating system. (b) Normalized M^2 profiles in an overexpanded jet exhausting from a nozzle with $h = 4.5$, $H/h = 2.7$ and $M^* = 3.1$.

x, characterizing the smaller size of the nozzle, was equal to zero (Fig. 10.1(a)).

The experimental setup is described in Alkhimov *et al.* (1997). The Schlieren method was used to study jet structure. Pressure probes with an outer diameter of 0.5 mm were used in these experiments. A thermocouple was used as a temperature probe.

10.2.2 Profiles of parameters in jets

As argued in Ginevskii (1969), Abramovich (1984), and Vulis and Kashkarov (1965), it is known that the profiles of velocity (v) and dynamic pressure (ρv^2) are self-similar at the initial and main parts of the jet. It is convenient to present the dynamic pressure in the form $\gamma p_c \mathrm{M}^2$; in the case of an isobaric flow, this leads to self-similarity of the profiles of M^2. Approximation formulae for velocity profiles are encountered in the literature, but it is rather difficult to find formulae for M^2 profiles. At the same time, it is the Mach number that is the most convenient quantity to be used in experiments because its determination requires only the knowledge of the pressure field (and not temperature fields). In addition, it is convenient to express the law of conservation of excess momentum in terms of the Mach number. Thus, one of our tasks was to verify self-similarity of the M^2 profiles and to find an approximation function for them.

Mach number profiles

The profiles of M^2 were reconstructed from the experimentally obtained profiles of the static (p_c) and Pitot (p_0') pressures. The data gained in studying jets with different initial (indicated by an asterisk) parameters (h = 1–4.5 mm, H/h = 2.7–8, M^* = 1.85–3.1 and T_0^* = 300–600 K, where h and H are the small and large transverse sizes of the jet at the nozzle exit, respectively) and plotted in the coordinates $(\mathrm{M}/\mathrm{M}_m)^2$, x/δ_M are well fitted by one curve (Fig. 10.1(b)). The nozzle shape and the coordinating system are shown in Fig. 10.1(a). With an insignificant scatter, this curve can be described by the function

$$\varphi_\mathrm{M} = \exp\left[-\left(\sqrt{\ln 2}\,\frac{x}{\delta_\mathrm{M}}\right)^2\right] \qquad [10.1]$$

where δ_M is the jet thickness along the smaller size (in the x direction), determined as the distance from the jet axis to the point where $\mathrm{M}^2(\delta_\mathrm{M}) = 0.5\mathrm{M}_m^2$.

It is worth noting that the stagnation temperature (T_0) and non-isobarity factor (n) (exit pressure divided by the ambient pressure) in the examined range (T_0 = 300–600 K, n = 0.5–1) do not exert any substantial effect on the M^2 profile.

Profiles of excess temperature

It is known from the jet theory that the profiles of the excess stagnation temperature ($\Delta T_0 = T_0 - T_a$) are also self-similar and admit the relation

$$(T_0 - T_a)/(T_{0m} - T_a) = \Delta T_0/\Delta T_{0m} = (v/v_m)^\sigma$$

where $\sigma = 0.5$ for plane jets, $\sigma = 0.75$ for symmetrical axis jets, v is the gas velocity and T_a is the ambient temperature. Assuming that the profiles are described by functions of the same form, we can find the relation between the velocity-profile thickness (δ_v) and the temperature-profile thickness (δ_T):

$$\delta_v = \delta_T \sqrt{\sigma}$$

A series of experiments was performed to validate self-similarity, to find the approximation function for excess stagnation temperature profiles and to find the relation of the thicknesses of the M^2 and ΔT_0 profiles. The stagnation temperature was determined by a thermometric probe utilizing a thermocouple. The profiles plotted in the coordinates $\Delta T_0/\Delta T_{0m}$, x/δ_T are fitted by the curve

$$\varphi_T = \exp\left[-\left(\sqrt{\ln 2}\,\frac{x}{\delta_T}\right)^2\right] \qquad [10.2]$$

with an insignificant scatter (Fig. 10.2). The experimentally found value of δ_T/δ_M for the examined range of parameters is close to 2.

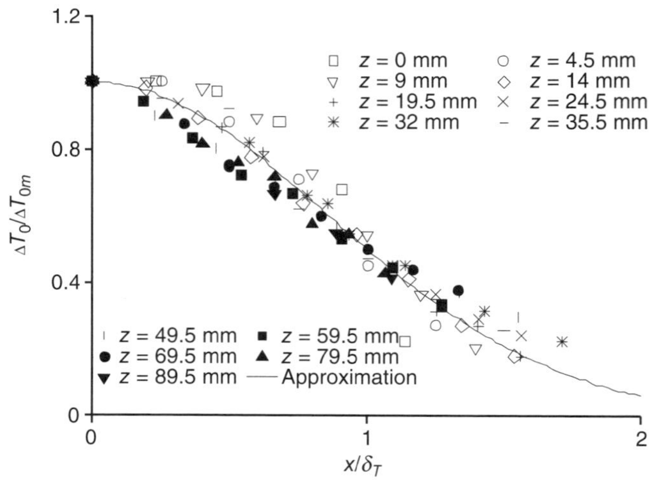

10.2 Normalized profiles of excess stagnation temperature in an overexpanded jet exhausting from a nozzle with $h = 4.5$, $H/h = 2.7$ and $M^* = 3.1$.

10.2.3 Streamwise distribution of axial parameters

One of the problems of the jet theory is finding the axial values of parameters denoted here by the subscript m. Thus, Ginevskii (1969) used the matching of two solutions for the initial and main parts; there was an inflection at the point of the matching (transitional part of the jet) which was not observed in our experiments. We made an attempt to find a smooth approximation function in the transitional region. Figure 10.3 shows the data borrowed from Abramovich (1984), and the results obtained in our experiments are plotted in Fig. 10.4. All data lie on one curve of the form

$$\left(\frac{M_m}{M_m^*}\right)=\left[1+3(z/z_{0.5}^{M})^4\right]^{-0.5} \qquad [10.3]$$

where z is the streamwise coordinate of the jet and $z_{0.5}^{M}$ is the coordinate defined from $M_m^2(z_{0.5}^{M}) = 0.5M_m^{*2}$. The greatest deviation from this curve is observed for data obtained in studying an overexpanded jet, but the upper peaks lie on the curve even in this case, whereas the lower peaks at $z/z_{0.5}^{M} < 1$ lie approximately at one level. It should be noted that this function also yields a correct asymptotic value, because we have $M_m^2 \sim 1/z^2$ for symmetrical axis jets in accordance with the equation of conservation of momentum (all jets at large distances can be presented as a symmetrical axis jet).

For moderate heating of the jets, the relation between the axial excess stagnation temperature and the axial value of M^2 should be close to the form

$$\Delta T_{0m}/\Delta T_0^* = (M_m^2/M_m^{*2})^{0.25}$$

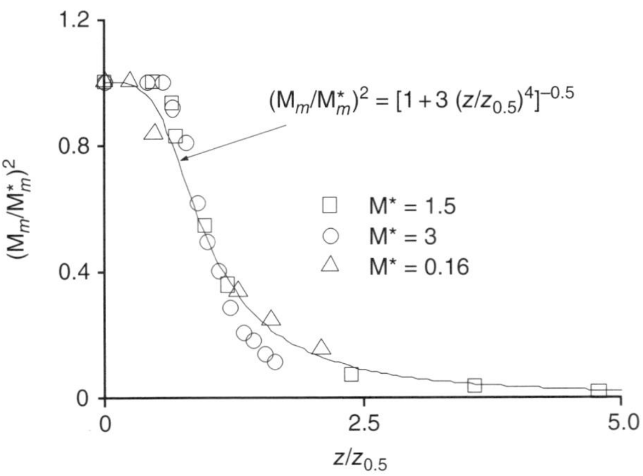

10.3 Generalized distribution of the axial values of M^2 versus the streamwise coordinate (Abramovich 1984).

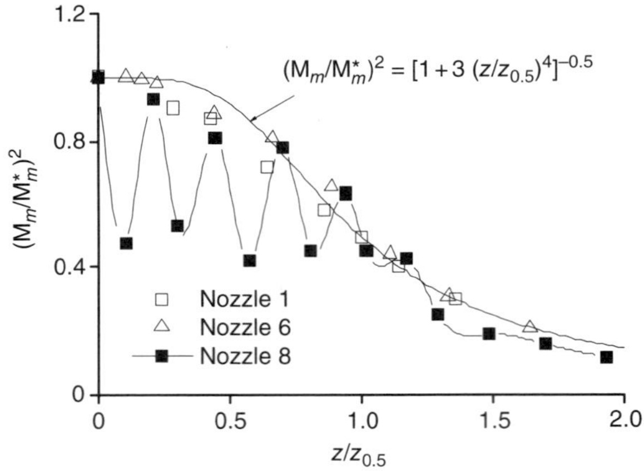

10.4 Generalized distribution of the axial values of M^2 versus the streamwise coordinate.

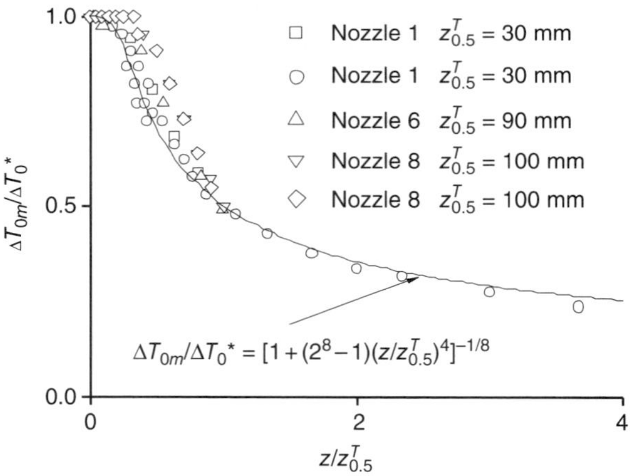

10.5 Generalized dependence of normalized excess stagnation temperature on the streamwise coordinate.

We used this circumstance to find the distribution function of the axial values of excess stagnation temperature (Fig. 10.5). A comparison with experimental data allows us to assume that

$$\Delta T_{0m} = \frac{\Delta T_0^*}{[1+255(\bar{z})^4]^{0.125}} \qquad [10.4]$$

$$\bar{z} = z / z^T_{0.5}, \qquad z^T_{0.5} \approx 2z^M_{0.5}$$

The ratio $z^{T}_{0.5}/z^{M}_{0.5}$ obtained for three jets from the experiment is approximately equal to 2.

10.2.4 Jet thickness

One important problem is to determine the jet-thickness growth as a function of the streamwise coordinate. It is known from the literature that the jet-thickness growth in the initial and main parts of the jet is linear but with different proportionality factors (Abramovich 1984). Thus, there is a transitional part where the jet-thickness growth is a non-linear function. We studied jets with a non-uniform initial profile because of a noticeable boundary layer on the nozzle walls, therefore we would expect that the potential core region is weakly expressed; the entire region under study can thus be considered as transitional, with the jet thickness approximated by a non-linear function. If we assume that the jet is planar, i.e. neglect its expansion in the direction of the larger size, we can obtain the relation between the jet thickness and the axial value of M^2 from the equation of conservation of momentum:

$$\delta_M = C^*h[1 + 3(z/z^{M}_{0.5})^4]^{0.5} \qquad [10.5]$$

Here, C^* is a coefficient that takes into account the jet thickness in the very beginning (i.e. at the nozzle exit). It differs from the nozzle-exit thickness because the jet thickness is conventionally determined by the ratio of the boundary value of the squared Mach number to the value at the jet axis (here, we used the value of 0.5). Moreover, this coefficient changes, depending on the profile fullness (i.e. its closeness to the rectangular profile). Clearly, this coefficient cannot be rigorously equal to unity.

If we assume further that expansion along the larger size is exactly the same as expansion along the smaller size (quasi-symmetrical axis case), i.e. $\delta_y/\delta_M = H/h$ (δ_y is the jet thickness along the larger size) and use the expression

$$(M/M_m)^2 = \exp\left[-\left(\sqrt{\ln 2}\,\frac{x}{\delta_M}\right)^2\right]\exp\left[-\left(\sqrt{\ln 2}\,\frac{y}{\delta_y}\right)^2\right] \qquad [10.6]$$

we obtain

$$\delta_M = C^*h[1 + 3(z/z^{M}_{0.5})^4]^{0.25} \qquad [10.7]$$

Based on experimental results, we can find a more exact curve (Fig. 10.6):

$$\delta_M = 0.75h[1 + 3(z/z^{M}_{0.5})^3]^{0.4} \qquad [10.8]$$

It can be seen in Fig. 10.6 that the proposed formulae yield approximately identical results in the region $z/z^{M}_{0.5} < 1$. Significant differences are observed

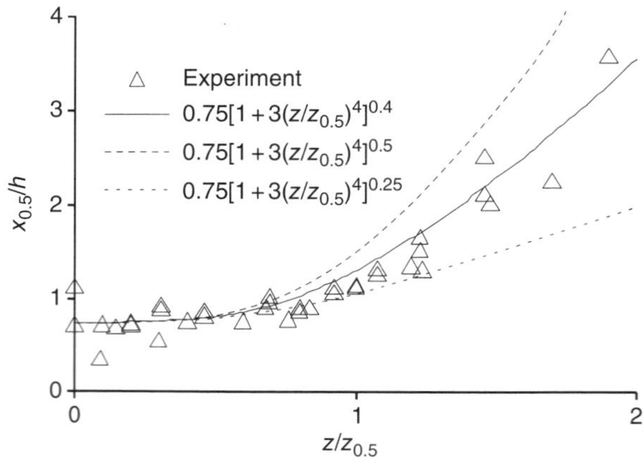

10.6 Generalized dependence of the jet thickness on the streamwise coordinate.

in the farther region $z/z_{0.5}^{M} > 1$. Thus, as it could be expected, the examined jets can be referred neither to the plane case nor to the symmetrical axis case.

10.2.5 Effect of the jet-pressure ratio

It is of interest to consider the dependence of the periodic structure of the non-isobaric jet on the jet-pressure ratio. It is convenient to use the distance from the nozzle exit to the shock wave as the characteristic distance. According to published data, this distance in the case of large pressure ratios ($n >> 1$) should be proportional to $n^{0.5}$. The jet was visualized by the Schlieren technique. Figure 10.7 shows photographs of the jet exhausting from a nozzle with $h = 3$ mm, $H/h = 3.3$ and $M^* = 1.5$ with different pressure ratios. An analysis of the photographs obtained confirmed the proportionality to $n^{0.5}$. It is seen that a paraboloid shock wave originates at $n \sim 2.5$ and becomes straight at $n \sim 4.5$–4.8. It is worth noting that the straight shock wave originates at pressure ratios much higher than those in the case of symmetrical axis jets ($n \approx 1.5$ (Avduevskii *et al.* 1985)).

It is known that the number of barrels in the jet decreases with increasing pressure ratio, because of faster equalization of pressure. Therefore, we had to find the effect of the jet-pressure ratio on gas parameters far from the initial part of the jet, where the static pressure is already equal to the atmospheric value. This can be done by measuring the length of the supersonic part of the jet l_s, because the jet flow reaches the velocity of sound prior to the point where the atmospheric pressure prevails, as was found

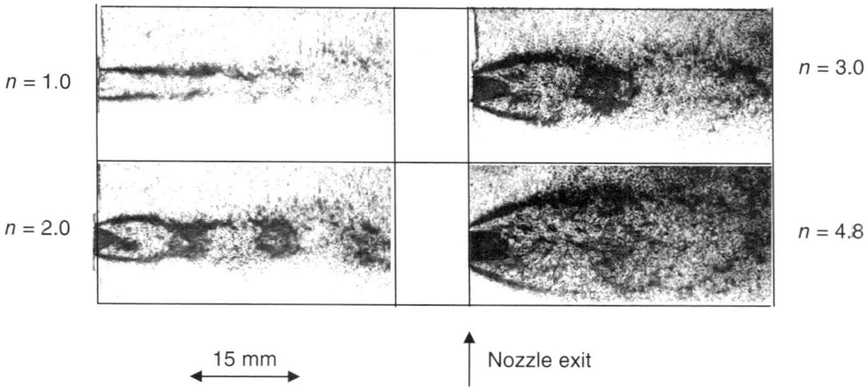

10.7 Schlieren pictures of the jet exhausting from a nozzle with $h = 3\,\text{mm}$, $H/h = 3.3$ and $M^* = 1.5$ with different pressure ratios.

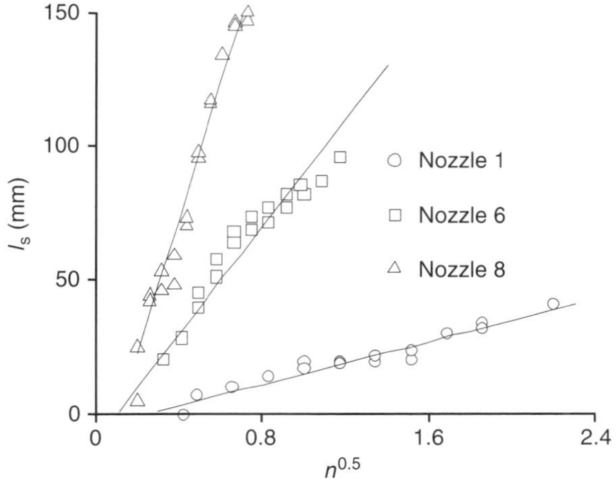

10.8 Length of the supersonic part of the jet versus the jet-pressure ratio $n^{0.5}$.

experimentally. It is seen from Fig. 10.8, which shows these dependences, that the same law of proportionality with $n^{0.5}$ is observed with insignificant deviations.

It was found in the present study that the initial non-uniformity of gas parameters at the nozzle exit makes the transition in the streamwise distribution of M^2 from the initial to the main part of the jet smoother. Observation of jet-thickness growth showed that the observed growth is smaller than that predicted for a plane jet, which seems to be attributed to tip effects arising in jets with a finite ratio of its sides. It was experimentally verified in the examined jets that the length of the element of the periodic jet structure and the supersonic length of the jet depend on the jet-pressure ratio as $\sqrt{n}$.

The study performed validated self-similarity of M^2, ΔT_0 and v profiles. The range of self-similarity starts at a certain distance from the nozzle exit and extends infinitely in the downstream direction. The transition through sonic lines has no effect on parameter profiles. Because of a significant thickness of the boundary layer formed on the nozzle walls, the initial profiles can hardly be distinguished from the self-similar profiles. For this reason, the range of self-similarity can be extended to the entire jet, beginning from the nozzle exit.

10.3 Impact of a supersonic jet on a substrate

This section describes the results for the interaction of supersonic air jets with a rectangular cross-section incident onto a flat infinite substrate at different impact angles (Alkhimov *et al.* 2000a; Papyrin *et al.* 2001). Although the shape of the coated part is not always a flat surface, the jet size is comparatively small, and the parameters of the gas, particles and surface at the impact moment can be determined in the first approximation by solving the problem of interaction of supersonic rectangular jets with a flat infinite substrate. Particle concentrations typical for cold spray are normally much lower than the values at which the particles begin to noticeably affect the gas parameters. Therefore, in the first approximation, we can ignore the presence of particles in the flow and, thus, substantially simplify the problem.

A specific feature of the studied nozzles is the large relative length, which leads to formation of a jet with a non-uniform profile at the nozzle exit (Alkhimov *et al.* 1997). Most results published in the literature refer to symmetrical axis or plane jets with a uniform velocity profile at the nozzle exit. Therefore, it was important to study the impact of plane jets, which are common enough for cold spray.

The experiments were performed on a setup including a gas heater and a pre-chamber with attached nozzles of different geometry. A steel plate with an orifice 0.2 mm in diameter was used to measure the pressure on the substrate surface. The plate was mounted on the coordinate table, which made it possible to change the distance from the nozzle exit z_0, to move in the substrate plane into mutually perpendicular directions x and y to the nozzle centerline, and to vary the angle of incline of the substrate plane φ_{im} to the nozzle centerline, corresponding to rotation of the substrate plane around the y axis.

The pressure profile on the substrate surface was measured in two mutually perpendicular directions x and y. The origin was located at the center of the projection of the rectangular nozzle-exit section onto the substrate surface. The coordinate axes x and y were aligned parallel to the sides of the nozzle exit; the x axis was parallel to the smaller side, the y axis was parallel to the larger side, and the z axis was normal to the substrate surface.

The values of the Mach number M in the region of the near-wall jet were determined by a Pitot tube with an inner diameter of 0.2 mm. The coordinate gear on which the Pitot tube was mounted allowed its motion in three mutually perpendicular directions.

10.3.1 The pressure distribution on the substrate surface and velocity gradient at the stagnation point

One of the tasks of experimental investigations of jet–substrate interaction was to determine the pressure distribution on the substrate surface. This information was necessary to reconstruct the character of the flow in the shock layer (based on the presence or absence of peripheral maximums) and to find the velocity distribution at the edge of the near-wall boundary layer with the use of the Bernoulli integral. Figure 10.9 shows the normalized pressure profiles on the substrate for different impact angles of the jet φ_{im} and different distances between the nozzle exit and the substrate. Here,

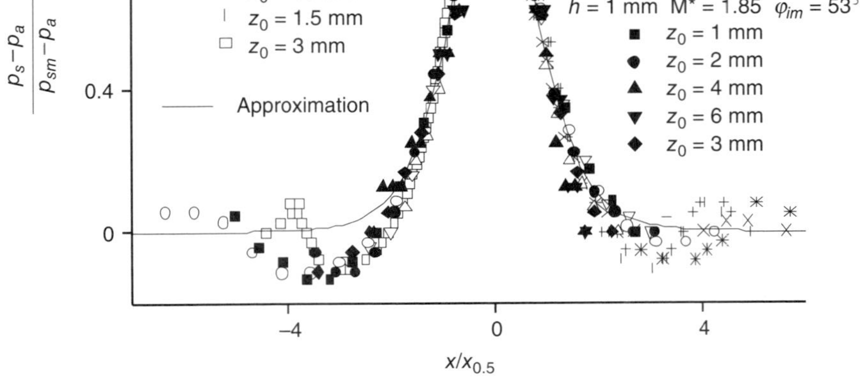

10.9 Pressure profiles on the substrate surface for an isobaric jet.

h is the smaller size of the nozzle exit, z_0 is the distance from the nozzle exit to the substrate, p_s is the pressure on the substrate surface, p_{sm} is the pressure on the substrate surface for $x = 0$, p_a is the ambient pressure and $x_{0.5}$ is the half-width of the pressure profile ($p_s(x_{0.5}) - p_a = 0.5(p_{sm} - p_a)$).

The points and the curve show the experimental data and the approximation. The experimental results are adequately approximated by the function

$$\frac{p_s - p_a}{p_{sm} - p_a} = \left[1 + (\sqrt[4]{2} - 1)\left(\frac{x}{x_{0.5}}\right)^2\right]^{-4} \qquad [10.9]$$

borrowed from Belov (1983), where it is claimed that the pressure distribution on the substrate surface is self-similar. For $z_0 \leq 4h$, the value of $x_{0.5}$ is approximately one half of h. Figure 10.9 shows that self-similarity is observed in our case along the smaller size of the jet for angles $\varphi_{im} = 50–90°$.

Figure 10.10 shows the pressure profiles along the larger size of the nozzle exit H. The distribution (especially for small z_0) has a segment of roughly constant values of pressures, and the pressure practically vanished only at the ends of the curve. Thus, we can state that, on the one hand, the pressure distribution along the y axis is not self-similar and, on the other hand, it can be assumed constant for rather large values of H/h.

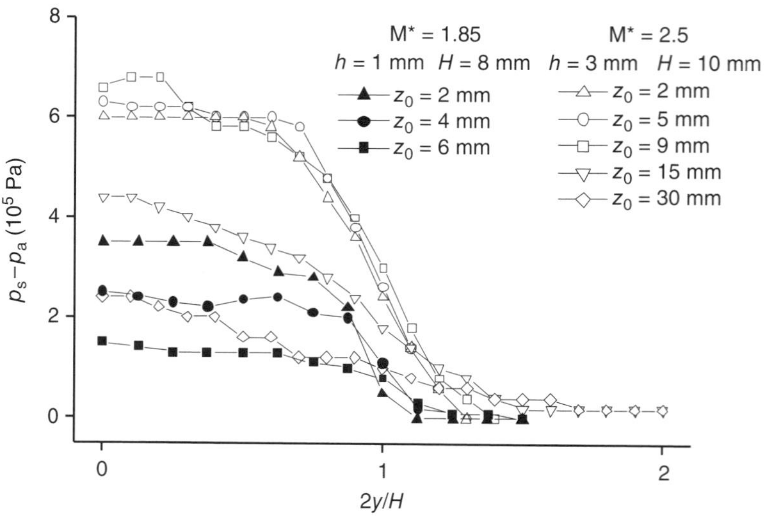

10.10 Pressure profile on the substrate surface for the larger size of an isobaric jet exhausting from the nozzle.

Velocity gradient at the stagnation point

Based on the pressure profiles on the substrate surface, we can find the velocity gradient at the stagnation point along the x axis. For this purpose, we use the expression for pressure (equation [10.9]) and the isentropic relation between velocity (Mach number) and pressure. Finally, we obtain the expression for the velocity gradient at the stagnation point

$$\frac{\mathrm{d}u}{\mathrm{d}x} = 2\frac{a_{cr}}{x_{0.5}}\sqrt{\frac{\gamma+1}{\gamma}\left(\sqrt[4]{2}-1\right)\left(1-\frac{p_a}{p_0'}\right)} \qquad [10.10]$$

where a_{cr} is the critical velocity of sound and p_0' is the pressure measured by the Pitot tube. The value of the root in the right-hand part of the equation for typical Mach numbers M = 1.8–3.1 and $\gamma = 1.4$ equals 0.5 within 5%. Hence, the velocity gradient can be estimated by the following expression with accuracy sufficient for practical applications:

$$\beta = \frac{\mathrm{d}u}{\mathrm{d}x} = \frac{a_{cr}}{x_{0.5}} \approx 2\frac{a_{cr}}{h} \qquad [10.11]$$

From equation [10.9], we find the ratio between the coordinate x_{cr} of the critical transition and the coordinate $x_{0.5}$. In the general case, this expression depends on the Mach number and jet-pressure ratio, but for an isobaric jet in the typical range M = 1.8–3.1, it can be estimated within 5% as

$$x_{0.5}/x_{cr} = 0.87$$

Thus, the value $x_{0.5}$ is fairly close to x_{cr}; hence, these two quantities can be estimated as $h/2$. The velocity in the accelerating flow region can be found by the formula (Yudaev *et al.* 1977a)

$$\frac{u}{u_{ac}} = 1.5\frac{x}{x_{ac}} - 0.5\left(\frac{x}{x_{ac}}\right)^3 \qquad [10.12]$$

where x_{ac} is the length of the acceleration region and u_{ac} is the velocity at the end of the acceleration region. Knowing the velocity gradient, we can find the relation between x_{ac} and u_{ac}:

$$x_{ac} = 1.5\frac{u_{ac}}{\beta}$$

The value of u_{ac} is found from the condition of isentropic expansion of the gas with a stagnation pressure p_0' to the atmospheric value:

$$u_{ac} = a_{cr}\sqrt{\frac{\gamma+1}{\gamma-1}}\left[1-\left(\frac{p_a}{p_0'}\right)^{\frac{\gamma-1}{\gamma}}\right]^{0.5} \qquad [10.13]$$

The value of p_a/p'_0 is determined by the Rayleigh formula. For typical conditions of an impact of an isobaric jet with $M^* = 2.5$ onto a substrate located in the region of the initial part of the jet, we obtain $M_{ac} = 2$, $u_{ac} = 1.62a_{cr}$ and $x_{ac} = 2.43x_{0.5}$, where $x_{0.5} \cong h/2$. Thus, the region of acceleration is slightly greater than the jet size h, starting from the stagnation point of the flow.

The above-made observations and estimates allow us to suggest a simple scheme of near-wall jet formation (Fig. 10.11(a)). The bow shock wave can be conventionally replaced by a rigid wall, and the gas can be assumed to be accelerated in a nozzle formed by this wall and the substrate surface. The critical parameters of the gas are reached near the boundary of the incoming jet, i.e. the region from the stagnation to the critical point corresponds to the confuser part of the Laval nozzle. After that, jet expansion and acceleration to supersonic velocities follow, which correspond to the diverging part of the Laval nozzle.

Comparison of pressure distributions in the jet and on the substrate surface

By comparing the pressure distributions in the free jet and on the substrate surface (Fig.10.11(b)), we can see that they are fairly close (z_0 is the distance from the nozzle exit to the substrate and to the input orifice of the Pitot tube). Thus, to find the pressure distribution over the substrate surface, we have to know the pressure at the center and the thickness $x_{0.5}$, which depend only on z_0. In this case, the problem becomes one-dimensional. It also follows from the data described above that the value of $x_{0.5}$ can be assumed to be equal to this value in the free jet. The same statement refers to the pressure at the substrate center: it can be estimated by the pressure at the centerline of the free jet behind the normal shock. We have to clarify the limits where these approximations can be used. For this purpose, we performed experiments whose results were compared with the distribution in the free jet.

10.3.2 Effect of the distance from the nozzle exit to the substrate on jet parameters: oscillations of the jet

Figure 10.12 shows the distributions of the maximum pressure behind the shock wave in the free jet and on the surface of a substrate mounted at an angle of 90° to the jet axis. The values of p'_0 and p_s initially coincide, but the pressure on the substrate decreases much more significantly with distance. Thus, at a distance $z_0 \approx 7h$, the pressure in the free jet does not yet decrease too much and equals, on average, the pressure near the nozzle exit, whereas its value under impact conditions at the same distance is approximately half

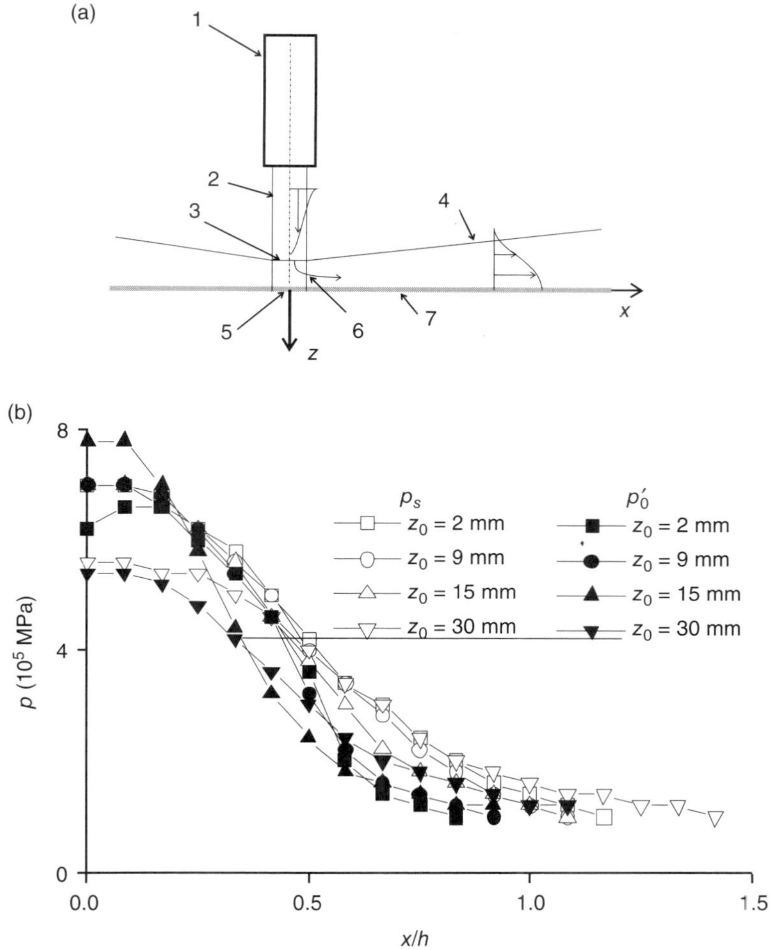

10.11 (a) Pattern of jet impingement. 1, nozzle; 2, jet; 3, bow shock; 4, near-wall jet; 5, stagnation point; 6, critical section of near-wall jet; 7, substrate surface. (b) Pressure profiles on the substrate surface p_s and behind the normal shock wave p'_0 along the smaller size of an isobaric air jet exhausting from the nozzle. $h = 3$ mm, $H/h = 3.3$, $M^* = 2.5$ and $\varphi_{im} = 90°$.

that at the nozzle exit. Such a pattern is typical not only for isobaric but also for non-isobaric jets. Thus, the presence of the substrate involves some changes in the flow structure, although the values of p'_0 and p_s coincide in the regions $z_0 \leq 4h$ and $z_0 \geq 15h$. If the impact angles differ from 90° (φ_{im} = 53–90°), the behavior of the curves remains qualitatively the same as in the case of an impact onto a normally located substrate, and the pressure decrease is faster than that in the free jet. Apparently, this decrease is

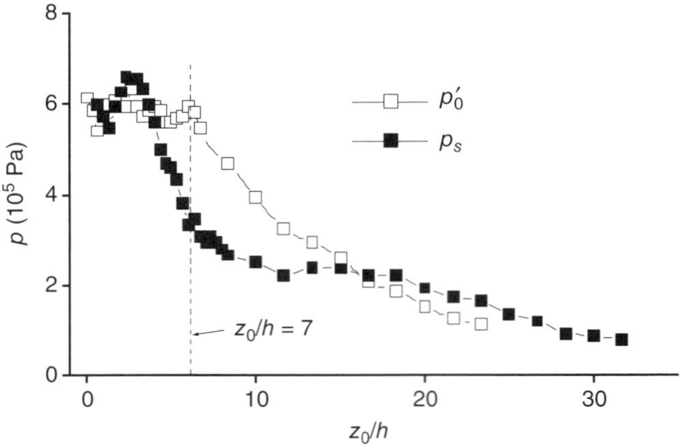

10.12 Maximum pressure on the surface of a substrate mounted at an angle of 90° to the jet axis and behind the normal shock versus the distance to the substrate for an isobaric air jet (T_0 = 300 K, p_0^* = 1.4 MPa) exhausting from the nozzle. h = 3 mm, H/h = 3.3 and M* = 2.5.

related to an emergence of oscillations of the jet incident onto the substrate, which increase the inflow of ambient air into the jet, and the total pressure at a certain part of the jet drastically decreases. Indeed, photographs taken using a ruby laser (Fig. 10.13) show that jet instability manifested in transverse oscillations is observed at certain distances from the nozzle exit to the substrate. The main parameters of oscillations gained from the photographs are summarized in Table 10.2. The data from Table 10.2 and the form of the photographs indicate that disturbances do not develop immediately when the jet leaves the nozzle but at a certain distance L_1 from the nozzle exit. It is seen in Table 10.2 that the mean length of this undisturbed section is about 2–6 nozzle thicknesses. The frequencies based on the wavelength and gas velocity lie in the ultrasonic range (~50–100 kHz). It should be noted that the amplitude of oscillations increases in the downstream direction and reaches 2–3 nozzle thicknesses. Stability improves as the pressure in the pre-chamber increases and the flow regime becomes non-isobaric with $n > 1$. In practice, the spraying distances are chosen such that instability does not have enough time to develop, in other words, from the condition $z_0 \leq L_1$.

10.3.3 Near-wall jet

One of the main goals of the present study was to obtain the M^2 profiles in the near-wall jet (Fig. 10.14). In the region of the near-wall jet, where the

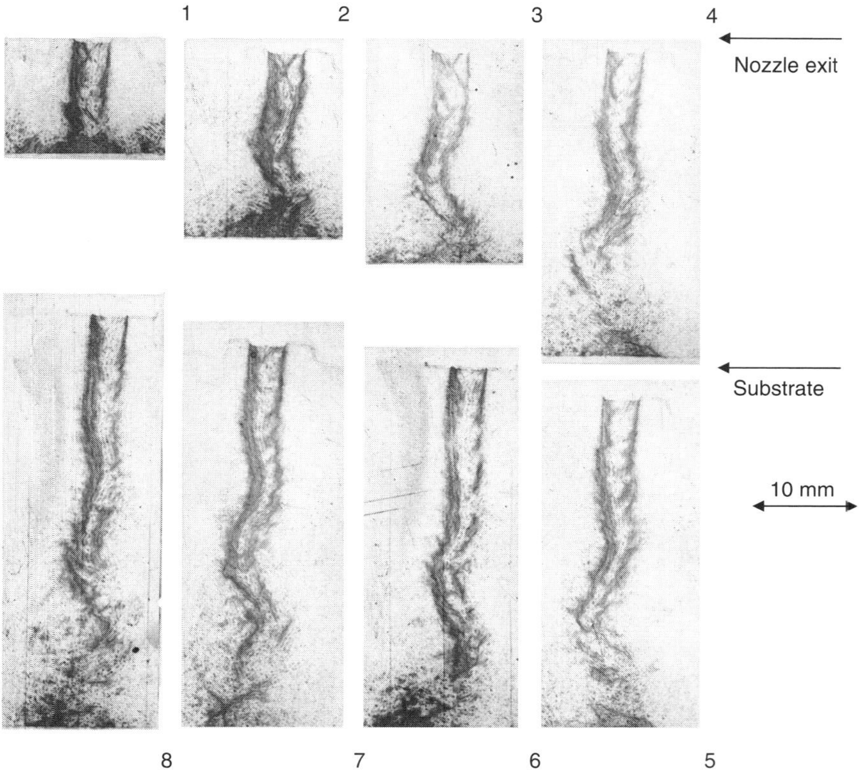

10.13 Oscillatory impact of a rectangular supersonic jet onto a normally located substrate. The exposure time is $\tau_{ex} = 30$ ns, $M^* = 2.25$, $h = 3$ mm and $z_0/h = 3$ (1), 5 (2), 5.7 (3), 8 (4), 9 (5), 9.7 (6), 10.3 (7) and 11 (8).

Table 10.2 Characteristics of supersonic jet oscillations

Parameters	z_0/h	L_1/h	λ (mm)
$h = 3$ mm, $v^* = 500$ m/s, $M^* = 2.5$, $n = 1$	7.7	1.4	9.9
	8.3	1.6–1.9	10.4–10.7
	9.3	1.3	11.7
	9.7	2.9	10.7
	10	1.7–2.6	11.1–11.6
	10.6	1.7	11.7
$h = 1$ mm, $v^* = 460$ m/s, $M^* = 1.9$, $n = 1$	11.3	2.3	6.5
	24	3.9	4.5
	31	3.9	5.2
	32	6.5	9.7
	$z_0 >> h$	4.8	6.5–8.4

Here, L_1 is the length of the initial undisturbed section and λ is the wavelength determined as the distance between two neighboring apices.

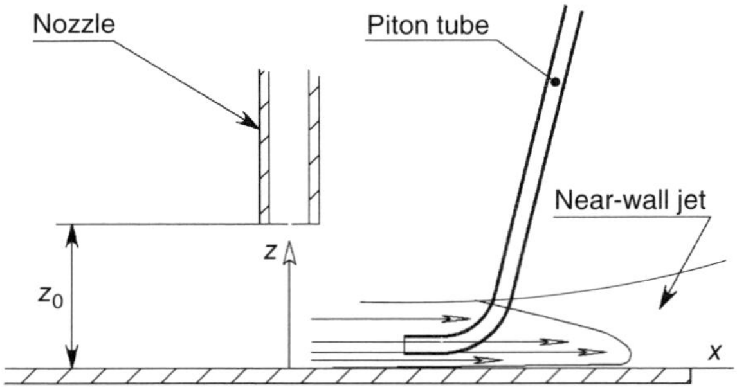

10.14 Schematic of experimental investigation of the near-wall jet.

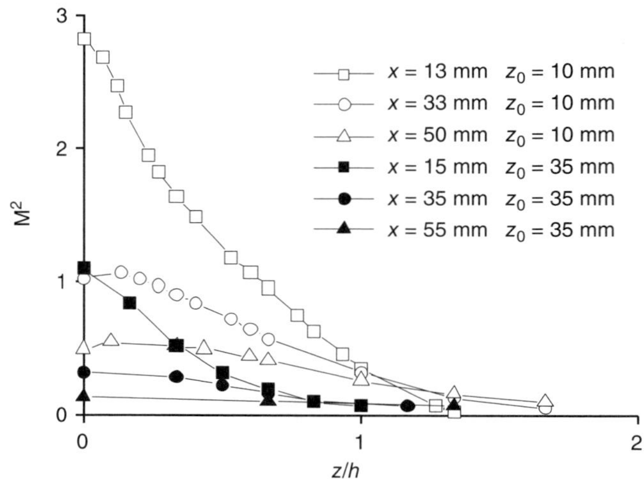

10.15 Mach number profiles in the near-wall jet far from the critical point for a jet impacting on a normally located substrate. $h = 3\,\text{mm}$, $H/h = 3.3$ and $M^* = 2.5$.

static pressure was close to the ambient pressure, the value of M^2 was determined by the Rayleigh formula

$$\frac{p_0'}{p_a} = \left(\frac{\gamma+1}{2}\right)^{\frac{\gamma+1}{\gamma-1}} M^2 \left(\frac{M^2}{\gamma M^2 - \frac{\gamma-1}{2}}\right)^{\frac{1}{\gamma-1}}$$

The value of M^2 reconstructed from the measured pressures is plotted in Fig. 10.15. Rather large values of the x coordinate were chosen to avoid the region of the accelerating gas flow and to reach the region of prevailing atmospheric pressure, which simplifies determination of M^2.

An analysis of Fig. 10.15 shows that the near-wall boundary layer on the substrate surface does not have enough time to develop (at least, it is thinner than the Pitot tube), and the boundary layer on the external side of the jet occupies the main part. To verify self-similarity, we plotted the normalized M^2 profile in Fig. 10.16. There is some difference between these profiles, but they are rather adequately approximated by the function

$$\frac{M^2}{M_m^2} = \exp\left[-\left(\sqrt{\ln 2}\,\frac{z}{z_{0.5}}\right)^2\right] \qquad [10.14]$$

To find the Mach number in the compressed layer (near the critical point), we used the values of pressure measured on the substrate surface p_s under the assumption of a constant total pressure, which allows us to use the isentropic formula

$$\frac{p_0}{p_s} = \left(1 + \frac{\gamma - 1}{2} M^2\right)^{\frac{\gamma}{\gamma-1}}$$

After these procedures, we obtained the distributions of M_m^2 along the x axis of the substrate surface, corresponding to the maximum velocity in the near-wall jet (Fig. 10.17). The gas is accelerated to supersonic velocities up to the distance $x_{ac} = (2 - 3)h$ and then decelerated.

Thus, it is shown that the pressure distribution over the substrate surface along the smaller size of the nozzle is self-similar in the case of the classical

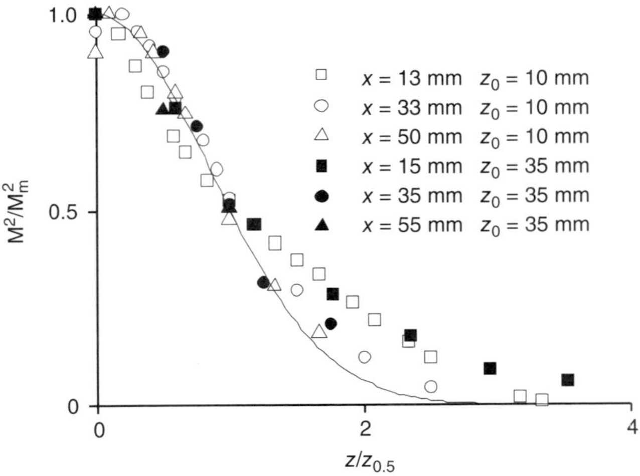

10.16 Normalized Mach number profiles in the near-wall jet far from the critical point for a jet impacting on a normally located substrate. $h = 3$ mm, $H/h = 3.3$ and $M^* = 2.5$.

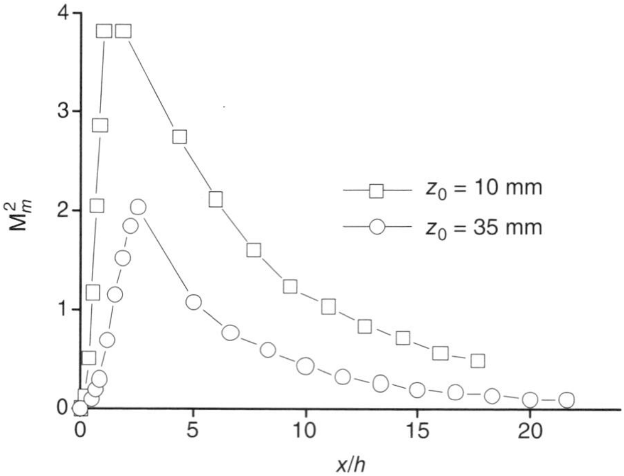

10.17 Distributions of $M_m^2(x/h)$ along the surface of a normally located substrate. $h = 3\,mm$, $H/h = 3.3$ and $M^* = 2.5$.

flow regime (i.e. in the absence of oscillations and circulation zones) and independent of the impact angle for φ_{im} = 50–90°. The critical parameters of the gas accelerated along the surface are reached in the vicinity of the boundary of the incident jet. The velocity gradient at the stagnation point can be determined by the formula $\beta = 2a_{cr}/h$. If the distance between the nozzle exit and the substrate is small ($z_0/h \leq 5$), the gas parameters can be assumed to be constant and equal to the parameters at the nozzle exit. The study of the near-wall jet showed that the velocity and Mach number profiles are self-similar, and the thickness of the near-wall boundary layer is negligibly small up to distances $x/h \approx 18$.

The results described above allow more detailed consideration of the processes of particle acceleration, including the specific features of particle motion within the jet, and the character of heat exchange between the jet and the substrate. Thus, these results ensure a better insight into the physical background of the cold spray method and more profound development of the spraying technology.

10.3.4 Thickness of the compressed layer

An important problem is to determine the compressed layer thickness as a function of the jet parameters and the distance z_0. We consider a supersonic jet impacting onto a normally located substrate (Fig. 10.18). Deceleration and deflection of the gas flow occur ahead of the substrate surface. The transition from the high-velocity supersonic flow to the low-

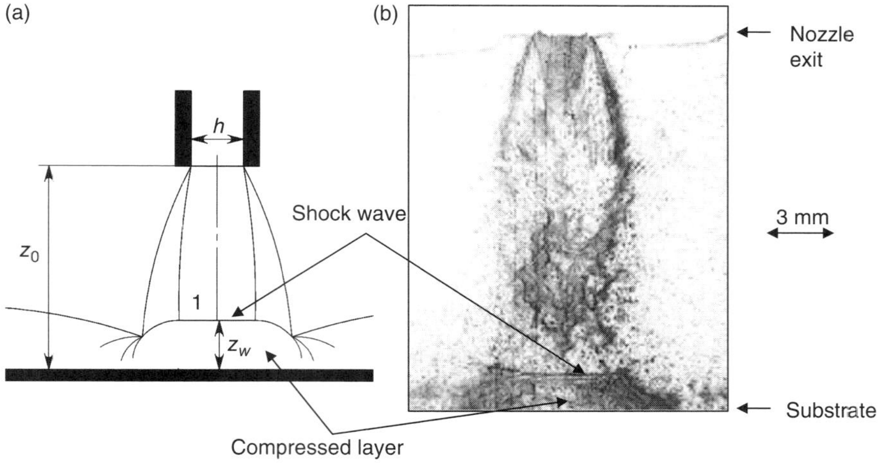

10.18 Schematic and instantaneous photograph ($\tau_{ex} \approx 30 \times 10^{-9}$ s, $n = 3$, $z_0 = 20$ mm) of the impact of a supersonic gas jet onto a flat infinite substrate. The detached shock wave is indicated by 1, *h* is the nozzle thickness and z_0 is the distance from the nozzle exit to the substrate.

velocity subsonic flow occurs on the shock wave located at a certain distance z_w from the substrate surface. A high-pressure, high-density gas layer is formed between the substrate surface and the shock wave. Obviously, fine particles of the deposited material passing through this layer are decelerated; the greater the compressed layer thickness, the greater the deceleration.

To determine the compressed layer thickness, we performed experiments with nozzles of different thicknesses and axial Mach numbers at the nozzle exit (Alkhimov *et al.* 2000a). We used an experimental setup including optical diagnostics for observation of the object. Photographs were taken, and the compressed layer thickness was estimated on the basis of the photographs obtained. These data are plotted in Fig. 10.19 and 10.20; in addition to the data obtained for an isobaric jet, Fig. 10.19 shows the points obtained for $n = 3$. The compressed layer thickness and the distance from the nozzle exit to the substrate are normalized to the nozzle width.

For an isobaric jet, the compressed layer thickness in the first approximation can be assumed to be constant and equal to one-half of the jet thickness. Figure 10.20 shows the data obtained for the non-isobaric flow mode ($n = 1.5$). Strong non-monotonicity within the first barrel can be noted. Such a behavior can be explained by interference of the wave structure of the jet with the bow shock wave arising on the body. This is particularly noticeable for large jet-pressure ratios. The compressed layer structure can be

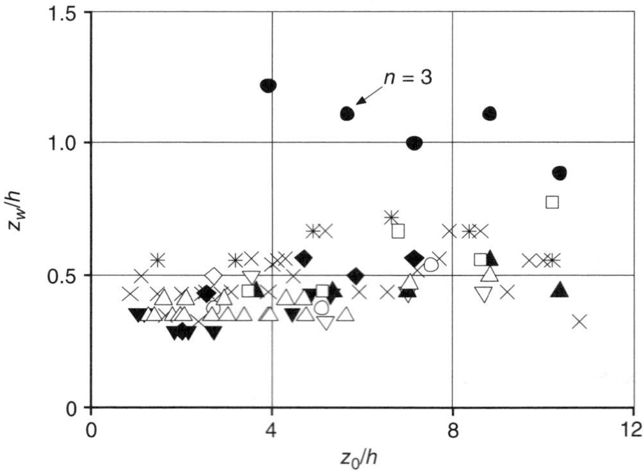

10.19 Compressed layer thickness versus the distance z_0. The parameters were varied in the following ranges: $h = 1\text{–}5$ mm, $H/h = 2.7\text{–}8$ and $M_{id} = 1.8\text{–}3.1$.

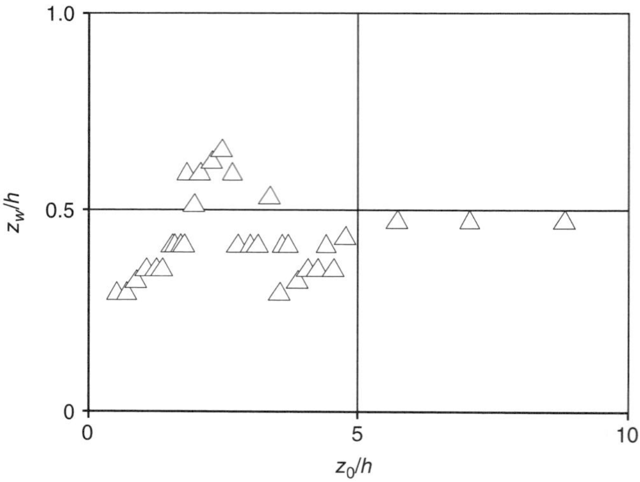

10.20 Compressed layer thickness versus the distance for a supersonic non-isobaric ($n = 1.5$) jet exhausting from the nozzle ($h = 5$, $H/h = 1.68$ and $M_{id} = 2.55$).

seen in the photographs in Fig. 10.21. We can give some simple considerations for estimating the compressed layer thickness. We write the equation of conservation of momentum of the jet with allowance for self-similarity of the dynamic pressure (or M^2) profiles

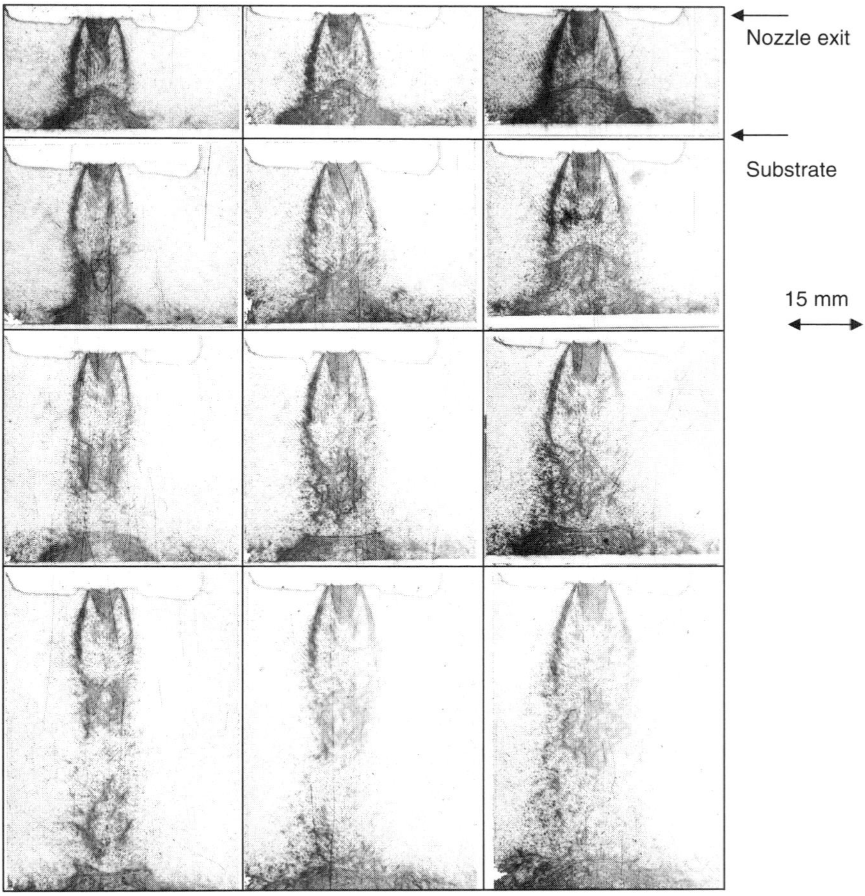

10.21 Compressed layer structure. The jet-pressure ratio is 2.5 (first column), 3 (second column) and 3.5 (third column). The distance is 10 mm (first row), 15 mm (second row), 20 mm (third row) and 30 mm (fourth row).

$$\frac{hH}{4}\rho_m v_m \int_0^1 \varphi(\xi)\mathrm{d}\xi \int_0^1 \varphi(\eta)\mathrm{d}\eta = z_w \frac{h+H}{2}\rho_k' u_m \int_0^1 \varphi(\zeta)\mathrm{d}\zeta \qquad [10.15]$$

where $\varphi(\xi) = \mathrm{M}^2/\mathrm{M}_m^2$, $\varphi(\eta) = \mathrm{M}^2/\mathrm{M}_m^2$ along the x and y axes, $\xi = x/h$, $\eta = y/h$, $\varphi(\zeta) = \mathrm{M}^2/\mathrm{M}_m^2$ along the z axis and $\zeta = z/z_w$. In the case of a uniform cross-sectional distribution of the gas parameters, we have

$$\int_0^1 \varphi(\xi)\mathrm{d}\xi = \int_0^1 \varphi(\eta)\mathrm{d}\eta = \int_0^1 \varphi(\bar{\zeta})\mathrm{d}\bar{\zeta} = C = 1$$

and the compressed layer thickness is

$$z_w = \frac{h}{2}\frac{1}{1+h/H}\left(\frac{\rho_m v_m}{\rho'_k u_m}\right) = \frac{1}{2}\frac{h}{1+h/H}\Psi \qquad [10.16]$$

$$\Psi = \frac{(1+a\mathrm{M}_m^2)^{0.5}(\gamma \mathrm{M}_m^2 - a)^{0.5/a}}{(c\mathrm{M}_m)^b}$$

$$a = \frac{\gamma - 1}{2}, \qquad b = \frac{\gamma + 1}{\gamma - 1}, \qquad c = \left(\frac{\gamma+1}{2}\right)^{0.5}$$

M_m is the Mach number at the axis of the incoming jet. The calculation of the function Ψ for Mach numbers of 1.8–3.1 yields 0.86–0.72, which can be approximated by a roughly constant value of ~0.8. Thus, we can assume that the ratio of the compressed layer thickness to the jet thickness is approximately 0.4 for the plane jet. For the rectangular jet, a correction in accordance with equation [10.16] should be used.

It is seen from equation [10.16] that the compressed layer thickness tends to be approximately one-half of the jet size for H much greater than h, i.e. for ideally plane nozzles and jets. After averaging all the data in Fig. 10.19 that refer to the isobaric air jet, we can estimate the compressed layer thickness as $\approx 0.45h$, independent of the distance from 0 to $10h$.

10.4 Heat transfer between a supersonic plane jet and a substrate under conditions of cold spray

Investigation of heat transfer between a two-phase jet and a substrate under conditions of cold spray (Alkhimov *et al.* 2000b; Kosarev *et al.* 2003) is important both for the theory and for applications. First, the processes of adhesive attachment of particles on the surface significantly depend on the surface temperature, as in the case of gas-thermal spraying (Shorshorov and Kharlamov 1978; Kudinov *et al.* 1990). In addition, it is important to properly control the temperature of the coated part in many technological processes to ensure the required mode of coating formation and the required state of the part surface.

Heat transfer can be divided into two components: heat transfer between the particles and the surface, and heat transfer between the carrier-phase gas jet and the surface. If the particle concentrations are low ($\varphi_p \leq 10^{-6}$), which is normally the case in cold spray, and the particle temperature is lower than the stagnation temperature of the jet, heat transfer between the particles and the surface is rather low, as compared with heat transfer between the surface and the gas flow. Therefore, in estimating the surface temperature, it is important to take into account heat transfer between the gas flow and substrate.

10.4.1 Method for measuring the heat-transfer coefficient

The heat-transfer coefficient was measured by a calorimetric probe, which was a copper washer with an imbedded thermocouple. The probe was flush-mounted into a plate made of a heat-insulating material (Fig. 10.22). A shield (steel plate) reflecting the gas flow from the nozzle was placed between the nozzle and the substrate. When the registering equipment was ready for operation, the shield was rapidly removed, and the jet was impinging on the substrate with the calorimetric probe. For correct operation of this system, the time needed for the washer temperature to reach a steady level t_0 should be much greater than the time needed for stabilization of the gas flow; the latter can be estimated with the help of the velocity of sound a and the characteristic scale of the problem (e.g. distance between the substrate and the nozzle exit z_0) as $\tau_r \approx z_0/a$. For air, under our test conditions, we have $\tau_r \approx 10^{-4}$ s. For chosen parameters of the calorimetric probe, we obtain $t_0 = 1$–3 s. Based on the resultant dependences of temperature on time, we determined the heat-transfer coefficient by solving the one-dimensional heat-conduction equation.

Let us consider heat transfer between the gas and an infinite plate of thickness δ_s. Figure 10.22 shows a schematic of the plane supersonic jet incident onto the substrate with indication of the coordinate axes and basic geometric parameters of the problem considered. We assume that the heat-transfer coefficient α and the stagnation temperature of the gas T_0 are constant. At the initial time $t = 0$, there begins heat transfer at the interface $z = 0$ between the plate with the initial temperature T_{s0} and the gas. The second side of the plate is assumed to be thermally insulated. We introduce the dimensionless quantities

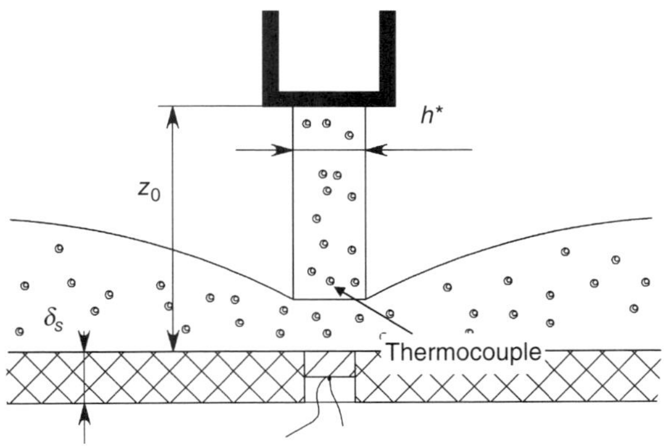

10.22 Schematic of heat-transfer coefficient measurement for a supersonic jet impacting on the substrate.

$$z = \xi\delta_S, \qquad t = \tau\frac{\delta_S^2}{\chi_S}, \qquad T_S = T_0(1-\theta)$$

to write the unsteady equation of heat conduction, initial and boundary conditions in the following form:

$$\begin{aligned}
&\frac{\partial\theta}{\partial\tau} = \frac{\partial^2\theta}{\partial\xi^2}\\
&\frac{\partial\theta}{\partial\xi} = \frac{\alpha\delta_S}{\lambda_S}\theta_w \qquad \text{at } \xi = 0\\
&\frac{\partial\theta}{\partial\xi} = 0 \qquad \text{at } \xi = -1\\
&\quad \theta = \theta_0 = 1 - \frac{T_{S0}}{T_0} \qquad \text{at } \tau = 0
\end{aligned} \qquad [10.17]$$

($\chi_s = \lambda_s/(c_s\rho_s)$ is the thermal diffusivity of the plate).

The solution is presented in the form of the series $\theta(\xi, \tau) = \sum_{i=1}^{\infty} f_i(\xi)\exp(-\varepsilon_i^2\tau)$, where $f_i(\xi)$ can be presented as the sum $A\cos(\varepsilon_i\xi) + B\sin(\varepsilon_i\xi)$. Using the boundary condition $f_i'(-1) = 0$, we obtain $B = A\text{tg}(\varepsilon_i)$. From the boundary condition:

$$f_i'(0) = -\frac{\alpha\delta_S}{\lambda_S}f_i(0)$$

we find the expression for determining ε_i:

$$\varepsilon_i\text{tg}(\varepsilon_i) = \frac{\alpha\delta_S}{\lambda_S} \qquad [10.18]$$

For

$$\text{Fo} = \frac{\chi_S\tau}{\delta_s^2} \geq 0.3$$

which corresponds to

$$\tau \geq \frac{0.3c_S\rho_S\delta_S^2}{\lambda_S} \approx 0.1\text{s}$$

we can cancel all terms of the series except for the first one, which yields a 1% error. In this case, it is convenient to present the temperature in a logarithmic form:

$$\ln\left(\frac{T_0 - T_S}{T_0}\right) = a + bt \qquad [10.19]$$

$$b = -\frac{\varepsilon_1^2\chi_S}{\delta_S^2} \qquad [10.20]$$

For experimental determination of α, we have to solve an inverse problem. Knowing the time evolution of temperature at a certain point of the plate, we can present it in the coordinates

$$\ln\left(\frac{T_0 - T_S}{T_0}\right), t$$

After that, we perform a root-mean-square approximation of all experimental points and construct a curve with coefficients a and b in accordance with equation [10.19]. The value of ε_1 is found from the known value of b, by inverting equation [10.20]:

$$\varepsilon_1 = \sqrt{(-b)\frac{\delta_S^2}{\chi_S}}$$

The heat-transfer coefficient is found from equation [10.18]:

$$\alpha = \frac{\lambda_S}{\delta_S}\varepsilon_1 \operatorname{tg}\varepsilon_1$$

As an example, Fig. 10.23 shows the results of experiments determining the coefficient of heat transfer between the jet and the substrate ($p_0 = 1.45$ MPa, $T_0^* = 330$ K, which is the stagnation temperature at the nozzle exit) at different distances x from the nozzle axis. Figure 10.23(a) shows the experimental points of the temperature dependence on time, which was registered by the thermocouple on the reverse side of the probe. In Fig. 10.23(b), the same data are plotted in the following coordinates:

$$\ln\frac{T_0(x) - T_S(x)}{T_0(x)}, t$$

together with the approximated curves passing through these experimental points. Heat-transfer coefficients corresponding to these cases and calculated on the basis of the curve slopes are also indicated.

A typical behavior of the normalized stagnation temperature $\overline{T}_0(x) = [T_0(x) - T_a]/(T_0^* - T_a)$ in the near-wall jet is shown in Fig. 10.24. In a wide range of z_0, the data are approximated by the function

$$f(x/x_{0.5}^T) = [1 + 15(x/x_{0.5}^T)^2]^{-0.25} \qquad [10.21]$$

It should be noted that the same distribution would be observed in the case of the jet impacting onto a non-heat-insulated surface (i.e. metallic surface) because heat transfer between the air jet and the substrate surface is only a small portion of the total amount of heat transferred by the jet (typical values of the Stanton number are St ~ 0.01). Thus, the decrease in stagnation temperature along the surface is mainly determined by the inflow of ambient air to the near-wall jet.

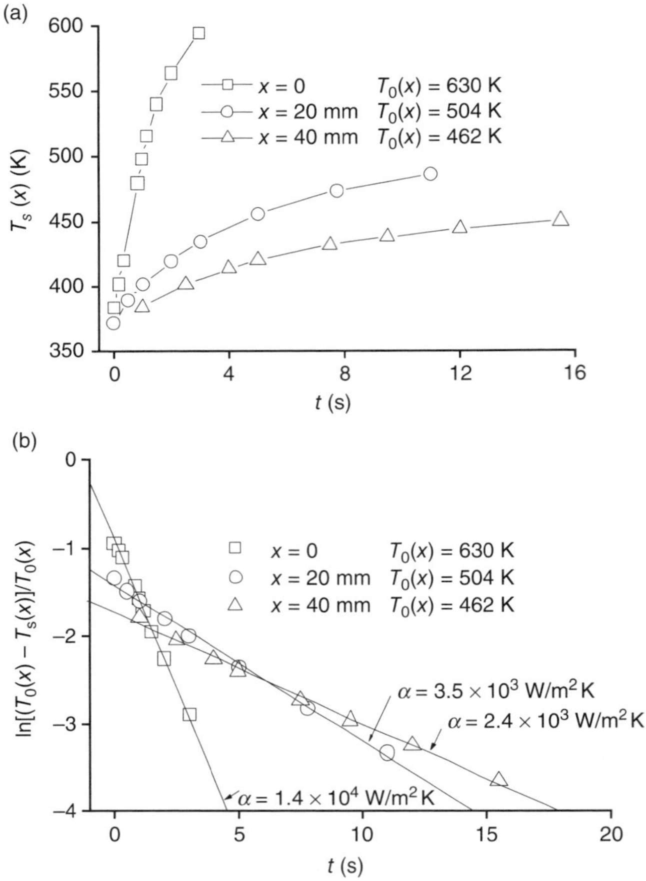

10.23 Time evolution of temperature for an isobaric air jet. p_0 = 1.45 MPa, T_0^* = 330 K, h^* = 3 mm (jet thickness at the nozzle exit) and z_0 = 15 mm.

10.4.2 Data on heat-transfer coefficient

Using the above-described procedure, we measured the heat-transfer coefficients for the jet impacting on the substrate. Figure 10.25 shows the heat-transfer coefficient $\alpha(0)$ at the stagnation point of the flow versus the distance from the nozzle exit to the substrate. The heat-transfer coefficient is almost independent (in the range considered) of the stagnation temperature of the jet at the nozzle exit; it reaches the maximum value at a certain distance z_0/h = 5–7 and then decreases.

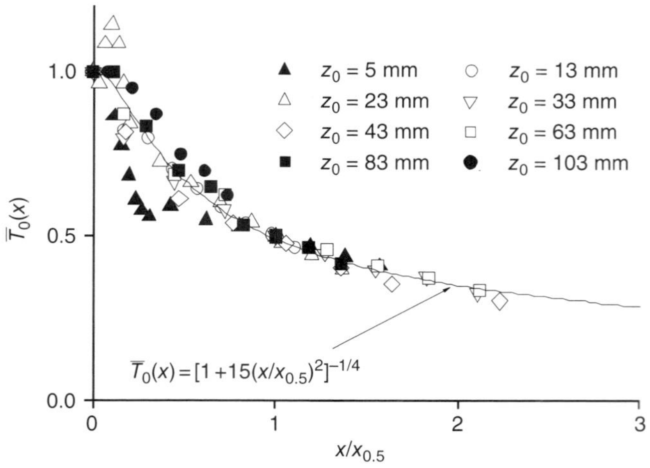

10.24 Stagnation temperature in a near-wall air jet impacting on a normally located substrate. $T_0^* = 550\,K$, $T_a = 300\,K$ (ambient air temperature).

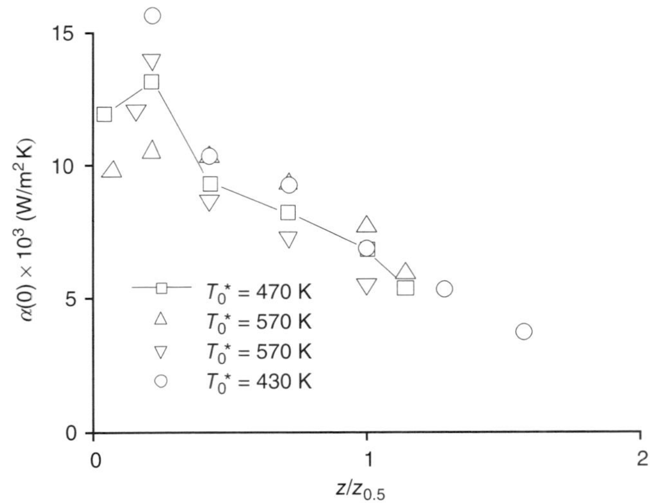

10.25 Heat-transfer coefficient versus distance for the normal impact of an isobaric air jet. $z_{0.5} = 70\,mm$, $\alpha_0 = 14 \times 10^3\,W/m^2 K$.

The experimentally measured Nusselt number Nu (represented in the graph (Fig. 10.26) as Δ) versus the Reynolds number Re_x, based on the distance from the stagnation point along the x coordinate, is shown in Fig. 10.26(a). The dependence of the Nusselt number proportional to the heat-transfer coefficient is approximated by the formula

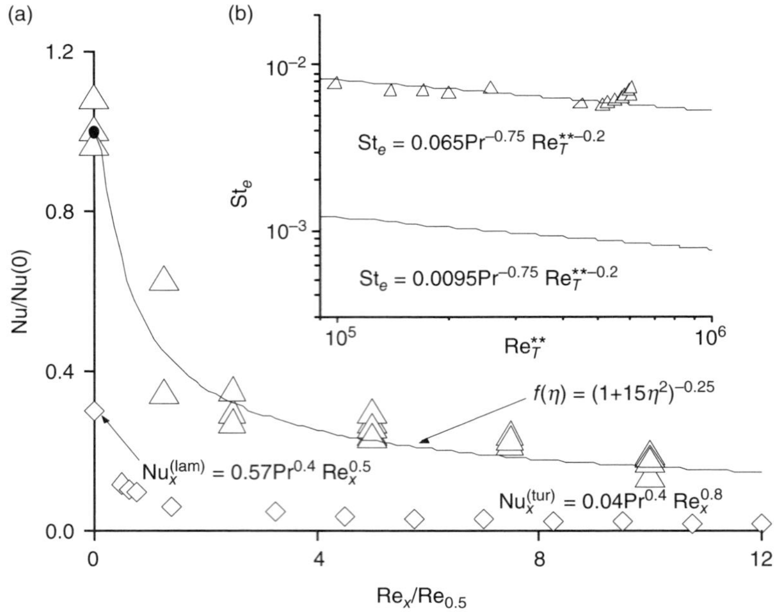

10.26 Nusselt number (a) and Stanton number (b) versus Reynolds number. $z_0/h^* = 5$, $Re_{0.5} = 6.8 \times 10^4$, $Nu(0) = \alpha(0)h^*/\lambda_0 = 980$, $\lambda_0 \approx 0.04\,W/m\,K$ and $\eta = Re_x/Re_{0.5}$. Superscripts (lam) and (tur) indicate laminar and turbulent, respectively.

$$\frac{Nu(x)}{Nu(0)} = \left[1 + 15\left(\frac{Re_x}{Re_{0.5}}\right)^2\right]^{-0.25} \qquad [10.22]$$

which covers the entire range of x, including the decay far from the impact point in the subsonic part of the near-wall jet, where $Nu \sim 1/x^{0.5}$ according to Yudaev *et al.* (1977b). For comparison, we can calculate the heat-transfer coefficient as a function of Re_x, using the experimentally measured distributions of parameters at the outer edge of the near-wall jet $M_e^2(x)$, $u_e(x)$ (Alkhimov *et al.* 2000b), and $T_{0e}(x)$ by the formula from Belov *et al.* (1969) and Belov and Pamadi (1970)

$$\alpha = 0.04 Pr^{-0.6} Re_{xe}^{-0.2} \rho_e u_e c_p (T_w/T_r)^{-0.16} \qquad [10.23]$$

where ρ_e and u_e are, respectively, the density and velocity at the outer edge of the near-wall jet. Since $(T_w/T_r)^{-0.16}$ (T_w and T_r are, respectively, the temperature of the substrate surface and the recovery temperature in the near-wall jet) has only a minor effect (for our test conditions, the maximum correction is about 1.1), it was replaced by unity. Equation [10.23] was derived on the basis of the theory of the turbulent boundary layer; a formula of the same form was derived in Gubanova *et al.* (1971).

The heat-transfer coefficient in the vicinity of the stagnation point was calculated on the basis of the laminar boundary-layer theory for the jet impacting on the substrate with allowance for the velocity distribution in the vicinity of the stagnation point $u = \beta x$ (Volchkov and Semenov 1994):

$$\alpha = 0.57 / \mathrm{Pr}^{0.6} \sqrt{\rho_0 \beta \mu_0 c_p^2} \qquad [10.24]$$

Typical values for test conditions obtained by equation [10.24] are (3.8–4.2) $\times\ 10^3$ W/m^{2}K at the stagnation point for $\beta = 2a_{cr}/h = (2.4\text{–}3) \times 10^5\ \mathrm{s}^{-1}$.

The data calculated by equations [10.23] and [10.24] are plotted by diamonds in Fig. 10.26(a). The calculated values are substantially lower than the experimental results. This difference can be attributed to the influence of velocity fluctuations in the vicinity of the critical point and in the near-wall jet. For instance, in Belov (1983), the influence of fluctuations of velocity and other parameters is taken into account by the formula $\mathrm{Nu}_t(0) = \mathrm{Nu}(0)(1 + 0.75b^{0.54})$, $b = 0.18\varepsilon v^* \sqrt{\rho_0/\beta\mu_0}$, where Nu(0) is the Nusselt number calculated with neglected turbulent oscillations, $\varepsilon = \sqrt{v'^2}/v^*$ is the degree of turbulence, ρ_0 is the density at the stagnation point and μ_0 is the viscosity based on the stagnation temperature of the flow. For the results calculated by this formula to match the experimental data, we should use $\varepsilon = 0.25$. This value is within the range (0.04–0.5) of ε measured experimentally by independent methods (Kalghatgi and Hunt 1976).

Figure 10.26(b) shows the Stanton numbers $\mathrm{St}_e = \alpha/\rho_e u_e c_p$ versus the Reynolds number based on the energy thickness δ_T^{**} calculated by the energy equation for a turbulent boundary layer, based on the postulate that the heat-transfer law in a turbulent flow is conservative (Volchkov and Semenov 1994):

$$\mathrm{St}_e = A\,\mathrm{Pr}^{-0.75}\mathrm{Re}_T^{**-m}, \qquad A = 0.0095, \qquad m = 0.2 \qquad [10.25]$$

Then, we obtain the following expression for Re_T^{**}:

$$\mathrm{Re}_T^{**} = \frac{1}{\Delta T}\left[\frac{A(1+m)}{\mathrm{Pr}^{0.75}}\int \Delta T^{1+m}\frac{\rho_e u_e \mathrm{d}x}{\mu_e}\right]^{1/(1+m)}, \qquad \Delta T = T_0(x) - T_w$$

Using the experimentally measured distributions of the parameters M_e^2, u_e and T_{0e} on the upper edge of the near-wall boundary layer, we can integrate this expression and obtain the values of Re_T^{**}. As it follows from Fig. 10.26(b), the experimental data are again higher than the calculated values. This suggests that heat transfer in the examined flow cannot be calculated by equations [10.23] and [10.24]; apparently, the heat-transfer calculation model developed in Gubanova *et al.* (1971) and Volchkov and Semenov (1994) is more accurate and promising. Nevertheless, there are some difficulties in calculating and measuring velocity fluctuations, and this problem, in many cases, is much more complicated than the determination of the heat-transfer coefficient itself.

10.4.3 Temperature of the substrate surface

The experimental results obtained allowed us to determine the temperature conditions on the substrate surface and inside the substrate. The temperature distributions in the substrate of length $2L_s$ and thickness δ_s were calculated by simultaneously solving the steady heat-conduction equation

$$\frac{\partial^2 T(x, z)}{\partial x^2} + \frac{\partial^2 T(x, z)}{\partial z^2} = 0$$

($T(x, z)$ is the temperature in the substrate) and the law of conservation of heat in the steady case

$$\int_0^{L_S} \alpha(x)[T_0(x) - T_S(x, 0)]\mathrm{d}x = 0$$

using the experimental value of stagnation temperature and heat-transfer coefficient in the near-wall jet. The solution was sought in the form $T(x, z) = T_s(x) + a(x)z + b(x)z^2$, which is justified in the steady case with $\delta_s \ll L_s$ and low temperature gradients in the substrate. The boundary conditions on the surfaces $z = 0$ and $z = -\delta_s$ yield the relations

$$\lambda_S \left.\frac{\partial T(x, z)}{\partial z}\right|_{z=0} = \alpha(x)[T_0(x) - T_S(x)] \Rightarrow a(x) = \frac{\alpha(x)}{\lambda_S}[T_0(x) - T_S(x)]$$

$$\lambda_S \left.\frac{\partial T(x, z)}{\partial z}\right|_{z=-\delta_S} = 0 \Rightarrow b(x) = \frac{a(x)}{2\delta_S} = \frac{\alpha(x)[T_0(x) - T_S(x)]}{2\delta_S \lambda_S}$$

Finally, we have

$$T(x, z) = T_S(x) + \frac{\alpha(x)}{\lambda_S}[T_0(x) - T_S(x)]\left(z + \frac{z^2}{2\delta_S}\right) \qquad [10.26]$$

Substituting equation [10.26] into the heat-conduction equation and making the necessary transformations, we obtain the equation for the substrate-surface temperature:

$$\frac{\partial^2 T_S(x)}{\partial x^2} = \frac{\alpha(x)}{\delta_S}[T_0(x) - T_S(x)] \qquad [10.27]$$

Solving it together with the integral equation of conservation of heat

$$\int_0^{L_S} \alpha(x)[T_0(x) - T_S(x,0)]\mathrm{d}x = 0$$

we obtain the temperature distribution on the substrate surface for $0 \le x \le L_s$. After that, from the known $T_s(x)$, we calculate the coefficients $a(x)$ and $b(x)$ and find the temperature distribution in the substrate. The results calculated for the case $L_s = 100 \times 10^{-3}$ m, $T_0^* = 1200$ K and $\delta_s = 3$ mm are plotted in Fig. 10.27. Figure 10.27(a) shows the surface temperature for substrates made of different materials.

The calculations performed show that the noticeable decrease in surface temperature (for materials with $\lambda_s \ge 40$ W/m K), as compared with the stagnation temperature of the incoming jet, is caused by heat redistribution inside the substrate. As is seen from Fig. 10.27(b), heat penetrates into the substrate at the beginning ($0 \le x \le 4h$), whereas the process is inverted at large distances: heat leaves the substrate and enters the near-wall jet. Based on the calculation results, we plotted the surface temperature in the deposition spot ($x = 0$) versus the substrate length for different substrate thicknesses δ_s (Fig. 10.28).

As the substrate length increases to (15–20)h, the surface temperature at the substrate surface decreases noticeably; the greater the value of δ_s, the greater the decrease. A further increase in the substrate length has practically no effect on the surface temperature at the substrate center.

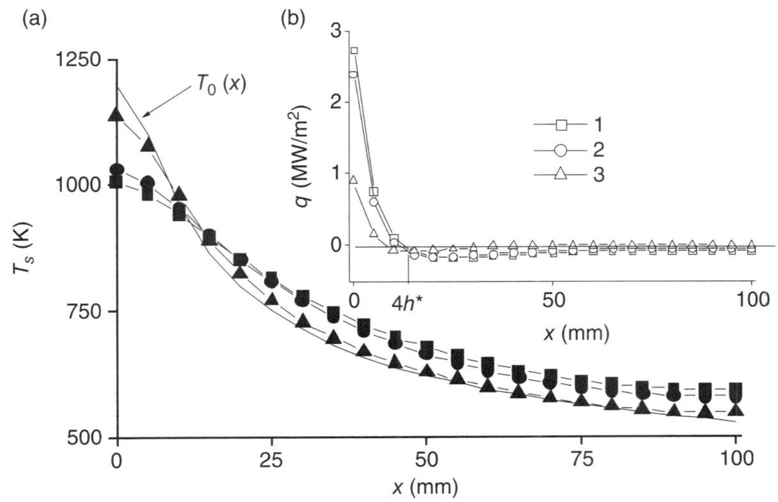

10.27 Distributions of the surface temperature (a) and heat flux (b) on different substrates. 1, Cu ($\lambda = 350$ W/m K); 2, Al ($\lambda = 250$ W/m K); 3, steel ($\lambda = 40$ W/m K).

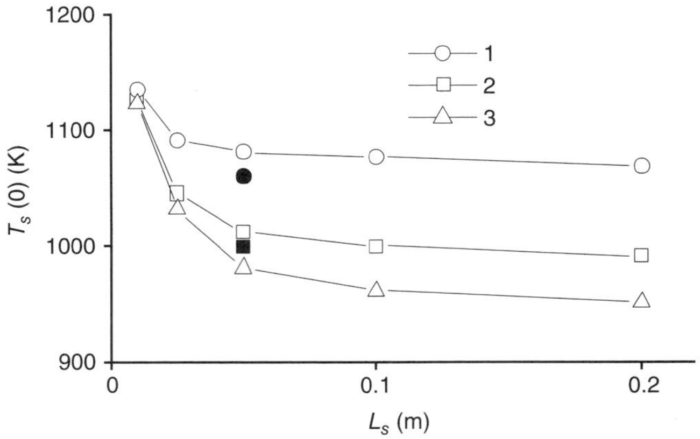

10.28 Surface temperature in the deposition spot ($x = 0$) on a copper substrate versus its size. 1, $\delta_s = 1\,\text{mm}$; 2, $\delta_s = 3\,\text{mm}$; 3, $\delta_s = 5\,\text{mm}$.

Experimental verification (symbols ● and ■) showed that the measured surface temperature near the critical point is in good agreement with the calculated values, which confirms the validity of the assumptions made and allows the use of this heat-transfer model for practical estimates.

Using equation [10.26], we can estimate ΔT_{max} (the maximum difference in surface temperatures for $z = 0$ and $z = -\delta_s$). We find

$$\Delta T_{max} = \frac{\alpha(0)\delta_S}{2\lambda_S}[T_0(0) - T_S(0)]$$

and $\Delta T_{max} \approx 20\,\text{K}$ for the calculation conditions in Fig. 10.27(a). Hence, the surface temperatures for $z = 0$ and $z = -\delta_s$ are almost identical.

Thus, the distributions of stagnation temperature and heat-transfer coefficient in the near-wall jet at different distances from the exit of a rectangular supersonic nozzle to the substrate were obtained experimentally. The experimental values of the heat-transfer coefficient are substantially higher than the calculated results, and this difference can be explained by velocity fluctuations in the vicinity of the critical point and in the near-wall jet. Using experimental data on stagnation temperature and heat-transfer coefficient, we calculated the substrate temperature in the steady case and showed that the surface temperature in the deposition spot is noticeably lower than the stagnation temperature. This is due to heat redistribution inside the substrate for heat-conducting materials ($\lambda_s \geq 40\,\text{W/m K}$).

10.5 List of symbols

a	Sound velocity
a_{cr}	Critical sound velocity
c_s	Substrate heat capacity
c_p	Gas heat capacity at constant pressure
Fo	Fourier number
H	Maximum exit dimension of nozzle
h	Minimum exit dimension of nozzle
L_1	Length of initial undisturbed part of jet
L_s	Length of the substrate
l_s	Length of supersonic part of jet
M	Mach number
M_{ac}	Mach number at the end of acceleration region
M_{id}	Mach number for ideal gas
M_m	Mach number at jet axis
M_m^*	Mach number at jet axis at nozzle exit
M^*	Mach number at the nozzle exit
$\mathrm{Nu} = \alpha h/\lambda_0$	Nusselt number
n	Non-isobarity factor
Pr	Prandtl number
p	Gas pressure
p_0	Stagnation pressure
p_0'	Stagnation pressure behind shock wave
p_a	Ambient pressure
p_c	Static pressure
p_s	Pressure at substrate surface
p_{sm}	Pressure at substrate surface at $x = 0$
$\mathrm{Re}_{0.5}$	Reynolds number based on the $x_{0.5}$ distance
Re_x	Reynolds number based on the distance from the stagnation point along the x coordinate
Re_{xe}	Reynolds number based on x coordinate and outer edge parameters of near-wall jet
Re_T^{**}	Reynolds number based on the energy thickness δ_T^{**}
$\mathrm{St}_e = \alpha/\rho_e u_e c_p$	Stanton number
T	Temperature in the substrate
T_0	Stagnation temperature
T_{0e}	Stagnation gas temperature in near-wall jet
T_{0m}	Stagnation temperature at jet axis
T_0^*	Stagnation temperature at jet axis at nozzle exit
T_a	Ambient temperature
T_m	Gas temperature at jet axis
T_{s0}	Initial temperature of plate

T_r	Temperature of restitution
T_s	Temperature of plate surface
T_w	Temperature of substrate surface
$\overline{T_0}(x) = [T_0(x) - T_a]/(T_0^* - T_a)$	Normalized stagnation temperature
t	Time
t_0	Time of probe temperature growth to steady state
u	Gas velocity along x axis
u_{ac}	Velocity at exit of acceleration part
u_e	Velocity of wall jet at external boundary
v	Gas velocity along z axis
v_{av}	Gas velocities averaged over the cross-section of the nozzle
v_m	Gas velocity on the axis
v^*	Gas velocity at the nozzle exit
v_m^*	Gas velocity at the nozzle exit on the axis
x	Coordination perpendicular to nozzle walls
$x_{0.5}$	Half-thickness of pressure profile ($p_s(x_{0.5}) - p_a = 0.5(p_{sm} - p_a)$)
$x_{0.5}^T$	Coordinate along near-wall jet where normalized stagnation temperature $\overline{T}_0$ is equal to 0.5.
x_{ac}	Length of acceleration part
x_{cr}	Coordinate of the critical transition
z	Longitudinal coordinate in jet
z_0	Distance from nozzle exit up to substrate surface
z_w	Distance from shock wave to substrate surface
$z_{0.5}^M$	Coordinate where $M_m^2(z_{0.5}^M) = 0.5M^{*2}$
$z_{0.5}^T$	Coordinate where $T_m(z_{0.5}^T) = 0.5T^*$
α	Heat exchange coefficient
$\alpha(0)$	Heat exchange at stagnation point
$\beta = 2a_{cr}/h$	Velocity gradient at stagnation point
γ	Ratio of specific heats
ΔT	$\Delta T_0 = T_0 - T_w$
ΔT_{max}	Maximal difference of surface at $z = 0$ and $z = -\delta_s$
ΔT_0	$\Delta T_0 = T_0 - T_a$
ΔT_{0m}	$\Delta T_{0m} = T_{0m} - T_a$
ΔT_0^*	$\Delta T_0^* = T_0^* - T_a$
δ_M	Jet thickness along minor size (in direction of x coordinate) defined as distance from jet axis up to point where $M^2(\delta_M) = 0.5\,M_m^2$
δ_s	Thickness of plate
δ_T	Thickness of temperature profile

δ_T^{**}	Energy thickness calculated by the energy equation for a turbulent boundary layer
δ_v	Thickness of velocity profile
δ_y	Jet thickness along major size
$\varepsilon = \sqrt{v'^2}/v_l$	Turbulence factor
μ	Gas viscosity
μ_0	Viscosity at stagnation temperature
λ_s	Heat conductivity of substrate
λ_0	Heat conductivity of gas
$v = \mu/\rho$	Dynamical viscosity
ρ_0	Stagnation gas density
ρ	Gas density
ρ_e	Density of wall jet at external boundary
ρ_k'	Gas density behind bow shock at critical point (where critical gas velocity is achieved)
ρ_m	Gas density at jet axis
ρ_s	Substrate density
τ	Dimensionless time
τ_{ex}	Exposure time
$\tau_r \approx z_0/a$	Time of relaxation of gas flow
φ_{im}	Angle of impingement
φ_p	Particle concentration
χ_s	Temperature conductivity of plate

10.6 References

Abramovich G I, 1984. *Theory of Turbulent Jets* [in Russian], Moscow: Nauka.

Alkhimov A P, Klinkov S V, Kosarev V F and Papyrin A N, 1997. Gas-dynamic spraying. Study of a plane supersonic two-phase jet. *Journal of Applied Mechanical and Technical Physics*, **38**(2), 324–330.

Alkhimov A P, Klinkov S V and Kosarev V F, 2000a. *Teplofizika Aeromekhanika*, **7**(2), 232–255.

Alkhimov A P, Klinkov S V and Kosarev V F, 2000b. *Teplofizika Aeromekhanika*, **7**(3), 389–396.

Alkhimov A P, Klinkov S V and Kosarev V F, 2001. The features of cold spray nozzle design. *Journal of Thermal Spray Technology*, **10**(2), 375–381.

Avduevskii V S, Ashratov E A, Ivanov A V and Pirumov U G, 1985. *Supersonic Non-Isobaric Gas Jets* [in Russian], Moscow: Mashinostroenie.

Belov I A, Ginzburg I G, Zazimko V A and Terpigor'ev V S, 1969. Effect of turbulence of the jet on its heat exchange with the target. *Heat and Mass Transfer Conference* [in Russian], 1969, Minsk: ITMO, Vol. 2, pp. 167–183.

Belov I A and Pamadi B N, 1970. Jet impingement upon a flat plate. *IIT-AERO-TN*, Bombay Institute of Technology, No. 3.

Belov I A, 1983. *Interaction of Non-uniform Flows with Targets* [in Russian], Moscow: Mashinostroenie.

Dykhuizen R C and Smith M F, 1998. Gas dynamic principles of cold spray. *Journal of Thermal Spray Technology*, **7**(2), 205–212.

Ginevskii A S, 1969. *Theory of Turbulent Jets and Wakes. Integral Methods of Calculation* [in Russian], Moscow: Mashinostroenie.

Grujicic M, Zhao C L, Tong C, Derosset W S and Helfritch D, 2004. Analysis of the impact velocity of powder particles in the cold-gas dynamic-spray process. *Materials Science and Engineering Journal*, **A368**(1–2), 222–230.

Gubanova O I, Lunev V V and Plastinina L I, 1971. *Izvestiya Akademii Nauk SSSR, Mekhanika Zhidkosti Gaza*, **2**, 135–138.

Jodoin B, Raletz F and Vardelle M, 2006. Cold spray modeling and validation using an optical diagnostic method. *Surface and Coating Technology Journal*, **200**(14–15), 4424–4432.

Kalghatgi G T and Hunt B L, 1976. The occurrence of stagnation bubbles in supersonic jet impingement flows. *The Aeronautical Quarterly*, **27**, 169–185.

Kosarev V F, Klinkov S V, Alkhimov A P and Papyrin A N, 2003. On some aspects of gas dynamics of the cold spray process. *Journal of Thermal Spray Technology*, **12**(2), 265–281.

Kudinov V V, Pekshev P Yu, Belashchenko V E, Solonenko O P and Safiulin V A, 1990. *Plasma Application of Coatings* [In Russian], Moscow: Nauka.

Kwon E H, Han J W, Lee C H and Kim H J, 2005. Computer simulation of injected particle behavior during cold spray process. In *International Thermal Spray Conference*, 2–4 May 2005, Basel, Swizerland (Ed. E. Lugscheider), Ohio: ASM International, pp. 1345–1348.

Papyrin A N, Alkhimov A P, Kosarev V F and Klinkov S V, 2001. Experimental study of interaction of supersonic two-phase jet with a substrate under cold spray. In *International Thermal Spray Conference*, 18–30 May 2001, Singapore (Ed. C. Berndt), Ohio: ASM International, pp. 423–431.

Pattison J, Morgan P, Selotto S, Khan A and O'neill B, 2005. Cold spray nozzle design and performance evoluation using particle image velocimetry. In *International Thermal Spray Conference*, 2–4 May 2005, Basel, Swizerland (Ed. E. Lugscheider), Ohio: ASM International.

Sakaki K, Huruhashi N, Tamaki K and Shimizu Y, 2002. Effect of nozzle geometry on cold spray process. In *Tagungsband Conference Proceedings*, 4–6 March 2002, Dusseldorf (Ed. E. Lugscheider), Germany: Deutscher Verband Für Schweißen, pp. 385–389.

Shorshorov M KH and Kharlamov Yu A, 1978. *Physical and Chemical Fundamentals of Detonation Gas Deposition of Coatings* [In Russian], Moscow: Nauka.

Shukla V and Elliott G, 2000. The fluid dynamics of cold gas dynamic spray. *ASM International Materials Solutions Conference*, 8–12 October 2000, St. Louis, Missouri, USA, Ohio: ASM International.

Stoltenhoff T, Voyer J and Kreye H, 2002. Cold Spraying: state of the art and applicability. In *Tagungsband Conference Proceedings*, 4–6 March 2002, Dusseldorf (Ed. E. Lugscheider), Germany: Deutscher Verband Für Schweißen, pp. 366–374.

Volchkov E P and Semenov S V, 1994. *Fundamentals of the Boundary-Layer Theory* [in Russian], Novosibirsk: ITP SB RAS.

Vulis L A and Kashkarov V P, 1965. *Theory of Viscous Fluid Jets* [in Russian], Moscow: Nauka.

Yudaev B N, Mikhailov M S and Savin V K, 1977a. *Heat Transfer during Interaction of Jets with Targets* [in Russian], Moscow: Mashinostroenie.

Yudaev B N, Mikhailov M S and Savin V K, 1977b. *Heat Transfer During Interaction of Jets with Targets* [in Russian], Moscow: Mashinostroenie.

11
Portable, low pressure cold spray systems for industrial applications

R. MAEV, V. LESHCHYNSKY and E. STRUMBAN,
University of Windsor, Canada

11.1 State-of-the-art cold spray systems

Widespread adoption of the cold spray (CS) technology implies, as the next logical step, its application in the industrial environment. Therefore, the development of a portable low pressure CS system is of fundamental importance. Of special practical interest is establishing the relationship between the operating parameters of a portable CS machine and the resultant microstructure of the formed spray coatings. To obtain high quality coatings, the configuration of a portable CS system and the spray process parameters have to be selected thoroughly. Due to the number of CS process parameters, it is inevitable that a certain degree of trial and error is required to optimize the spray process for each specific coating–substrate combination. Presented below are some specific characteristics related to a portable, low pressure system. These characteristics are subsequently compared with those of established high pressure CS systems.

The stage related to the particle acceleration in low pressure CS process is considered to consist of three subsequent processes:

- gas and powder mixing;
- acceleration of the particles within the divergent section of the spray nozzle;
- particle movement within the free jet area (before and after contact with the substrate).

Each of these processes has its own characteristics and may be described by distinct parameters. Both the gas and particle velocities change during each phase of acceleration, as shown in Fig. 11.1. One can observe a substantial difference between the velocities of the gas and particle – this is due to the significant disparity in the drag forces that each encounters. This variation is taken into account when the parameters of the powder-laden jet are calculated.

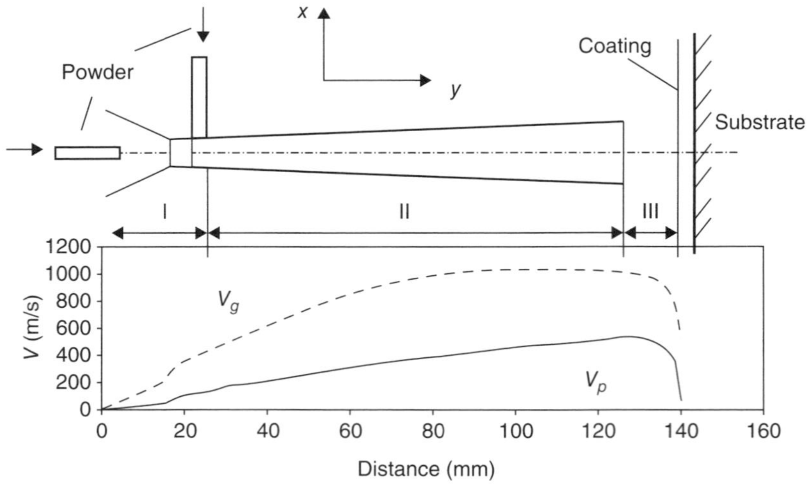

11.1 Schematic of a cold spray system with axial and radial injection of the powder.

There are two fundamental aspects that serve to distinguish the low pressure spray process from the established high pressure, or stationary, CS process:

1 The utilization of low pressure gas (0.5–1 MPa instead of 2.5–3 MPa).
2 Radial injection of powder instead of axial injection (in most cases).

Through close examination of these differences it can be noted that the particle content in the gas–particle jet is considerably higher in low pressure systems due to a decrease in the gas mass flow rate. In general, the dependency of particle volume concentration on the powder mass flow rate (as shown in Fig. 11.2) reveals that the particle concentration is proportional to both the air flow rate and the powder mass flow rate. In a low pressure spray process both of these variables can be compared with those of the traditional axial injection high pressure CS. In Fig. 11.3, the powder concentration and deposition efficiency of low pressure CS and of different variations of spray systems developed by Alkhimov *et al.* (1994) and Van Steenkiste *et al.* (2000) are presented. The decrease in the gas flow rate due to the diminishing gas pressure can be observed to cause a sharp increase in the particle concentration of the gas–particle jet. It is known (Alkhimov *et al.* 1994) that the solid phase content in the gas–particle jet in the traditional CS is in the range of 10^{-5}–10^{-6} (by volume), while the content for a low pressure CS system is in the range of 10^{-4}–10^{-5}. For this reason the characteristics of gas–powder flow vary considerably. Hence, the particle impact parameters depend strongly on the type of CS system that is utilized.

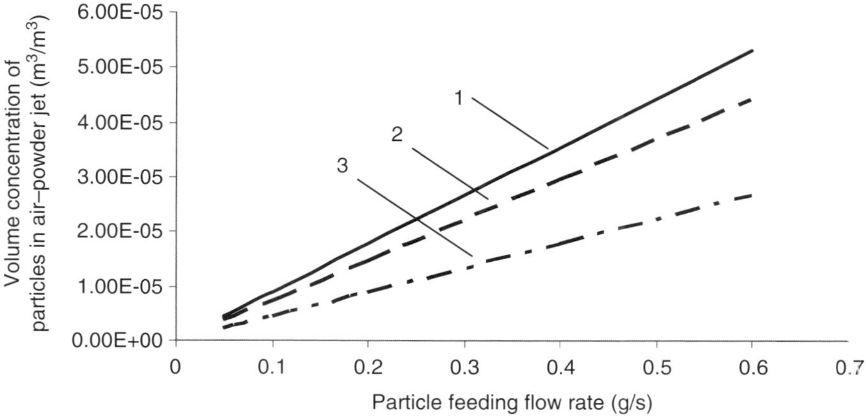

11.2 The particle volume concentration in the powder-laden jet for low pressure gas dynamic spray. 1, air flow rate 0.25 m^3/min; 2, air flow rate 0.3 m^3/min; 3, air flow rate 0.5 m^3/min.

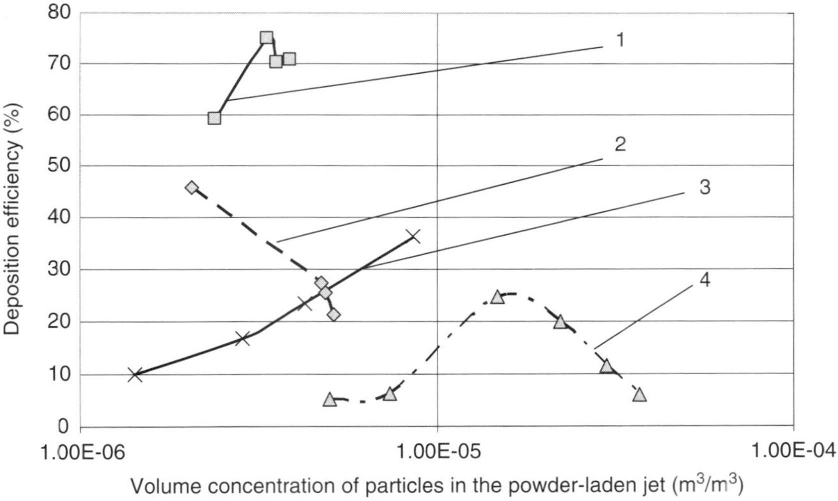

11.3 Deposition efficiency of high pressure gas dynamic spray (1, 2, 3) and low pressure cold spray (4) processes. 1, copper, particle size 63–106 μm; 2, copper, particle size <45 μm (Van Steenkiste *et al.* 2000); 3, aluminum, particle size 45 μm (Alkhimov *et al.* 1994); 4, aluminum, air temperature 5500 °C, air pressure 7 bar, particle size 45 μm (Kashirin *et al.* 2002).

As illustrated by Papyrin (2001), Van Steenkiste (2003), and Kay and Karthikeyan (2003), there are various features that characterize typical CS systems. The main trait of the CS system presented by Alkhimov *et al.* (1994) is that the gas and particles form a supersonic jet – with particle velocities ranging from 300 to 1200 m/s – whose temperature remains insufficient to bring about thermal softening. Although the particles are introduced into the gas stream at subsonic velocities, the gas – particle jet is brought to a supersonic speed via the de Laval-type nozzle.

The initial design of the CS system put forth by Alkhimov *et al.* (1994) was rather complex. However, Kay and Karthikeyan (2003, 2004) made significant improvements to this model by redesigning the nozzle. This design change allowed the heater mechanism to be removed from the main body of the gun. In this way, the spray gun applications proved to be considerably more flexible, permitting its use within deposition geometries that would otherwise have been physically precluded in the previous version of the CS system (Alkhimov *et al.* 1994). As a consequence, a bulkier and heavier electrical heater is required in order to heat the large volume of processing gas. Furthermore, moving the gun and heater assembly requires a heavy-duty mechanical manipulator and, as a result, restricts the movement of the spray beam. Thus, the flexibility of the entire spray operation is highly restricted in this arrangement. Also, as a heater may be required to heat the gas to as high as 700 °C, it may operate at a high pressure and temperature. During the spraying of some materials such as aluminum, the powder particles may get deposited inside the nozzle on the walls, blocking the gas flow path. When this happens, the gas flow may be greatly reduced or even stopped, causing an abnormal increase in the temperature and pressure of both the heating element and the gun. Such sudden increase in temperature and pressure can damage the gun and the heater, and also affect the safety of the operator.

The high pressure CS system (so-called 'kinetic spray' system) developed by Van Steenkiste (2003) is based on the Alkhimov *et al.* (1994) model and is able to produce spray coatings using particle sizes of greater than 50 μm and up to about 106 μm. A modified version of this device is capable of producing coatings using particles of larger size (up to 120 μm) and is designed to heat a high pressure air flow up to about 650 °C. After introduction of the particles, the heated air–particle jet is directed through a de Laval-type nozzle to achieve particle exit velocities between 300 m/s and 1000 m/s. The primary drawback of this system is that the spray material is injected into the heated gas stream prior to its passage through the de Laval nozzle. There is an inherent tendency for clogging, resulting in backpressure and spray gun malfunction. The second difficulty in CS devices can be attributed to the low durability of the convergent throat portions of nozzle. Because the heated main gas stream is under high pressure, injection of the powder itself requires a high pressure powder delivery system. Unfortunately, these are quite expensive and are not conducive to the design of portable cold spraying devices. In the Van Steenkiste (2004) design, in an effort to overcome this shortcoming, a specially developed radial method is used for powder injection. This system involves injecting the particles by means of a positive pressure that exceeds that of the main gas pressure. Radial injection takes place within the diverging region of the nozzle, just after the throat. However, the solution of Van Steenkiste (2004) does not eliminate the use of a high pressure powder delivery system. Furthermore, it introduces an additional difficulty – the necessity to permanently maintain a specific pressure difference between the positive injecting pressure and the main gas pressure. Moreover, the value of this difference is strongly dependent on the location of the injection point. Thus, in general, an additional adjustment of pressure is needed for different injection points along the nozzle, adding certain complications to the spray system.

The first portable, low pressure CS system was developed by Kashirin *et al.* (2002). The apparatus was called the Gas Dynamic Spray (GDS) system. A GDS device consists of a compressed air source that is connected by a gas passage to a heating unit, and in turn fed into a supersonic nozzle. The divergent section of the nozzle is then connected by a pipe to a powder feeder just before the throat. When in operation, the compressed air is delivered to the heating unit where it is heated to approximately 600 °C and subsequently accelerated by the supersonic portion of the nozzle. The powder material is then transferred via the feeder to the supersonic nozzle where it is mixes with the air flow and is accelerated towards the injection point on the nozzle. Within the nozzle the static pressure is maintained below atmosphere, ensuring that powder is effectively drawn in from the powder feeder. This pressure can be maintained only if the cross-sectional area of the supersonic nozzle in this portion exceeds that of the throat. The main difficulty associated with the system of Kashirin *et al.* (2002) is that the spatial uniformity of powder–air flow is quite low due to particle movement in the long powder passage. Both the particle-wall friction and the inter-particle friction result in particle accumulation in various places within the powder feeder conduit. The second difficulty is that precise control of the powder feeding rate is nearly impossible because particle movement within the powder feeder is strongly dependent on the length of the passage pipe, and the fluidity and specific weight of the powder or powder mixture. Additionally, the application of a heating unit, the outlet of which is connected to the convergent part of the nozzle, does not facilitate intensive gas heating in a small coil heating unit.

Therefore, in order to improve the mechanism of gas heating as well as the precision with which powder injection may be controlled, the low pressure portable GDS system needs to be modified. Only after the above-mentioned shortcomings are addressed, can a really practical portable GDS system become a reality.

11.2 State-of-the-art powder feeding systems

One of the primary goals of low pressure portable CS systems is the development of a powder feeding and metering device for use in delivering powder mixtures to the divergent part of the spray nozzle. A variety of vibratory powder feed systems have been known for many years. Such feeding systems, dating back to the 1950s and 1960s, have included both rotationally vibrated bowls and linearly vibrated channels or troughs. In the field of CS, a distinct need exists for similar portable feed systems.

Not only must this system be portable, it must be reasonably accommodating in that it must facilitate the use of numerous metal and ceramic powders in the spray process. Naturally, each of the powder mixtures has its own particle size distribution, specific gravity, coefficient of friction, or other properties that affect its flow characteristics. Therefore, the feed system must ensure a consistent particle flow of powders having significantly different powder flow rates.

Typical designs of powder feed systems may be divided into two groups: (a) fluidized bed hoppers and (b) dispensing hoppers. A powder feeding system comprising an enclosed hopper and a carrier gas conduit was suggested by Fabel (1976). A schematic of this device is shown in Fig. 11.4(a). Here the carrier gas supply is connected to a conduit that extends through the lower point of a hopper and continues to the point of powder-carrier gas utilization. In order to regulate the amount of fluidizing gas that is supplied to the hopper, the pressure at a point in the carrier gas line is monitored, as per the method suggested by Dalley (1970). As the pressure within the gas line is dependent on the mass flow rate of solids, the change in pressure may be used to regulate its flow.

Although the feeding system designed by Fabel (1976) has good repeatability, various problems have recently been suggested, particularly with respect to the control and uniformity of the powder feed rate. One such problem is pulsation; this shortcoming, apparently due to a pressure oscillation, results in uneven coating layers. A modified system by Rotolico *et al.* (1983) (Fig. 11.4(b)) has eliminated this difficulty inserting an additional fluidizing powder near the conveying gas line.

The use of the fluidized bed hoppers in the high pressure CS systems has been proven effective in academic laboratories and in the industrial environment. The obtained results have been summarized in a review by Karthikeyan (2005). However, despite this degree of success, dispensing hoppers with powder feeding rate controls have not yet been applied on a large scale.

The powder feeder designs of Schinella (1971) and Tapphorn and Gabel (2004) use a vibrating structure to move powder from the receiving surface to the discharge channel (Fig. 11.5). Schinella (1971) describes the use of a

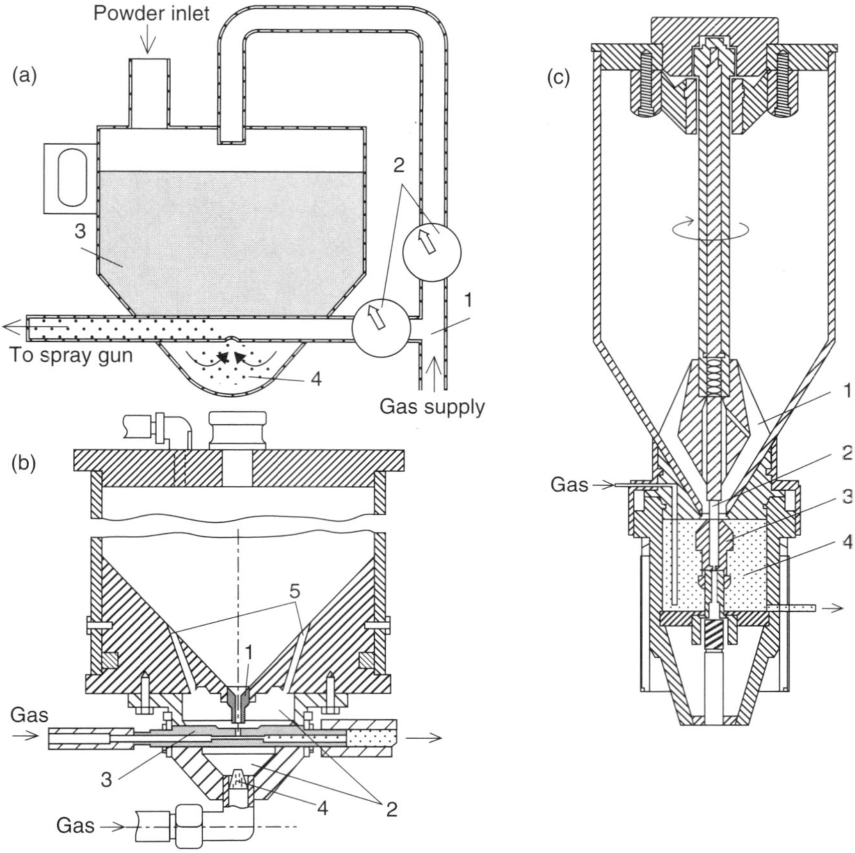

11.4 Fluidizing powder feeding systems design. (a) US Patent 3,976,332 (Fabel 1976): 1, carrier gas supply system; 2, pressure sensors and valves; 3, hopper; 4, fluidizing chamber. (b) US Patent 4,381,898 (Rotolico *et al.* 1983): 1, powder discharge fitting; 2, fluidizing chamber; 3, powder distribution system; 4, gas porous insert; 5, fluidizing orifices. (c) US Patent 3,501,097 (Dalley 1970): 1, powder feed passages; 2, tie rod; 3, powder chamber valve; 4, fluidizing chamber.

hopper with an outlet channel and hemispherical cup for metering powder and powder flow control. The primary limitation of this powder feeder is that the consistency of powder metering is affected by the agglomeration of powder in the hemispherical cup. In some cases it may even become plugged.

An improvement in the design of the powder metering device has been made by Tapphorn and Gabel (2004). In this system, sieve plates are mounted within the hopper for precise metering of the powder as it moves

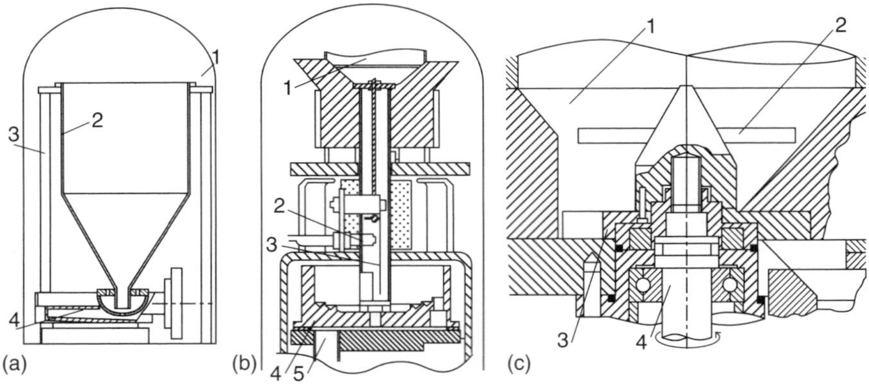

11.5 Powder metering devices. (a) US Patent 3,618,828 (Schinella 1971): 1, pressurized chamber; 2, feed hopper; 3, hopper frame; 4, vibratory powder feeder-metering device with outwardly spiral path. (b) US Patent 6,715,640 (Tapphorn and Gabel 2004): 1, powder hopper; 2, powder feeding rate sensor; 3, flexible metal vane; 4, vibrating spiral bowl; 5, powder discharge outlet. (c) US Patent 4,808,042 (Muehlberger and de la Vega 1989): 1, powder hopper; 2, powder mixer; 3, powder feeder-metering disc; 4, powder feeder shaft.

into the vibrating bowl. The powder is metered through the sieve plates by a hopper vibrator that is controlled by a level sensor mounted in the vibrating bowl. Unfortunately, the limitation of this design lies in its repeatability and accuracy. In fact, powder feeding rate measurement by a level sensor is strongly dependent on the agglomeration and fluidity of both powder and, in particular, powder mixtures. Moreover, plugging of funnels with flexible metal vanes occurs; this results in failure to control the amount of powder that flows from the funnel tube to receiving surface, and from the receiving surface to the discharge channel.

11.3 Modification of the low pressure portable gas dynamic spray system

A portable gas dynamic CS gun developed by the authors (Fig. 11.6) eliminates many of the inherent limitations described above by minimizing the scatter of operating parameters and, thus, improving the spray gun efficiency. One advantage of this new system is that the powder flow rate is continuously measured. In this way, the powder flow rate and/or the flow rate of the pressurized gas can be adjusted in real time, leading to more controlled and efficient deposition by the spray gun. In this design a gas passageway extends through the spray gun. A gas supply port supplies pres-

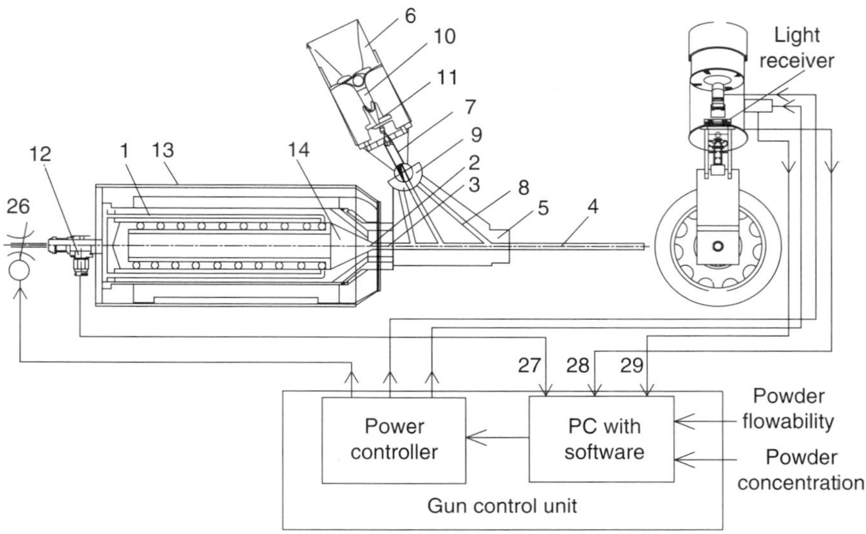

11.6 Low pressure cold spray gun with programmed controller.

surized air (or an alternative gas) to the passageway inlet. A nozzle in the passageway forms the pressurized air into a supersonic jet stream. A powder feed passage leads to the passageway and supplies powder at a controlled rate to the passageway, where it is entrained in the gas and exits the spray gun in the supersonic jet stream. The spray gun further includes a powder flow rate sensor which measures the powder flow rate of the powder. The powder flow rate sensor includes a light emitter transmitting light across a duct through which the powder travels. A light receiver mounted opposite the light emitter determines the flow rate of the powder based upon the amount of light received from the light emitter. The signal from the light receiver is processed by a specially developed controller, which adjusts the gas flow rate and/or the powder flow rate based upon the measured powder flow rate and based upon a set powder flow rate or a stored desired powder flow rate.

The portable GDS gun generally is connected to a pressurized gas source that supplies high pressure air or other gas to a heat chamber (1) (numbers in brackets refer to Fig. 11.6 and 11.7). The converging part (2) and the throat (3) of the nozzle are made of ceramic, while the diverging part (4) of the nozzle is made of steel. The diverging section of the nozzle extends through an outer housing (5) from which it is supplied with powder from a powder container (6). As the pressurized air or other gas passes through the nozzle, it reaches supersonic velocities and draws powder from the powder feeding hose (7) into the diverging part of the nozzle. The outer housing has multiple passages (8) each leading to the diverging part of the nozzle. Powder is selectively supplied to one of the multiple passages in the outer housing using a switch (9). The switch supplies the powder based upon the value of negative pressure at certain points of the air jet. The switch may be set manually, or automatically by a controller programmed to the expected negative pressure points along the diverging part of the nozzle. Depending upon the pressure from the pressurized gas source, the location of the negative pressure point along the diverging part of the nozzle may vary. The switch is set so that the selected passage coincides with the negative pressure point.

The powder from the container is fed to the switch through a vibrating bowl (10) and funneled (11) into the partial-vacuumed powder passages of the outer housing. The powder then mixes with the jet of conveyance air and then jointly with it flows through the nozzle extension conjugated with the diverging part of the nozzle, to impart supersonic velocities to the air and entrained powder.

A jet of conveyance air from the pressurized air source is heated in the heat chamber. The compressed-air line contains a variable throttle (12) by which the flow impedance (e.g. the flow cross-section) is regulated from a controller as a function of a setpoint value of the volumetric flow of conveyance air and/or of a setpoint value for the volume concentration of the particles in the powder-laden jet. The controller may be a computer having a processor, memory, and other storage, and being suitably programmed to perform the operations described herein.

The heat chamber includes a serpentine or helical coil heating element mounted on a ceramic support and an insulation chamber that is located in an internal chamber housing (13). The air flows along the helical path defined by the helical coil heating element. The heated air exits the heater via a tapered chamber (14), which together with ceramic insert (2) forms the convergent portion of the nozzle.

The powder supply system is shown in more detail in Fig. 11.7. The powder supply system includes a powder container with powder to be sprayed in loose particulate form, a bowl vibration unit (15) (such as a motorized vibration unit) for control of the powder flow rate, and a funnel (11) connected to a flexible hose (7). Additionally, two powder container vibration units (bowls) (16 and 10) positioned one below the other along the powder flow are incorporated into the upper portion of the powder supply system. Simultaneous control of the two vibration units provides precise and constant control of the powder feeding rate. Powder is fed into the lower vibration unit (10) so that a certain level of powder is maintained by means of a sensor (17) which controls the operation of a main powder hopper (not shown). The rate of powder dispensing (powder flow rate) is additionally controlled by the removable lower bowl nose (18) with variable

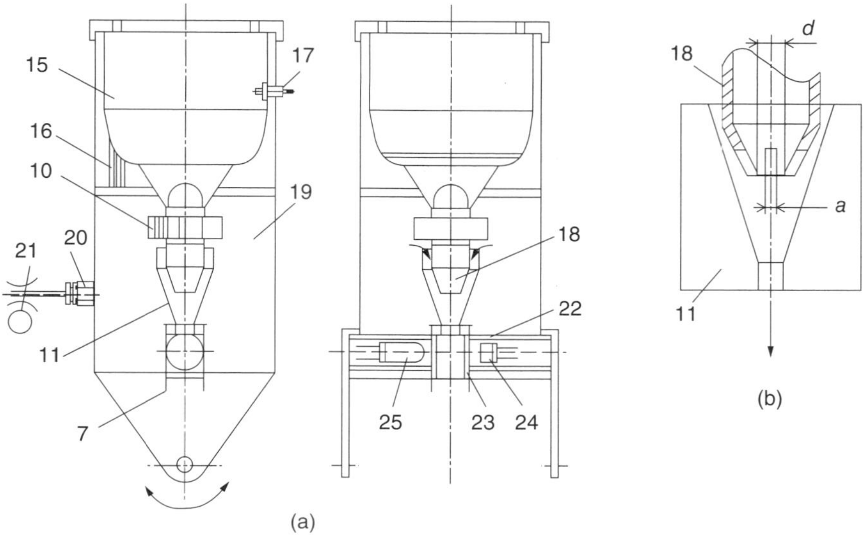

11.7 Low pressure cold spray gun. The powder supply system: (a) pick-up device; (b) removable nose and funnel. The powder/air duct is composed of a powder transportation tube (23) connecting funnel 11 to a powder aspirating hose (7).

hole diameter, *d*, and slot size, *a*. The rate of powder dispensing is defined by the flowability of the powder.

The partial vacuum existing in the partial-vacuum zone in the lower portion (19) of the pick-up housing (chamber) aspirates air from the atmosphere while being strongly throttled by a flow throttle when passing into the partial-vacuum zone of the chamber. The chamber is fitted with a flow sensor (20) generating a measurement signal as a function of the air flowing from the atmosphere through a throttle (21) into the partial-vacuum zone of the chamber, i.e. the quantity per unit time, or rate, of air passing through the throttle, and hence also being a control of the rate of powder passing through the powder passages (8).

The pick-up device includes a powder metering unit (24) detecting a flow of powder particles in a measurement duct, which can be a simple glass powder transportation tube (23) connecting the funnel (11) to a powder aspirating hose (7) attached to the powder switch of the outer housing. The powder-metering unit includes an infrared sensor (24) and an infrared emitter or light source (25) disposed within the channel made in the pick-up bottom plate. The infrared sensor can determine the mass flow of powder through the glass tube based upon the amount of light from the light source that is able to pass through the glass tube to the infrared sensor. Although an infrared light source and infrared sensor are preferred, other wavelengths of light or other waves could also be used. Optionally, an additional powder metering unit can be mounted in the pick-up housing on opposite sides of the funnel (11). The additional powder metering unit is similar to the described power metering unit and includes an infrared sensor (or light sensor) and an infrared emitter (or light source). This powder metering unit measures the powder dispensing rate ω_d from the lower vibrating bowl (10). The powder dispensing rate ω_d can then be compared with the conveyed powder rate ω_p. The amplitudes of the vibration units can be adjusted relative to one another in order to ensure that the powder dispensing rate ω_d is equal (over some short period of time) to the conveyed powder rate ω_p. This prevents clogging of the funnel.

The particle volume concentration significantly affects the deposition efficiency. The particle volume concentration in a powder-laden jet greatly influences the effectiveness of the GDS process particularly in the case of radial injection of powder by conveyance air of the partial-vacuum zone. Control of the volume concentration of particles is achieved by regulation of two parameters: the rate of conveyed powder and the rate of conveyance air. The rate of conveyed powder ω_p is substantially dependent on the powder dispensing rate ω_d and the rate of conveyance air. The powder rate is approximately proportional to the rate of conveyance air of the partial-vacuum zone of the pick-up housing chamber. Therefore, the conveyance air must be adjusted to achieve the desired particle volume concentration of the powder-laden jet. Thereupon the controller can automatically set the rate of conveyance air by means of the adjustment motor and a throttle (26) (Fig. 11.6) in such a way that the volumetric flow remains at the setpoint. From another side of the system, the controller will automatically set the powder dispensing rate ω_d by means of the adjustment of amplitudes of vibration units on the basis of measurements of the rate of conveyed powder ω_p in order to achieve the permanent balance $\omega_d = \omega_p$. Additionally, the rate of conveyance air is regulated by changing the injection point location, using the outer housing switch either manually or automatically. The controller (Fig. 11.6) regulates the powder feeding flow rate, the carrier air flow rate, and the feed of powder conveyance air in the partial-vacuum zone of the pick-up chamber as a function of the measurement signals of the measurement lines (27, 28, and 29), and as a function of the setpoint value of the volume concentration of particles in the air–powder jet, by means of the vibration units and two throttles. The controller comprises two inputs: (a) an input for the powder flowability setpoint value (i.e. a manual or automatic, fixed or variable setpoint of the powder dispensing flow rate, ω_d) and (b) an input for the powder volume concentration setpoint value, C_v, allowing the carrier air flow rate to be determined for the air passing through the powder/air duct according to:

$$C_V = \frac{\omega_p}{\rho_p \omega_{air}} \quad [11.1]$$

where ω_p is the particle feeding flow rate from the funnel (Fig. 11.7(b)), ρ_p is the material density, and ω_{air} is the carrier air flow rate controlled by air pressure and throttle.

Alternative heat chamber designs are shown in Fig. 11.8. The heat chamber includes the helical coil heating element mounted on a ceramic tube within a carrier air transportation pipeline (Fig. 11.8). The carrier air transportation pipeline is mounted inside the internal chamber housing to define a hollow cylindrical passageway. The air flows in from the entry line in a forward direction between the internal chamber housing and the pipeline. The air then enters the forward end of the heat chamber pipeline and flows rearwardly within the helical coil heating element. At the rearward end of the heat chamber pipeline, the air enters the ceramic tube and then travels in a forward direction through the ceramic tube, the tapered chamber, and the ceramic insert forming the converging part of the nozzle. Thus, the air gathers heat from the helical coil heating element on three serpentine passes. This increase in the heating surface intensifies the heating of the air and increases the temperature of the carrier air up to 650–700 °C in the portable heating chamber. The system incorporates safety features for the protection of both the system and the operator. The control system switches off the power supply and sends a signal out in cases of abnormal increases in the temperature of the gas above a set value.

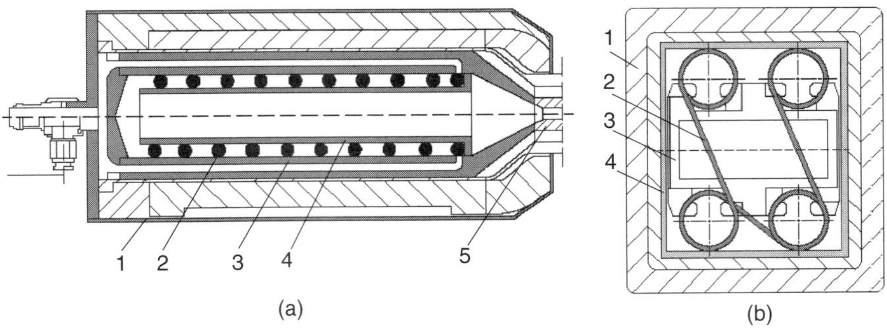

11.8 Low pressure cold spray gun. Heating chamber design. (a) Helical coil-heating design: 1, housing; 2, helical coil-heating element; 3 and 4, ceramic tubes; 5, nozzle insert. (b) Square chamber design: 1, multilayer insulation; 2, heating element; 3, heating element support; 4, ceramic housing.

11.4 Industrial low pressure, portable gas dynamic spray system

A reliable, rugged model of a portable GDS machine for industrial applications was developed by Centerline Windsor Ltd (Windsor, Ontario, Canada). The all-metal enclosure provides outstanding durability and it can be moved from worksite to worksite easily and quickly. A touch-screen display panel provides convenient and precise control while monitoring the machine's settings and diagnostics. Through the use of electronics, the digital controller ensures peak system performance by monitoring air pressure, air temperature, powder feed rates, etc. The SST Spray gun (Supersonic Spray Technologies (SST) is a division of Centerline Windsor Ltd) incorporates an ergonomic and durable design intended to promote operator control and feel. With two mode selection buttons and an easy to operate trigger, it is comfortable for both right- and left-handed operators. The hopper and powder delivery system are equipped with programmable feed control and are sure to provide a long period of trouble-free performance. Dual 400 ml hoppers feed custom powder and abrasive compounds which are supplied in convenient hopper refills that can be easily replaced when replenishment of material is required. An integrated 'explosion-proof rated' system is used to provide quick and easy worksite clean-up. An 'easy disposal' feature allows for the disposal of collected residual material safely and quickly. The touch-screen control provides precise control of the vacuum system.

11.5 References

Alkhimov A P, Papyrin A N, Kosarev V F, Nesterovich N I and Shushpanov M M (1994), Gas-dynamic spraying method for applying a coating, US Patent 5,302,414.

Dalley H S (1970), Powder feed device for flame spray guns, US Patent 3,501,097.

Fabel A (1976), Powder feed device for flame spray guns, US Patent 3,976,332.

Karthikeyan J (2005), Cold spray technology, *Advanced Materials & Processes*, **163**(3), 33–35.

Kashirin A I, Klyuev O F and Buzdygar T V (2002), Apparatus for gas-dynamic coating, US Patent 6,402,050.

Kay A and Karthikeyan J (2003), Advanced cold spray system, US Patent 6,502,767.

Kay A and Karthikeyan J (2004), Cold spray system nozzle, US Patent 6,722,584.

Muehlberger E and de la Vega P (1989), Powder feeder, US Patent 4,808,042.

Papyrin A (2001), Cold spray technology, *Advanced Materials & Processes*, **159**(9), 49–51.

Rotolico A J, Romero E and Lyons J E (1983), Device for the controlled feeding of powder material, US Patent 4,381,898.

Schinella A A (1971), Powder feeder, US Patent 3,618,828.

Tapphorn R M and Gabel H (2004), Powder fluidizing devices and portable powder-deposition apparatus for coating and spray forming, US Patent 6,715,640.

Van Steenkiste T H, Smith J R, Teets R E, Moleski J J and Gorkiewicz D W (2000), Kinetic spray coating method and apparatus, US Patent 6,139,913.
Van Steenkiste T H (2003), Method of producing a coating using a kinetic spray process with large particles and nozzles for the same, US Patent 6,623,796.
Van Steenkiste T H (2004), Low pressure powder injection method and system for a kinetic spray process, US Patent 6,811,812.

12

Stationary cold spray systems for industrial applications

H. HÖLL, Cold Spray Technology GmbH, Germany

12.1 Introduction

The principle of cold spray systems, which is to expand heated gas with injected metal powders through a de Laval nozzle, is generally simple. However, the severe pressure and temperature environment demands highly sophisticated materials and methods for an industrial, long-lasting system and for a secure and stable process. A complete system, shown in Fig. 12.1, will be described in the following sections.

12.1.1 Gas consumption

The gas consumption is basically limited by the throughput of the de Laval nozzle; however, the more important value is the weight of material that is sprayed. With a standard deposition rate of 2 to 4 kg/h and a deposition efficiency of 60 to 90% this relates to a mass of ~6 kg which must be accelerated to a velocity of Mach 2 to 3. In order to obtain a stable process the mass of the process gas should be ten times higher than the powder. For this reason the Kinetiks® 4000 is designed to control 60 to 120 m^3/h of nitrogen gas which equals roughly 75 to 150 kg of gas per hour. When helium gas is used, the values are up to 250 m^3/h which equals roughly 45 kg of gas. These values are based on a nozzle throat diameter of 2.7 mm.

12.1.2 Pressure regulation

The regulation circuit of a cold spray system has to control two gas lines: the process gas and the feeder gas. Both lines must be controlled separately. The process gas line of the Kinetiks® 4000 system has a volume of ~1 to 2 dm^3. The use of a PID regulator requires a slow-acting setup in order to obtain stable pressures. Distortions such as switching on the powder line during operation can cause variations in the loop which must be avoided. The solution for this problem is cascade regulation with a 'rate of

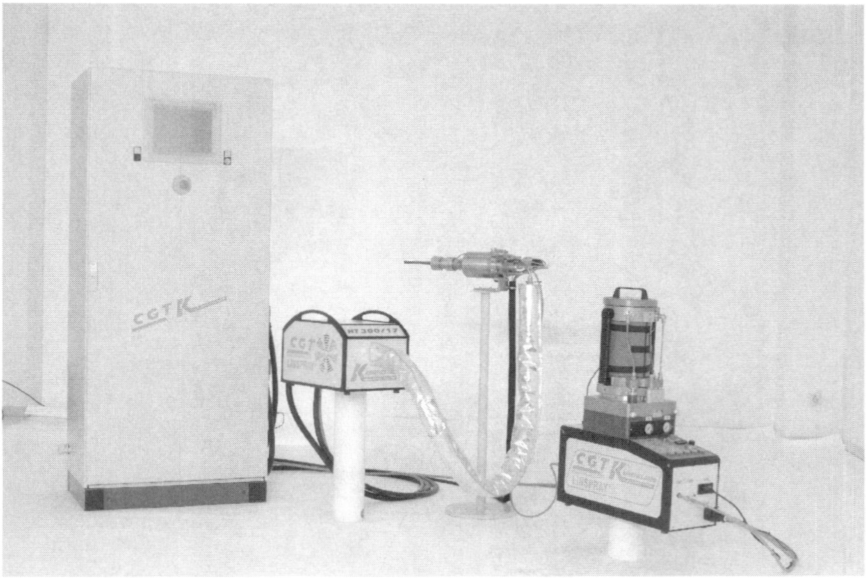

12.1 The Kinetiks 4000/34.

change' pressure regulator and a fast-acting mass flow regulator for the inner loop. This method has proven to yield the most stable pressure behaviour for the process.

The feeder gas line has a volume of ~3 to 5 dm^3 and is basically controlled by the pressure generated by the process gas flow in front of the de Laval nozzle. As the powder distribution inside the gas feeder line is homogeneous, any variation in the pressure will result in a variation of the feeder gas velocity and will respectively upset the amount of powder that is fed into the process gas. Therefore, any variation of the process gas (or feeder gas) pressure results in a variation of the powder feed rate and this will result in a variation of the coating thickness. Stable pressures and flow rates are the key for a stable and reliable process.

12.1.3 Gas heating

The heating of the process gas can be realized with a coil or a filament. The heating energy is proportional to the consumption of gas. Heating 80 m^3/h of nitrogen up to 800 °C needs about 24 kW of energy. Considering the loss of energy as 10 to 15%, a total energy of about 28 kW is required. The Kinetiks® 4000 accomplishes this with a two-stage heater system, shown in Fig. 12.2. The first stage heats the gas up to ~420 °C, and the second stage

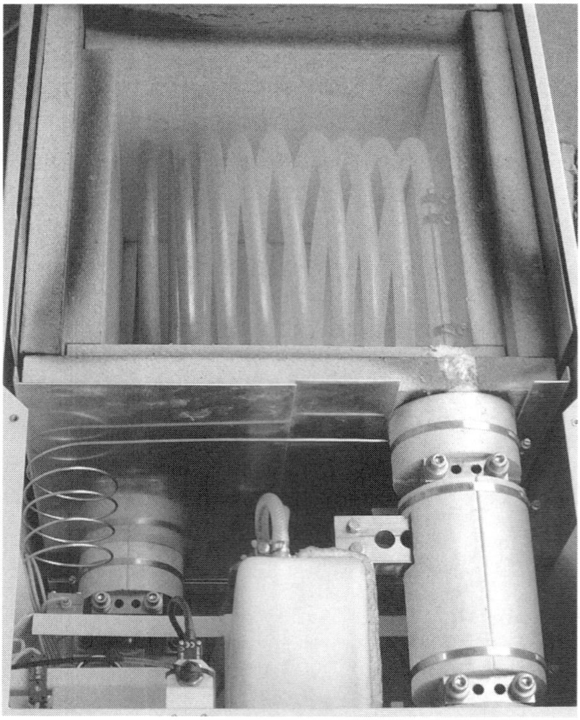

12.2 Linspray® heater.

increases the temperature up to 800 °C. The energy values required for helium heating are modified by the high thermal capacity of helium. Heating up 160 m^3/h of helium to 800 °C requires about 33 kW of energy.

12.2 Industrial cold gas spray equipment: system layout, gas supply and control cabinet

Figure 12.3 shows the Kinetiks® 4000 industrial cold spray system, with 47 kW of heating energy. The Linspray® heater and the ActiveJet® have a total of 47 kW heating power. This would be the unit for spraying materials with nitrogen and helium gas at a temperature up to 800 °C. Other configurations starting with 17 kW and 34 kW have a similar design. The correct system for the application is always defined by the maximum temperature at which the unit must be operated. An additional feature of the concept is to run a second powder feeder in-parallel. With this option it is possible to mix powders. This opens up new possibilities for coating in terms of which materials can be sprayed.

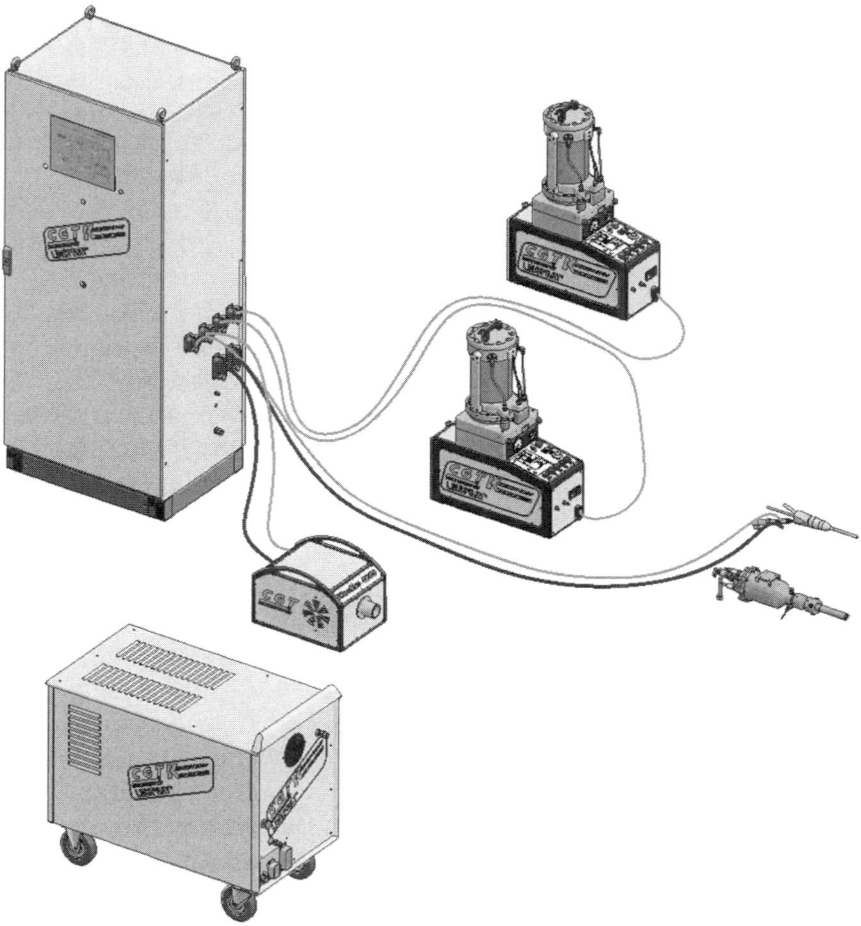

12.3 System layout.

12.2.1 Gas supply system

The gas supply system for the Kinetiks® 4000 is one of the key points for stable process operation. The supply must be able to deliver $120\,m^3/h$ of nitrogen gas at a stable pressure of 5 MPa. Using a dual gas system, the helium volume is $250\,m^3/h$. Pressure variations or oscillations have an effect on the process stability through the pressure in the cold spray gun. An increasing pressure in the gun decreases the feeder gas flow and vice versa. This results in a fluctuation of the powder fed through the gun and dimensional changes in the coating thickness.

The gases can be supplied in cylinder bundles; however, for productions purposes, more and more installations are equipped with a liquid nitrogen supply. The Linde Gas Company has developed a supply system that meets the requirements and has proved its reliability in several installations. In this case a standard tank is used with a DESY® pressure increasing system. This system is set to serve pressures from 5 to 20 MPa. In combination with the right pressure regulator upstream of the system, fluctuations are close to zero. The Linde system is shown in Fig. 12.4.

12.2.2 Control cabinet

The control cabinet is the second key factor for a stable process. Basically the pressure and the temperature in the gun have to be controlled. However, there are additional values which must be regulated in-parallel. The core of the regulation is based on three volume controllers from which the pressure regulation is set. In total there are five regulation circuits for the two major lines. The second function for an industrial system is the process data recording feature. Eighteen process values (temperatures, pressures, flow

12.4 Liquid nitrogen supply by the Linde Gas Company.

12.5 Control cabinet.

rates and regulation values) can be recorded via a LAN interface. These values can be used to examine the stability and the production. This built-in feature is a quality management tool. The third functional block is built in for personal safety and process reliability. All valve positions (open–closed) and measurement values are checked for correct operation. Leaking pipes can be detected and set up to shut down the unit if necessary. The operation is fully automated. An engineer sets the parameters. The operator must merely start and stop the unit. All this can easily be done via the 13-inch touch-screen. The control cabinet is shown in Fig. 12.5.

12.3 Industrial cold gas spray equipment: gas heater, powder feeder and cold spray gun

The Linspray® gas heater (Fig. 12.6) is designed for a maximum material temperature of 800 °C at 3.5 MPa. With a maximum power of 30 kW, this can lead to a gas temperature over 700 °C at the outlet of the heater. The

12.6 Linspray® gas heater.

maximum temperature of the heating coil is limited by the material properties. Using the heating system with the Kinetiks® 4000 system the temperature is automatically limited to 650 °C at 4.5 MPa. For the safety of the system there are four sensors installed: material, gas and contact temperature, as well as gas pressure. These values are monitored and lead to a shutdown if any irregular situation occurs.

A second option is the HT300/17 gas heater, shown in Fig. 12.7. This heater has a maximum of 17 kW power and can be used in the same way as the Linspray® heater. The heating technology is based on a filament which yields good reaction time; however, for the use of helium the 30 kW Linspray® heater is preferred.

12.3.1 Powder feeder

The process parameters are the important values for good coating properties. In addition, the powder feed rate is important for the uniform thickness

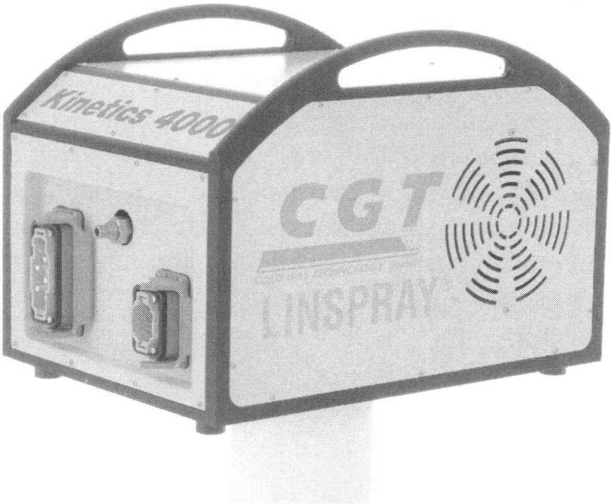

12.7 HT300/17 gas heater.

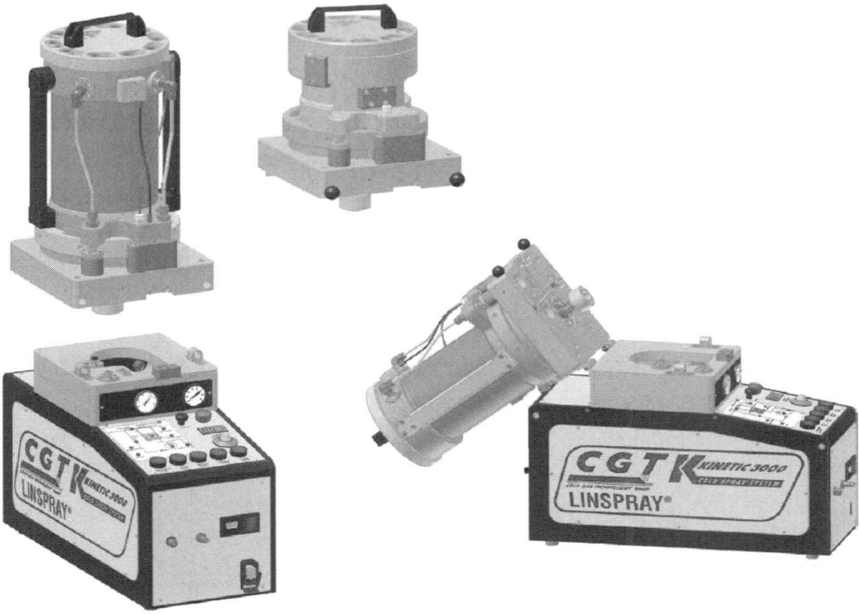

12.8 The PF 4000-2 Comfort powder feeder.

of each coating pass. The powder feeder PF 4000-2 was designed to meet the special process requirements for cold spraying. The second design issue was disassembling and assembling without tools.

Other features of the system, such as a bypass line and the separate de-ventilation system, make it possible to refill the powder during operation. The downtime in order to do this is less than 2 minutes. The major parameter for the feeder is the maximum approved pressure. The feeder is designed according to the European pressure regulations and can be operated up to 4.4 MPa. For the newest cold spray systems, this pressure has to be tolerated by the feeder. The feeder components are shown in Fig. 12.8.

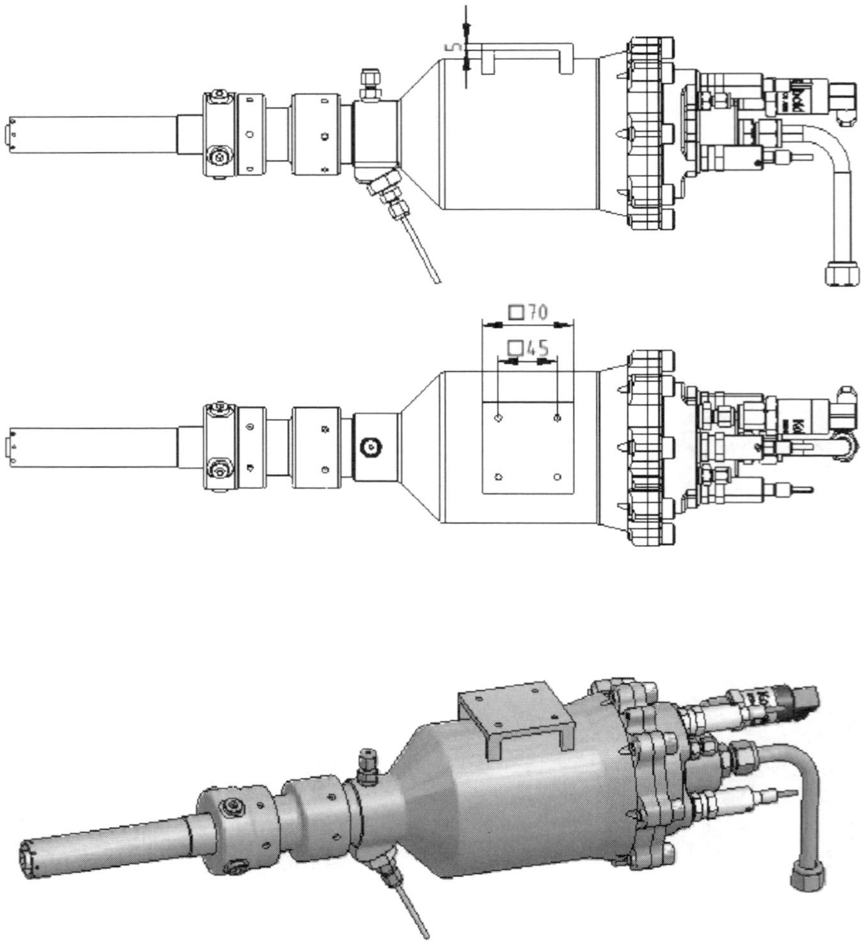

12.9 Nozzle assembly.

12.3.2 Cold spray gun

The cold spray gun and nozzle is the last module in the system. The Active-Jet® 4000 has a 17 kW heating stage within the nozzle assembly, shown in Fig. 12.9. The material properties and 4.0 MPa pressure have been the limiting factors for maximum allowable operating temperatures. As the second heating stage is in front of the nozzle it was possible to heat shield the pressure enclosure, which makes it possible to operate the system at 800 °C. The nozzles are made from a one-piece tungsten carbide sinter material. These nozzles have been available since 2004 and have proven their reliability and extremely good wear resistance in several installations. The nozzle limits the maximum temperature of the process to 550 °C. Purge air cooling is needed above this temperature.

Many new nozzle materials are under investigation, e.g. polymer materials with anti-sticking properties for low melting metals and ceramics for higher temperatures. The most common geometry is the 'Type 24 MOC' which gives superb results for spraying copper. However, very good results have also been achieved when spraying tantalum, nickel or powder mixtures.

12.4 References

Krömmer W, Heinrich P and Linde Gas AG, 'Cold Spraying – Equipment and Application Trends', Proceedings of the 2003 International Thermal Spray Conference, Orlando, Florida, USA, Vol. 1, pp. 97–102.

Richter P and Höll H, 'Latest Technology for Commercially Available Cold Spray Systems', Proceedings of the 2006 International Thermal Spray Conference, Seattle, Washington, USA.

Part III

Applications of cold spray coatings

13
Mechanical, thermal and electrical properties of cold sprayed coatings

A. GOULDSTONE, W. B. CHOI, W. CHI, Y. WU and S. SAMPATH, State University of New York at Stony Brook, USA

13.1 Introduction

Previous chapters in this publication have described the general characteristics and bonding mechanisms of cold spray (CS) coatings. In addition, conventional quality-control parameters for sprayed coatings, such as bond strength and hardness, have been addressed. In addition, previous efforts in the literature (Van Steenkiste, 2002; Hall, 2005) have involved direct tensile measurements of CS materials. In this chapter, we present and address results from a series of integrated property measurements on a range of CS metallic materials, in both the as-deposited and heat-treated state. Results are discussed in the context of microstructural formation during high velocity impact of particles, and the general characteristics of interparticle bonding. Interestingly, each of the experiments presented in this chapter by itself would not be sufficient to draw the conclusions found, and it is one goal of this writing to profess the value of such an integrated study, not only for CS but for sprayed materials in general.

When discussing the properties of CS or sprayed materials in general it is important to recognize the context in which they are taken, and what information is yielded. Sprayed materials require a multi-scale consideration, as they contain particles of tens of microns in diameter, micro- and nano-scale interfaces that may or may not be continuous. In addition, the highly directional nature of the process (i.e. high velocity impact) subsequently leads to highly anisotropic behavior in mechanical, thermal and electrical properties (Choi, 2007). Macroscopically, the measured properties of sprayed materials can be in the range of 10–80% of their respective bulk counterparts, and are strongly dependent upon process conditions (Prchlik, 2003; Sampath, 2004; Vaidya, 2005). Mechanical behavior may be quite nonlinear or hysteretic (Liu, 2006). Many property measurements in the literature are directly connected to performance; thermal barrier coating (TBC) thermal conductivity is widely reported, as is wear resistance of many hardface materials. On initial approximation, the properties described in this chapter (elastic, plastic, thermal, electrical) are not related to performance *per se*, but it will be shown that combined they provide a powerful and insightful supplement to microstructural characterization, as well as potential reliability assessments. Some of the results and discussions here are adapted from previous investigations (Sampath, 2004; Choi, 2007). While the primary focus of this work will be on CS Al, other studies of Ni-based alloys and composites will be included in some discussions.

13.2 Materials used for analysis

The materials described here were fabricated at Sandia National Laboratories (SNL; Al, Ni–5%wt. Al) and Army Research Laboratories (ARL; Al, Ni, Ni–30%wt. Al). Full details of processes are described elsewhere (Sampath, 2004; Choi, 2007), but general characteristics are presented here.

The coatings studied in this chapter were deposited at ARL and SNL under established conditions as listed in Table 13.1. The CS nozzles in both ARL and SNL were a de Laval type, with converging–diverging geometry.

13.2.1 Aluminum

Al coatings were prepared at both ARL and SNL with different He gas temperatures. The Al coating from SNL was deposited with He gas at 325 °C and 250 psi, denoted as Al-I, whereas the ARL coating was deposited with room temperature He gas, denoted as Al-II. The Al feedstock of both

Table 13.1 Spray parameters for Al, Ni, Ni–5%wt. Al and Ni–30%wt. Al coatings

	Al-I*	Al-II**	Ni	Ni–5%wt. Al	Ni–30%wt. Al
Feedstock size (μm)	15~45	15~45	4~8	11~45	Pure Al (15~45) coated with 70%wt. Ni
Feed rate (g/min)	11	5	15	–	20
He gas temperature (°C)	325	Room temperature	350	–	400
He gas pressure (psi)	250	380	380	–	380

* CS Al-I coatings were obtained from Sandia National Laboratories, Albuquerque, NM, USA.
** CS Al-II coatings were obtained from Army Research Laboratories, Aberdeen Proving Ground, MD, USA.

Al-II and Al-I was gas-atomized (Valimet H-20, Stockton, California) with a spherical morphology and a similar particle size distribution of 15~45 μm. The Al powder of SNL and ARL was fed into the nozzle at 11 and 5 g/min, respectively, and sprayed onto 6061-T6 Al substrate. Some Al-I specimens were annealed at 300 °C in air and Ar for 22 h (Choi, 2007).

13.2.2 Ni, Ni–5%wt. Al and Ni–30%wt. Al composite

The Ni feedstock was 4~8 μm, high density, spherical pure Ni powder, and sprayed onto 6061-T6 Al substrate. The Ni–30%wt. Al composite coating was prepared with the pure Al powder (Valimet H-40) coated with 70%wt. Ni, sprayed onto 6061-T6 Al substrate. Both coatings shared the same He gas pressure at 380 psi, but different He gas temperature of 350 and 400 °C and feed rates of 15 and 20 g/min, respectively. The Ni–5%wt. Al coating was obtained from SNL with a particle size of 11~45 μm (Sampath, 2004).

13.3 Mechanical property measurements

13.3.1 Elastic modulus

Continuous stiffness measurement (CSM) or the instrumented indentation method was used to extract elastic moduli using the microindentation instrument Nanotest 600 (Micro Materials Limited, Wrexham Technology Park, Wrexham, UK). The instrument obtained load–displacement (p–h) curves, from which the elastic moduli were extracted from the stiffness of the unloading curve based on the Oliver–Pharr method (Oliver, 1992). Each indentation procedure employed at least 15 loading/unloading cycles (15 indents) of the Berkovich indenter with a maximum load of 5 N.

The elastic moduli of the samples were measured using Hertzian indentation methods by Choi *et al.* (Choi, 2007), adapted from those developed for bulk ceramics (Pajares, 1996a, 1996b; Wei, 1996; Lee, 2000; Lawn, 2002). A brief summary of the procedure is presented here. Under frictionless spherical indentation (indenter radius, R = 4.76 mm), the mean contact pressure p_m is given by

$$p_m = \frac{2}{3}\left(\frac{6PE^{*2}}{\pi^3 R^2}\right)^{\frac{1}{3}} = \frac{P}{\pi a^2} \qquad [13.1]$$

where P is the applied load, a is contact radius, and E^* is reduced modulus); for a linear elastic specimen p_m is related to modulus by (Johnson, 1985)

$$\frac{1}{E^*} = \frac{1-v_s^2}{E_s} + \frac{1-v_i^2}{E_i} \qquad [13.2]$$

where E_s and E_i, are specimen and indenter moduli, respectively, and v_s and v_i are Poisson's ratio of specimen and indenter, respectively. It is clear from equation [13.1] that a plot of p_m versus P, for a given indenter tip, can be fit to reveal a value of E. As specimen deformation is elastic, optical measurement can be achieved via the observation of a crushed thin, soft film deposited on the surface before indentation. Sample observations of indents can be found in the literature (Choi, 2007).

13.3.2 Plastic properties

To explore the high-strain plastic behavior of coatings, Hall (2005) and Van Steenkiste (2005) used in-plane tensile tests, whereas Choi (2007) employed the 'Tabor method' which has been previously used for bulk ceramics and coatings (Padture, 1995; Pajares, 1996a, 1996b; Wei, 1996; Lawn, 2002). If an indentation load induces plasticity (and hence leaves an impression), a characteristic strain (ε) can be defined as $0.2a/R$. Although a characteristic flow stress was empirically determined to be $p_m/2.8$ for bulk metals, it is not obvious that this relation is valid for sprayed materials. Thus, a plot of characteristic flow stress versus strain provides a comparative measurement of yielding and hardening properties, analogous to stress–strain curves in compression. Indentation experiments were similar to the elastic tests, but with additional tips having radii 0.04, 0.08, 0.12 and 0.24 cm, in a Mitutoyo AVK-C2 indenter with load range of 1–50 kg. Sub-surface deformation was imaged using a bonded interface technique (Pajares, 1996a, 1996b; Wei, 1996; Lawn, 2002).

13.3.3 Measurement of coefficient of thermal expansion

Coefficient of thermal expansion (CTE) measurements (Choi, 2007) were performed using a Netzsch Thermische Analyse DIL 402C Al_2O_3 tube dilatometer with a TASC 414/4 controller. Correction files were generated using sapphire standard NBS SRM 732, DIN 51045. The temperature was programmed from room temperature to 300 °C at a heating rate of 5 °C/min. The push rod load was set to 25 cN. The CTE was determined using standard methods. Tests were repeated up to 10 times on each specimen, to differentiate microstructural changes (e.g. internal oxidation) from reversible thermal expansion. Most tests were performed in laboratory air. All tests were performed on free-standing coatings, with bulk Al used for comparison.

13.3.4 Thermal conductivity measurement

Measurement and understanding of the thermal properties of cold sprayed coatings is critical not only from design and application points of view but as a gage to assess the process–microstructure–property relationships. In a seminal paper, McPherson (1984) used through-thickness measurement of thermal conductivity as a method to quantify the nature of intersplat interfaces in thermal sprayed ceramic coatings. Recent studies at the Center for Thermal Spray Research (CTSR), State University of New York, have focused on developing correlations among process conditions, microstructure and thermal conductivity for wide ranging thermal sprayed and cold sprayed coatings. The results point to a powerful characterization approach to highlight the microstructure defects in these complex layered systems. Experiments presented here were performed at CTSR specifically for this chapter.

Thermal conductivity measurements were performed on a Holometrix Laser Flash thermal diffusivity measurement system. Measurements were made on free-standing coating samples devoid of substrate. The free-standing sample was rounded to a disk with diameter of 12.7 mm and coated with graphite for thermal conductivity measurement (the graphite acts as a laser absorber). The optimum sample thickness depends on the diffusivity of materials. The time taken for the back surface to reach half its maximum temperature ($t_{1/2}$) should be at least three times longer than the length of the laser pulse (0.33 ms) and no longer than 3 s. The suggested thickness is 0.4–1 mm for ceramic coatings and 1–2 mm for metal coatings.

In our studies, thermal conductivity was measured in one step by obtaining thermal diffusivity and specific heat in a single test. The thermal diffusivity was determined by subjecting one side to a short laser pulse and recording the temperature rise on the opposite side. The specific heat was calculated from the same data of temperature rise recorded in thermal diffusivity measurements with a suitable calibration by comparing with the temperature rise of a reference sample of known specific heat. With an additional measurement of bulk density, thermal conductivity was calculated using the formula $K = \alpha \rho Cp$ where K is the thermal conductivity, ρ is the bulk density and Cp is the specific heat.

13.3.5 Electrical measurement

Resistivity measurements (Choi, 2007) were taken in two orthogonal directions: in-plane and through-thickness, using a four-point probe method on a microprobe station (Karl Suss, model PSM6). Specimens were sliced from free-standing coatings, and typical dimensions were 2 mm × 2 mm × 150 μm. Measurements were performed on as-sprayed and annealed materials; in addition, the in-plane resistivity of (pre-oxidized) specimens polished to different thicknesses was measured – further experimental details are presented in Sharma (2006) and Choi (2007).

13.3.6 Residual stresses

Residual stress measurement on CS materials has been reported in the literature, using neutron scattering and substrate curvature methods (Matejicek, 1999, 2002, 2003a, 2003b; Sampath, 2003, 2004). In all cases, stresses were found to be negligible, and of the same order as surface peening effects. Thus, in this chapter the details of the measurement are not described extensively and appropriate references are provided.

13.4 Microstructural characterization of coatings

Figure 13.1 (Choi, 2007) displays cross-sectional scanning electron microscope images of as-sprayed and annealed coating microstructure, for both powders. To clearly distinguish interfaces, after polishing (0.05 micron silica) all coatings were etched (50 g NaOH/1 L H_2O, 5 min at room temperature); this is the reason for the dark regions in the images. Similar to thermal spray

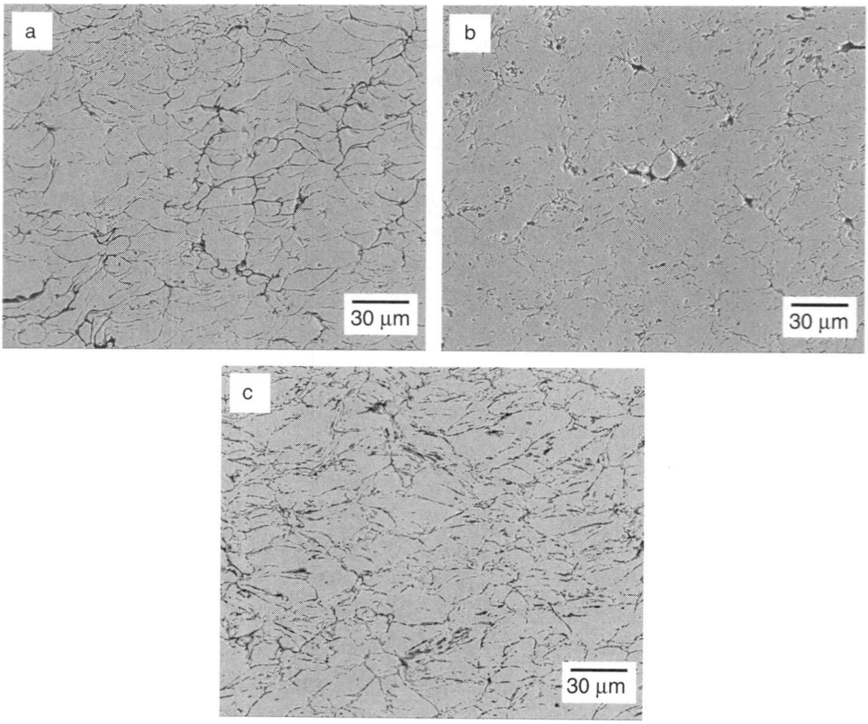

13.1 Cross-sectional polished and etched scanning electron micrographs of cold spray Al-I: (a) as-sprayed; (b) annealed in air; (c) annealed in Ar.

(TS), CS coatings display a network of interlocking splats, boundaries and larger pores. CS coating appears to have a flattening ratio far lower than that seen during TS, presumably due to the impact of solid, rather than molten particles. Coating density was measured via mercury intrusion porosity and was found to be ~99%. Interesting features appear after heat treatment (HT). 'Vertical' interfaces are far more apparent than 'horizontal' interfaces after HT in air, indicating an improvement in bonding of the latter, and a deterioration of the former. HT in Ar seems to cause deterioration of all interfaces. Specific comments on microstructure are made in the context of properties, in Section 13.6.2.

13.5 Properties of coatings

13.5.1 Elastic properties

Raw data from indentation measurements may be found elsewhere (Sampath, 2004; Choi, 2007); here we present resulting modulus values for CS materials. Table 13.2 gives values for a series of coatings, as sprayed and heat treated, and Table 13.3 compares them with bulk. A number of salient features arise. First, note that all modulus values are significantly lower than bulk values. Given the typical behavior of TS materials, this is not surprising. However, if one were to attempt to predict modulus values from, for example, cross-sectional micrographs or density measurements, it is clear this would lead to a qualitative overestimation. Second, for the case of CS Al, values of modulus *decrease* with heat treatment. This may be surprising, as annealing of TS coatings is typically associated with sintering or

Table 13.2 Elastic moduli, thermal conductivity and electrical resistivity of cold spray Al, Ni, Ni–5%wt. Al and Ni–30%wt. Al composite

	Elastic modulus (GPa)		Thermal conductivity (W/m K)	Electrical resistivity (μΩ cm)
	Instrumented indenter	Elastic indent method		
CS Al-I	61	45	–	10
CS Al-I HT-air	51	23	–	15
CS Al-II	66	–	173	3.9
HVOF Al	65	–	96	37
APS Al	51	–	–	–
CS Ni	135	125	91	8.8
CS Ni–5%wt. Al	110	–	11	140
CS Ni–30%wt. Al	96	–	65	6.2

HVOF, high velocity oxygen fuel.

Table 13.3 Ratio of coating elastic moduli (E), thermal conductivity (K) and electrical resistivity (ρ) of cold spray Al, Ni, Ni–5%wt. Al and Ni–30%wt. Al composite to known bulk or similar composition material values

	$E_{TS\text{-}coating}$*/E_{Bulk}	$K_{TS\text{-}coating}/K_{Bulk}$	$\rho_{TS\text{-}Coating}/\rho_{Bulk}$
CS Al-I	0.88	–	3.3
CS Al-I HT-air	0.74	–	5.0
CS Al-II	0.95	0.79	1.4
HVOF Al	0.94	0.44	12
APS Al	0.74	–	–
CS Ni	0.65	0.71	1.3
CS Ni–5%wt. Al**	0.58	0.17	4
CS Ni–30%wt. Al***	0.78	0.37	1.1

* Values of $E_{TS\text{-}Coating}$ are based on the instrumented indentation.
** Bulk values are calculated based on the rule of mixture of intermetallic compounds.
*** Bulk values are calculated based on the rule of mixture.

coagulation of interfaces, and thus an *increase* in such properties. Further, the drop in modulus cannot be accounted for by any kind of recovery mechanism in the metallic particles, as this would only affect plastic behavior; elastic properties in this case can only be attributed to micro-scale damage. It should be noted that such a decrease was also observed in tensile testing experiments (Hall, 2005). Further evidence for this and discussion is provided in Section 13.6.2. Finally, note that the relative elastic modulus (ratio of CS to bulk) is higher in the case of Al than Ni. This indicates perhaps that the softer, lower melting temperature material bonds more effectively. Another reason is perhaps somewhat higher porosity in the cases of the stiffer Ni coating.

One other important note is that the modulus measurements can be dependent upon the measurement technique. For example, a modulus calculated by the Oliver–Pharr method is quite a bit higher (in these cases) than that calculated from the Hertzian technique. This is probably due to the lower sampling volume in the former case, and the fact that the contribution from interfaces is not well-represented. The Hertzian technique probes a larger volume, is a true 'elastic' measurement, and is thus more correct.

13.5.2 Plastic properties

Figure 13.2 displays the plastic behavior of the CS Al specimens on stress–strain axes. Post-heat-treatment behavior seems to provide the most insight into coating structure – after annealing, the coatings display significantly

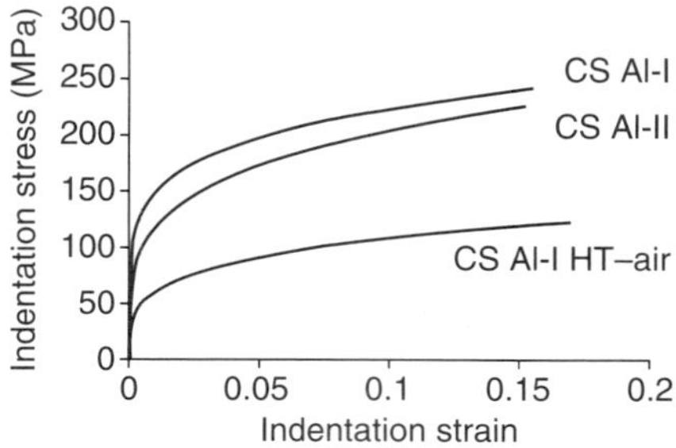

13.2 Indentation stress–strain curve of cold spray Al.

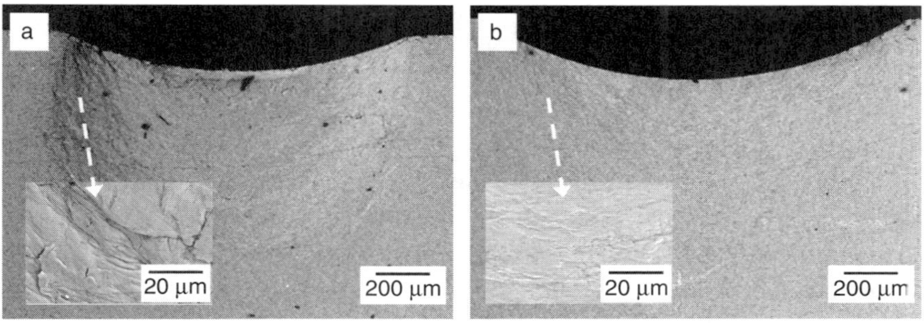

13.3 Scanning electron micrograph of sub-surface damage before and after annealing on CS Al-I coating using bonded specimens with the spherical indenter aligned diagonally to the bonded interface: (a) as-sprayed; (b) annealed.

lower values of flow stress than in the as-sprayed condition. Hardening slopes are similar before and after annealing. These trends are in agreement with tensile data, although the indentation tests span a much larger strain range (Van Steenkiste, 2002; Hall, 2005). In contrast to elastic behavior, plastic properties may be interpreted (at least in part) in the context of microstructural recovery of heavily cold-worked particles. This issue, that is to say, the intrinsic particle properties, was examined via a high-resolution transmission electron microscope (TEM) in the literature (Balani, 2005). In addition, sub-surface deformation images of CS Al-I provide interesting supplemental information on the nature of plastic deformation in Fig. 13.3; they clearly show different flow patterns before and after annealing; the former case exhibits significant micro-cracking, whereas the latter does not. In addition, the decrease in flow stress after annealing is similar to that seen for, for example, cold-worked metals and metallic coatings deposited by high velocity oxygen fuel (HVOF) but is opposite to that seen for coatings deposited via plasma spraying (Matejicek, 1999). Mechanisms for this, as well as the dependence of properties on the initial powder characteristic, are discussed further in Section 13.6.

Figure 13.4 shows plastic stress–strain data from Ni-based alloys and composites, some of which were reported in Sampath (2004). Perhaps not surprisingly, the Ni-based materials exhibit far higher flow stress than pure Al. Also, within with the Ni-based results, it is pure Ni that exhibits the highest strength, and the effects of mixing Al are increasingly deleterious (Fig. 13.5). That is to say, as the percentage of Al increases, flow stress decreases. Thus on initial approximation, it can be concluded that the flow properties of these sprayed coatings are strongly dependent on the individual starting materials. This needs further investigation but the results are nevertheless important for design and performance considerations of cold sprayed materials.

13.5.3 Thermal expansion

When considering the CTE of CS Al-I, or for that matter any sprayed material, it is useful to consider such a property over a well-defined temperature/time history range, rather than, for example, a near room temperature measurement. When tabulated in this way, the data transcend from a simple 'property' to a robust descriptor of coating structure and bonding character. As an example, Fig. 13.6 (adapted from Choi 2007) displays dilatometer CTE measurements of CS Al-I from room temperature to 300 °C, in air. Also included are data for bulk Al. During the first cycle, below a temperature of approximately 150 °C, the coating CTE values agree well with handbook values for bulk Al. However, above this temperature, they deviate strongly, as the coating CTE increases to 4.2×10^{-5}/°C at the maximum temperature tested. Repeated cycling reduces this deviation, and coating expansion behavior approaches that for bulk Al. Interestingly, the CTE is

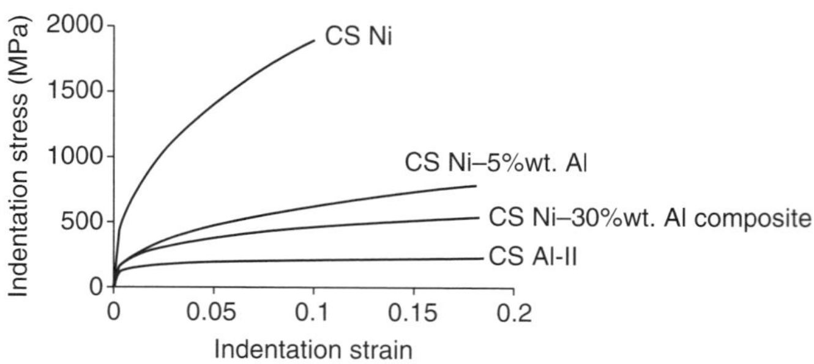

13.4 Indentation stress–strain curve of cold sprayed Al, Ni, Ni–5%wt. Al and Ni–30%wt. Al composite coatings.

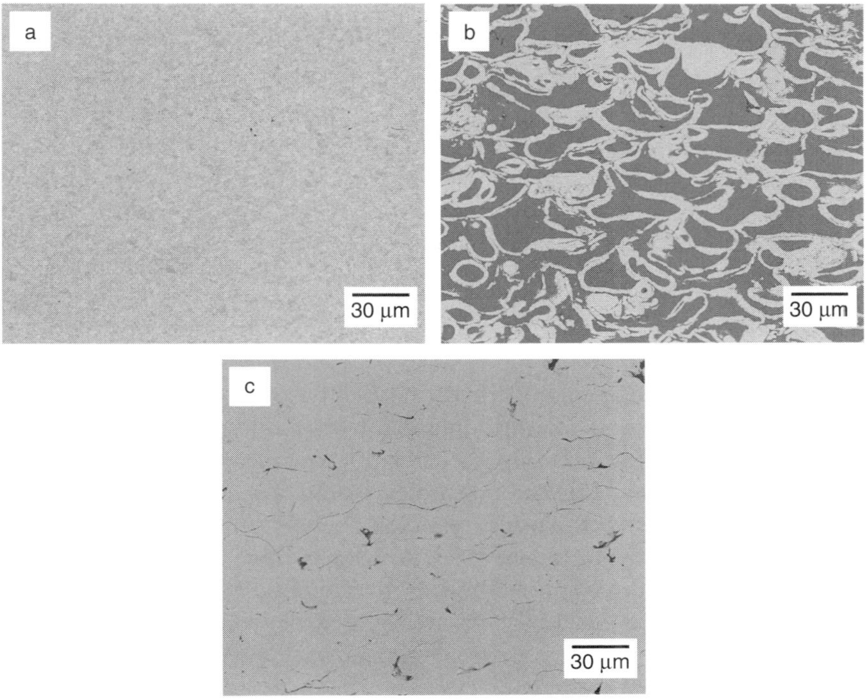

13.5 Cross-sectional polished scanning electron micrographs of cold sprayed materials: (a) Ni; (b) Ni–30%wt. Al composite; (c) Ni–5%wt. Al.

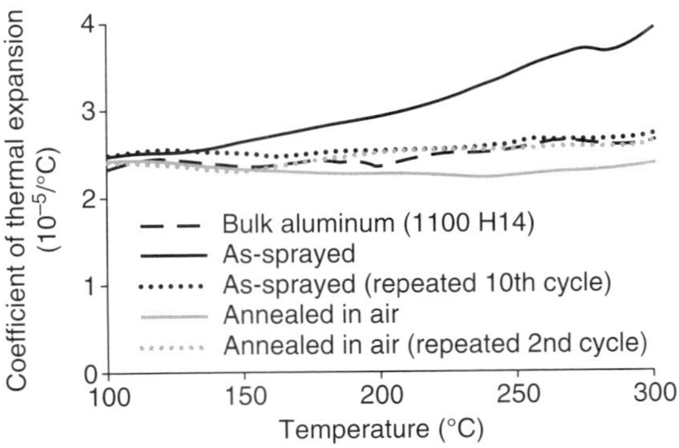

13.6 Thermal expansion measurement of CS Al-I coating.

lower than that of bulk Al after a large number of cycles, or after annealing in air. In Choi (2007), the underlying interpretation was that oxidation played an important role in expansion behavior. The porous structure of coatings allows more oxygen penetration to the interior of the sample along the splat boundaries so that a larger linear expansion would be exhibited as oxidation occurs, but then a smaller linear expansion would be shown by the resulting 'composite' specimen. As described by the author, such a measurement aided in the interpretation of mechanical measurements.

13.5.4 Thermal properties

Tables 13.2 and 13.3 compare the through-thickness thermal conductivity of various cold sprayed coatings obtained from the laser flash instrument. Both absolute results and those normalized with respect to the bulk counterpart are represented. It should be noted that in the case of alloys, thermal conductivity was calculated using a simple rule of mixtures. Several interesting observations can be made. The thermal conductivity values of the various coatings are substantially lower than bulk despite the coatings displaying very high compaction density in the as-deposited state. However, this result becomes less surprising when one considers the phonon scattering that may occur at the non-epitaxial interfaces (Kapitza, 1930) between particles, and provides a further argument for their importance in design. Furthermore, it is also clear that the mechanical properties of the starting material system play an important role in the cold sprayed coating thermal conductivity. On initial approximation, it may be asserted that the harder and higher melting temperature material (Ni) does not bond as effectively as its softer, lower melting temperature counterpart (Al), as discussed in Section 13.5.1 on elastic properties.

13.5.5 Electrical properties

Literature studies of electrical properties of CS metallic coatings provide an interesting statement about these materials and the highly directional nature of the interparticle bonding. Table 13.2 shows one example of resistivity data for CS Al coatings (adapted from Choi, 2007), in as-sprayed and annealed conditions, along with experimental and literature values for bulk Al. As-sprayed specimens show fairly strong anisotropy between in-plane and through-thickness properties. This is notable in that if one examines the microstructure of a typical CS material (e.g. Fig. 13.1), the structural anisotropy is not very apparent, as the particle flattening ratio is not very high compared with TS materials (Zhang, 2003; Li, 2004). Thus, this suggests a very different nature of bonding between interfaces that are parallel to the spray direction and those that are perpendicular to the spray direction. Further discussion is provided in Section 13.6.

13.6 Comparison of cold spray coatings with traditional thermal spray coatings

Although it can be argued that CS is an entirely different material deposition process, it originated and was developed through the TS community. It is often considered an extreme extension of TS processes and referred to as a logical transition of TS from a regime of high temperature–low velocity to one of high velocity–low temperature. As such it is appropriate to compare the mechanical, thermal and electrical properties of CS materials with those of their thermal sprayed counterparts. Table 13.3 compares the elastic modulus, thermal conductivity and electrical resistivity of Al and Ni–5%wt. Al alloy produced by CS and Al produced by other TS processes. The results are presented in the form of representative property ratios compared with the bulk material of similar composition. It is important to note that the thermal sprayed materials often experience oxidation and other thermal degradation and as such are compositionally somewhat different.

The plastic properties of CS Al and Ni–5%wt. Al coatings are compared with those of traditional TS coatings in Fig. 13.7(a) and (b), respectively. The results indicate that in comparison to atmospheric plasma spray (APS), CS Al has higher overall yield and tensile strengths with similar strain hardening coefficients. However, the HVOF sprayed Al shows a substantially larger yield and ultimate tensile strength, and a notably larger strain hardening response. Similar results have also been observed in the case of the Ni–5%wt. Al material (Sampath, 2004). However, note that in the case of the (higher melting temperature) Ni alloys, the difference between the

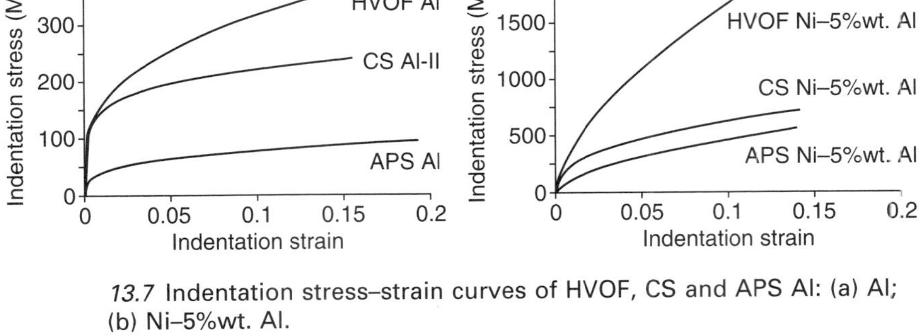

13.7 Indentation stress–strain curves of HVOF, CS and APS Al: (a) Al; (b) Ni–5%wt. Al.

HVOF and CS data is far more significant than that for the Al. The results from both Al and Ni–5%wt. Al suggest that the thermal input is perhaps an important parameter in creating a strong interparticle bond in these spray-deposited systems. This is particularly noteworthy given that the thermal sprayed materials exhibit significant oxidation at the interparticle interfaces.

13.6.1 As-sprayed properties

All literature studies, including those presented here, have shown significantly different properties between CS materials and their bulk counterparts (Van Steenkiste, 2002; Morgan, 2004; Balani, 2005; Choi, 2007). As stated above, this may be surprising if one attempts to predict behavior by recourse to density measurements or microstructural observations. However, if one considers the imperfect nature of the bonding between particles, such results logically follow.

The lower elastic moduli in these coatings can be attributed to the above effects; nano- or micro-scale cracks and gaps may close under compression or allow sliding under shear; this is similar to the origin of lower modulus in other TS coatings. The non-trivial contribution of such interparticle defects is further evidenced by the lower thermal conductivity results described above. In addition, although not discussed here, it is highly likely that some sort of elastic anisotropy exists in these materials – this is a subject for further study. Beyond the elastic limit, behavior is highly brittle in some of the coatings, as evidenced by observation of sub-surface deformation after indentation. Choi suggested, in particular for CS Al, that because of the higher yield strength and strain hardening properties of the coatings, and the similarity between coatings made from different powders, one could suggest that breaking/sliding of oxide/oxide interfaces is the primary mechanism of inelastic deformation (Choi, 2007). He also pointed out that in the as-sprayed condition, individual particles were quite hard for the following two reasons: (a) the sprayed metallic particles are small-volume structures (20 μm) and (b) they were highly work-hardened during deposition (supported via TEM observations in Balani, 2005).

Regarding the CTE, the deviation from bulk behavior in the Al coatings was probably due to oxidation at interfaces between splats. This is supported by the results from repeated cycling, as CTE behavior approached bulk behavior with repeated testing. The oxygen source in this case could be the dilatometer atmosphere, or trapped gas within the coating. This can be inferred from the data from Choi (2007) which show different CTE values during measurements in air (the CTE under the inert gas deviates from bulk less strongly, but non-trivially). In the discussion below, we describe how an observation of this type can have *profound* implications for the metastable nature of these materials; perhaps far from the near-bulk behavior predicted, expected or advertised upon their early usage (Papyrin, 2001).

Electrical resistivities of CS Al were higher than the bulk Al sample measured, and anisotropic. These characteristics can be attributed to intrinsic defects within Al splats (e.g. high dislocation density) and oxidized interfaces between splats. Anisotropy, on initial approximation, could be attributed to the quasi-lamellar microstructure containing more interfaces per unit length through the thickness of the coating versus in-plane. Nevertheless, examining the data from Choi (2007) on the thickness dependence of resistivity (as specimen thickness approaches splat size this effectively confines electrons to a single interrupted conductive path). This concept was fully explored in Sharma (2006), the imperfect interfaces between splats (interface normal perpendicular to substrate normal) have a non-trivial effect on anisotropy. It could further be argued that these interfaces are less well-bonded than those above and below particles, but this assertion must be certified with direct measurements.

Finally, comparison of mechanical, electrical and thermal properties of CS materials with identical or similar traditional TS materials revealed several interesting features. A notable feature is that the availability of thermal component in HVOF significantly enhances the interparticle adhesion and as such enhance plastic, thermal and electrical properties.

13.6.2 Annealed properties

The post-annealing behavior of CS metallic coatings may deviate strongly from many other TS materials. On initial approximation, heat treatment of plasma sprayed coatings leads to sintering and densification, which may either be advantageous (in the case of MCrAlY bondcoats for example; Matejicek, 1999) or deleterious (consequent sintering of yttria-stabilized zirconium (YSZ) TBCs at operating temperatures leading to increased thermal conductivity; (Sampath, 1999; Kulkarni, 2003)). In short, coatings tend to get more dense, stiffer (higher *E*) and stronger (higher hardness). In the case of HVOF coatings, the persistence of semi-molten particles, coupled with high velocity impact leads to relatively hard particles/splats. This usually leads to an increase in ductility upon heat treatment, but rarely decreases elastic modulus, as interface coagulation tends to occur (Hall, 2005; Choi, 2007). Through the studies presented in this chapter, we see this is not always the case for CS materials. Here we discuss characteristics of CS Al arising after annealing in both air and Ar. One notable result is a significant decrease in elastic modulus. Choi explained this by considering behavior at splat interfaces (see Fig. 13.1). The CTE values of Al and Al_2O_3 (at room temperature) are approximately $23 \times 10^{-6}/°C$ and $6 \times 10^{-6}/°C$, respectively. If one considers the oxide interphases between splats as thin films, a temperature increase of 250 °C would impart a mismatch strain of approximately 0.004, which would probably be sufficient to cause opening of any cracks present. They asserted that such damage could account for the decreased modulus. They also described another possible mechanism involving the migration of vacancies to the splat interfaces and subsequently opening gaps – this issue was left open. Interestingly, annealing in Ar causes a larger decrease in modulus than in air. Choi explained this by asserting that the increased oxidation in air imposes a compressive stress at the interfaces (i.e. 'filling the gaps'); this does not occur under the noble atmosphere. In addition, a contribution could arise from the higher amount of oxide in the coating.

Similarly and not surprisingly, annealing decreases flow stress of the CS Al coatings. Choi, Hall and Balani rationalized this by considering the microstructural recovery of the hardened as-sprayed splats (Balani, 2005; Hall, 2005; Choi, 2007), diverting plastic strain from the interfaces during indentation. In addition, Choi supported this hypothesis with images of sub-surface deformation (Fig. 13.3), in which (a) few cracks are seen and (b) deformation extends far below the contact region. It was agreed by all authors that annealing alters inelastic deformation from primarily brittle to ductile (Balani, 2005; Hall, 2005; Choi, 2007). The evolution of this transition is interesting and requires further systematic study. Finally, the flow stress of annealed CS Al is lower than that of bulk 1100 Al, illustrating continued imperfect contact between splats. This follows from the interpretation of elastic behavior as well.

Thermal expansion after different annealing conditions as well as subsequent cycling was interpreted by Choi as follows (Choi, 2007).

> Annealing or thermal cycling in air allows oxidation at splat interfaces, thus driving irreversible expansion. As oxidation progresses (meaning with increasing cycle #, not with increasing temperature), CTE decreases overall, for three possible reasons: 1) oxide film becomes thicker, decreasing the reaction rate by limiting oxygen diffusion to Al, 2) cracks and interfaces begin to fill and close, holding splats more tightly (this agrees with interpretations of elastic modulus change with annealing) and 3) effective CTE can be described by a rule-of-mixtures and is thus decreased by the presence of oxide. Annealing in Ar produces a CTE similar to that of bulk Al, likely due to sintering of small pores and interfaces. However, when examined together with mechanical results above, this suggests that while open porosity is greatly reduced, isolated defects remain, but do not contribute to oxidation.

Electrical resistivity after annealing of CS Al-I remains anisotropic. To describe the post-annealing behavior, Choi made the assumption that annealing would presumably cause an overall *decrease* in resistivity within splats and that any increases may be attributed to inter-splat effects.

13.7 Conclusions

In this chapter, we have discussed a series of integrated property measurements for metallic CS coatings, in particular Al- and Ni-based alloys. Consideration of as-sprayed and heat-treated behavior under all tests provides a powerful supplement to stand-alone microstructural characterization. In addition, results for CS coatings were compared with those for HVOF and APS coatings, displaying some effect of thermal input on macroscopic behavior, particularly important for the higher-melting-point materials. All data were contrasted with bulk material behavior. At this point, further high resolution studies of CS interfaces, as well as *in-situ* macroscopic properties during heat treatment, are recommended for a more complete understanding.

13.8 Acknowledgments

The authors gratefully acknowledge support from the National Science Foundation through the MRSEC program (DMR 0080021) and the GOALI FRG program (CMMI 0605704). This work was made possible through collaborative efforts in CS sample preparation at Sandia National Laboratory (Dr. Richard Neiser and Dr. Aaron Hall) and through the Army Research Laboratory (Victor Champagne and Philip Leyman). Their respective contributions in enabling this characterization study are acknowledged. Finally, the authors are grateful to Atin Sharma for their assistance in electrical and thermal property measurements.

13.9 References

Balani K (2005), 'Effect of carrier gases on microstructural and electrochemical behavior of cold-sprayed 1100 aluminum coating', *Surface & Coatings Technology*, **195** (2/3), 272–279.

Choi W B (2007), 'Integrated characterization of cold spray aluminum coatings', *Acta Materialia*, **55** (3), 857–866.

Hall A (2005), 'The effect of a simple annealing heat treatment on the mechanical properties of cold-sprayed aluminium', *International Thermal Spray Conference*, Basel, Switzerland, ASM International, Materials Park, Ohio, 2005.

Johnson K (1985), *Contact Mechanics*, Cambridge University Press, New York.

Kapitza P (1930), 'The changes of the resistance from gold crystallisation in a magnetic field in very low temperatures and supraleite capability. Critical observations on the new work from Meissner and Scheffers', *Physikalische Zeitschrift*, **31**, 713–720.

Kulkarni A (2003), 'Processing effects on porosity–property correlations in plasma sprayed yttria-stabilized zirconia coatings', *Materials Science and Engineering A*, **359** (1/2), 100–111.

Lawn B R (2002), 'Overview: damage in brittle layer structures from concentrated loads', *Journal of Materials Research*, **17** (12), 3019–3036.

Lee K S (2000), 'Model for cyclic fatigue of quasi-plastic ceramics in contact with spheres', *Journal of the American Ceramic Society*, **83** (9), 2255–2262.

Li L (2004), 'Substrate melting during thermal spray splat quenching', *Thin Solid Films*, **468** (1/2), 113–119.

Liu Y (2006), 'Optimizing nonlinear properties of thermal spray coatings through process parameters', *MRS Fall*, Boston, Materials Research Society, Warrendale, Pennsylvania, 2006.

Matejicek J (1999), 'Quenching, thermal and residual stress in plasma sprayed deposits: NiCrAlY and YSZ coatings', *Acta Materialia*, **47** (2), 607–617.

Matejicek J (2002), 'Residual stress in sprayed Ni+5%Al coatings determined by neutron diffraction', *Applied Physics A: Materials Science & Processing*, **74**, S1692–S1694.

Matejicek J (2003a), '*In situ* measurement of residual stresses and elastic moduli in thermal sprayed coatings – Part 1: apparatus and analysis', *Acta Materialia*, **51** (3), 863–872.

Matejicek J (2003b), '*In situ* measurement of residual stresses and elastic moduli in thermal sprayed coatings – Part 2: processing effects on properties of Mo coatings', *Acta Materialia*, **51** (3), 873–885.

McPherson R (1984), 'Model for the thermal-conductivity of plasma-sprayed ceramic coatings', *Thin Solid Films*, **112** (1), 89–95.

Morgan R (2004), 'Analysis of cold gas dynamically sprayed aluminium deposits', *Materials Letters*, **58** (7/8), 1317–1320.

Oliver W C (1992), 'An improved technique for determining hardness and elastic-modulus using load and displacement sensing indentation experiments', *Journal of Materials Research*, **7** (6), 1564–1583.

Padture N P (1995), 'Fatigue in ceramics with interconnecting weak interfaces – a study using cyclic hertzian contacts', *Acta Metallurgica Materialia*, **43** (4), 1609–1617.

Pajares A (1996a), 'Contact damage in plasma-sprayed alumina-based coatings', *Journal of the American Ceramic Society*, **79** (7), 1907–1914.

Pajares A (1996b), 'Mechanical characterization of plasma sprayed ceramic coatings on metal substrates by contact testing', *Materials Science and Engineering A*, **208** (2), 158–165.

Papyrin A (2001), 'Cold spray technology', *Advanced Materials & Processes*, **159** (9), 49–51.

Prchlik L (2003), 'Deformation and strain distribution in plasma sprayed nickel-aluminum coating loaded by a spherical indenter', *Materials Science and Engineering A*, **360** (1/2), 264–274.

Sampath S (1999), 'Substrate temperature effects on splat formation, microstructure development and properties of plasma sprayed coatings. Part I: Case study for partially stabilized zirconia', *Materials Science and Engineering A*, **272** (1), 181–188.

Sampath S (2003), 'Development of process maps for plasma spray: case study for molybdenum', *Materials Science and Engineering A*, **348** (1/2), 54–66.

Sampath S (2004), 'Role of thermal spray processing method on the microstructure, residual stress and properties of coatings: an integrated study for Ni–5 wt.%Al bond coats', *Materials Science and Engineering A*, **364** (1/2), 216–231.

Sharma A (2006), 'Anisotropic electrical properties in thermal spray metallic coatings', *Acta Materialia*, **54** (1), 59–65.

Vaidya A (2005), 'An integrated study of thermal spray process–structure–property correlations: a case study for plasma sprayed molybdenum coatings', *Materials Science and Engineering A*, **403** (1/2), 191–204.

Van Steenkiste T H (2002), 'Aluminum coatings via kinetic spray with relatively large powder particles', *Surface & Coatings Technology*, **154** (2/3), 237–252.

Van Steenkiste T H (2005), 'Fracture study of aluminum composite coatings produced by the kinetic spray method', *Surface & Coatings Technology*, **194** (1), 103–110.

Wei L H (1996), 'Effect of mechanical damage on thermal conduction of plasma-sprayed coatings', *Journal of Materials Research*, **11** (6), 1329–1332.

Zhang H (2003), 'Numerical and experimental studies of substrate melting and resolidification during thermal spraying', *Journal of Materials Science & Technology*, **19**, 137–140.

14

Cold spray particle deposition for improved wear resistance

D. WOLFE and T. EDEN, Penn State University, USA

14.1 Introduction

This chapter discusses hard coating materials applied by the cold spray process with an emphasis on Cr_3C_2–25wt.%NiCr and Cr_3C_2–25wt.%Ni blended coatings applied on 4140 alloy for wear-resistant applications. Select material properties of various carbides and cold spray deposition processing are also presented. The chapter discusses the improvements in hard, wear-resistant C_3C_2-based coating properties and microstructure through changes in nozzle design, powder characteristics, stand-off distance, powder feed rate, and traverse speed which resulted in an improved average Vickers hardness number (VHN) and wear resistance comparable with some thermal spray processes. The cold spray process optimization of the C_3C_2-based coatings resulted in increased hardness and improved wear characteristics with lower friction coefficients. The improvement in hardness is directly associated with higher particle velocities and increased densities of the C_3C_2-based coatings deposited on 4140 alloy at ambient temperature. Select coating microstructure and properties using optical microscopy, scanning electron microscopy, hardness, and ball-on-disc tribology experiments for friction coefficient and wear determination will be discussed. The presented results strongly suggest that cold spray is a versatile coating technique capable of tailoring the hardness of C_3C_2-based wear-resistant coatings on temperature-sensitive substrates.

14.2 Current state of the art

The current state of the art is rapidly changing. To date, most cold spray depositions have been made with nitrogen and have focused mainly on aluminum, copper, zinc, and other pure materials due to the limitations in the powder material's ability to plastically deform under certain deposition conditions.[1–10] Ceramics typically do not plastically deform to the extent that metallic materials do, making deposition of 100% ceramic materials at low temperature very difficult. Presently no commercial carbide wear-resistant cold spray coatings are available, but this should change in the next few years with the rapid advancements being made in the cold spray technology areas. The authors believe that cold spray processes will eventually excel with niche coating applications, such as corrosion-resistant coatings, sealants, wear and repair technology, in addition to microelectronic and biomedical applications. The Applied Research Laboratory of the Penn State University (ARL-PSU) has worked extensively on chromium carbide-containing blend materials for tailoring the wear resistance of the material for a given application using nitrogen as the precursor gas. Further enhancements and improvements are expected by using helium and WC–Co as it becomes more cost effective and efficient.

14.3 Cold spray wear-resistant material properties

In choosing the proper coating materials, one should also look at the feasibility of meeting application goals with regards to wear resistance, erosion, adhesion, ease of deposition, and deposition method feasibility. In addition, a materials selection should be performed weighting the parameters of interest, whether it is intrinsic hardness for improved abrasion resistance or low density for aerospace applications. Select material properties are listed in Table 14.1 for some carbide materials.

Table 14.1 Select properties of refractory metal carbides and binder materials[1,2,11]

Material	Hardness (VHN)	Crystal structure	Melting temperature (°C)	CTE (μm/m-C)	Elastic modulus (GPa)	Density (g/cm^3)
Carbide						
Cr_3C_2	1400	Orthorhombic	1800	10.3	373	6.66
TiC	3000	Cubic	3100	7.7	451	4.94
VC	2900	Cubic	2700	7.2	422	5.71
HfC	2600	Cubic	3900	6.6	352	12.76
NbC	2000	Cubic	3600	6.7	338	7.80
ZrC	2700	Cubic	3400	6.7	348	6.56
WC	2200 (0001) 1300 (1010)	Hexagonal	2800	5.2 (0001) 7.3 (1010)	696	15.63
TaC	1800	Cubic	3800	6.3	285	14.50
Mo_2C	1500	Hexagonal	2500	7.8	533	9.18
Binder						
Ni	<100	Cubic	1455	15.0	207	8.9
Co	<100	Cubic/ hexagonal	1495	16.0	207	8.9

VHN, Vickers hardness number; CTE, coefficient of thermal expansion.

14.3.1 Carbide properties[12–22]

1. Carbides generally have hardness values of 2000–3000 VHN.
2. Transition metal carbides have high melting temperatures, with the highest being HfC (3890 °C). Generally, the carbides of transition metal elements have higher melting temperatures than the parent transition metal elements, the exceptions being the group VI carbides (Cr_3C_2, MoC, and WC).
3. Probably the most important aspect of transition metal carbides lies in the fact that they have defect structures. Stoichiometry within these groups is rare. More common are deviations from stoichiometry. Carbide phases often exist over a wide range of compositions with appreciable vacancy concentrations. Up to 50 at.% vacancies on the non-metal lattice have been reported with fewer vacancies on the metal lattice sites. It should be noted that even though a carbide phase may be stoichiometric, a large vacancy concentration of the order of a few atomic percent could exist. In addition to the compositional variations that affect the physical properties, these vacancy concentrations also play an important role in determining the magnetic, mechanical, physical, and thermal properties of the materials.
4. Most carbides have high chemical stability at room temperature; the exception being VC which slowly decomposes in air. Around 450 °C, TiC oxidation starts. Oxygen reacts with TiC and forms a non-protective TiO_2 layer of anatase. Between 700 and 1000 °C, a protective rutile phase forms resulting in a parabolic oxidation rate that is determined by the diffusion of oxygen through the oxide layer. The oxidation rates of titanium and carbon are nearly the same at 1000 °C.
5. Transition metal carbides generally possess high values of Young's modulus (276–600 MPa). Most transition metals have values of 138–276 MPa. Carbides are generally very brittle at low temperatures. However, at temperatures greater than or equal to 1000 °C, they undergo a brittle-to-ductile transition, above which they plastically deform. As a result, these brittle polycrystalline carbides can become ductile. This phenomenon is reported to be the underlying factor which gives TiC one of the highest compressive yield strengths (248 MPa) of any material.
6. Transition metal carbides also exhibit metallic optical, magnetic, and electrical properties, even though they are considered ceramics. Generally, these properties differ only slightly from the parent transition metal element. However, due to the defect structure, large variations can occur in the electrical and thermal conductivities as a function of temperature depending on the processing.
7. Bonding in the transition metal carbides consists of a complex mixture of the metal-to-non-metal and metal-to-metal interactions found in

covalent and metallic bonding. It should be noted that a small amount of ionic bonding also occurs. The arrangements of the atoms are generally octahedrally coordinated with the non-metal atom surrounded by six metal atoms in an octahedral arrangement. This atom orientation also results in increased metal-to-metal bonding.

14.3.2 Titanium carbide

As with most carbides, TiC has a wide range of properties similar to both metallic and ceramic materials.[23–33] For example, TiC has very high hardness and strength, similar to ceramics, but still maintains the very good electrical and thermal conductivity associated with the parent metal. The unique properties of TiC are derived from its complex bonding nature. There is still controversy over the exact bonding mechanisms in TiC, but general agreement exists on the fact that it is complex.[23–28] The generally accepted bonding scheme for TiC is that it is a combination of metallic, covalent, and ionic bonding. Since TiC is a defect structure, the bonding schemes change depending on structure and composition. This is often why such a large range of values is found in the literature for TiC.

Because of the wide range of properties for TiC, it is used extensively in industry as a secondary carbide in WC–Co cutting inserts,[29] coatings for fusion-reactor applications,[30] blow plungers for the glass industry,[31] coatings for stamping and extrusion, pump shafts in the chemical industry for wear and corrosion resistance,[27–30] armor wall tiles,[32] and wear-resistant coatings for cutting tools.[33–39]

TiC has a B1 NaCl structure as shown in Fig. 14.1. Each titanium atom is surrounded by six carbon atoms and each carbon atom is surrounded by six titanium atoms in a perfect lattice. TiC is often discussed as an interstitial

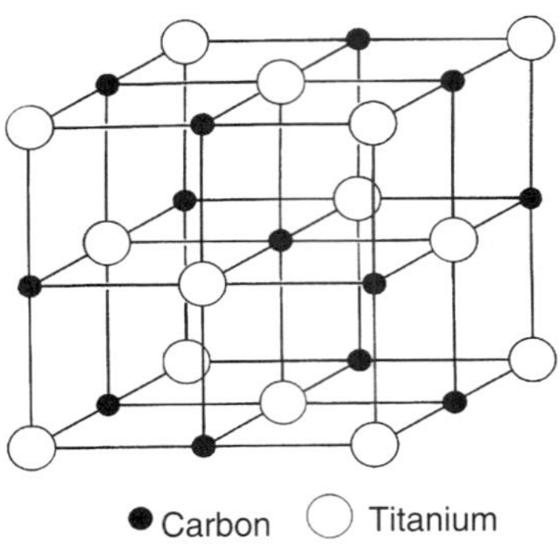

14.1 Structure of titanium carbide (cubic) showing the positions of the carbon and titanium atoms.

carbide where titanium occupies a close-packed structure and the carbon is located on specific interstitial sites (i.e. a titanium metallic lattice with carbon interstitials). As a result, the interatomic spacing between the titanium atoms and the carbon atoms determines the location of the interstitials and the stoichiometry is determined by the population density on the metallic lattice sites and the carbon interstitial sites, both of which dictate the material properties. It is this interstitial atom arrangement that gives rise to the combination of the metallic, covalent, and ionic bonding. As a result, TiC is brittle and hard whereas titanium metal is soft and malleable. The melting temperature for TiC is much higher than for pure titanium metal. These major differences between the carbides and the parent metal suggests that the bonding is much stronger in the carbide phase, particularly the metal-to-carbon bond.

14.3.3 Chromium carbide

Unlike TiC, chromium carbide has been much less studied.[40–51] Very little information regarding the various phases and properties of chromium carbides has been characterized and reported in the literature. Similar to β-titanium, chromium metal has the bcc (body centered cubic) structure. As a result, the same arguments for interstitial carbon apply, i.e. a change to a close-packed structure. There are three main phases for chromium carbide which are Cr_3C_2, Cr_7C_3, and $Cr_{23}C_6$ (as shown in Fig. 14.2).[39] The most common reported phase for coatings is Cr_3C_2, which is orthorhombic (a = 0.283, b = 0.554, and c = 1.1470). Cr_7C_3 has an hcp (hexagonal close-packed) structure with lattice parameters of a = 1.398 and c = 0.4523. $Cr_{23}C_6$ has a complex face centered cubic structure with 116 atoms per unit cell, which results in its large lattice parameter of 1.066 nm, and thus the lower melting temperature as compared with pure chromium (1875 °C). A schematic drawing of the Cr_3C_2 and $Cr_{23}C_6$ structures is illustrated in Fig. 14.2, showing the complexity of the atomic arrangement. The hardness values for Cr_3C_2, Cr_7C_3, and $Cr_{23}C_6$ are 1300, 1600, and 1000 kg/mm^2, respectively.[1,2]

Unlike TiC, which has a higher melting point than Ti, all of the chromium carbides have lower melting temperatures than pure chromium (1865 °C). Within the chromium carbide structure, both Cr–Cr and Cr–C bonds give rise to the unique properties of chromium carbide. However, unlike TiC, Cr_3C_2 has very weak Cr–C bonding and very strong Cr–Cr bonding. The electronegativity difference for C–Cr is ~0.9, which suggests a lower degree of ionic bonding than in TiC.

One of the advantages of cold spraying chromium carbide-containing components is that no melting or dissolution of the carbide phase occurs resulting in improved high temperature performance, as other chromium phases do not form.

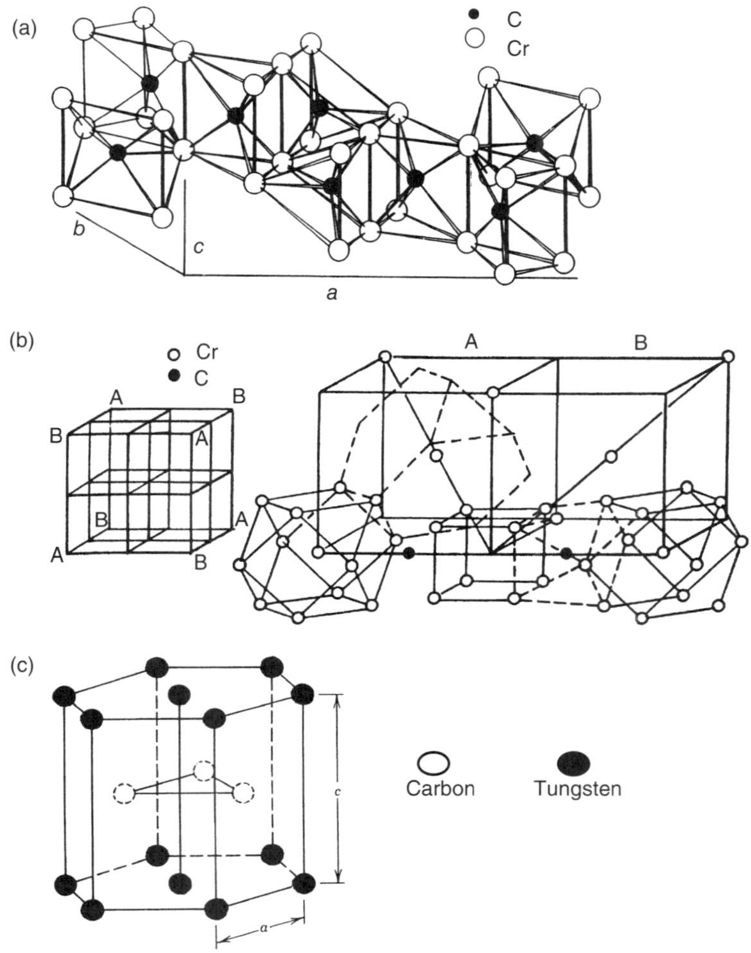

14.2 Illustration of (a) Cr_3C_2 (orthorhombic), (b) $Cr_{23}C_6$ (cubic), and (c) WC hexagonal crystal structures.

14.3.4 Tungsten carbide

WC has a hexagonal crystal structure (Fig. 14.2(c)) with a high melting point of 2800 °C, but has a relatively high density of 15.63 g/cm^3. The primary use of WC is in the manufacture of cemented carbides for the cutting tool industry or for wear-resistant applications. A cemented carbide material is fabricated by 'cementing' hard carbides, primarily WC, in a binder matrix using cobalt metal via liquid phase sintering. The excellent wetting characteristic of Co to WC and the high solubility of WC in Co at elevated temperatures results in excellent densification resulting in a high toughness, high strength, and high hardness material. Adding TiC and TaC to WC results in increased high temperature wear resistance and hot hardness and oxidation stability, and these materials are superior to high-strength steels in most machining applications in the cutting tool industry. In addition, the WC grain size is very important in determining the hardness, strength, thermal conductivity, toughness, and abrasion resistance. Ultrafine submicron WC-based systems are becoming more and more important for cutting tools, chipless forming, and wood cutting.

In addition, properties of representative cobalt-bonded cemented carbides are given in Table 14.1, showing a wider range of material properties as a function of nominal composition and grain size; these properties are discussed in more detail elsewhere.[11] The binder content can have a significant role in the performance in the coating material, especially in cold spray technology. One of the anticipated future thrusts in powder development will be the metallic coating of hard carbide particles for use and incorporation into cold spray processes. The more uniformly distributed metallic material around the particle will further increase the volume fraction and density of these coating systems.

Recently, WC–Co-based coatings have been applied by the cold spray process using both nitrogen and helium gas. The added advantage of using cold spray for WC–Co is that the cold spray process can eliminate decarburization, reduction, and oxidation reactions with H_2 (plasma spray) and the dissolution of WC and Co typically observed in the thermal spray and high-velocity oxygen fuel (HVOF) techniques that result in $Co_xW_yC_z$, W, WO_3, and W_2C phases. Kim *et al.*[52] showed that WC–Co nanostructured coatings with low porosity and average VHN greater than 2000 VHN could be deposited using cold spray.

14.4 Cold spray wear-resistant processing

Using the unique, flexible cold spray facility at the ARL-PSU, various Cr_3C_2-based coatings were applied on 25.4 mm × 25.4 mm × 6.35 mm 4140 alloy steel. The feedstock powder was baked out in an oven at 65 °C to remove moisture and improve powder flowability. The powder was then weighed, and placed in a Praxair (model number 1264 HPHV) powder feeder. Prior to spraying the coatings, the 4140 substrates were grit blasted (Trinco Dry Blast) using 16 to 20 grit aluminum oxide powder (Trinco) to remove surface oxides and create a minimum surface finish of $R_a \simeq 5.08\,\mu m$. In order to increase the mechanical bonding between the cold sprayed coating and the substrate, a minimum surface roughness is desired but depends on substrate hardness. The 4140 alloy was then ultrasonically cleaned in ethyl alcohol to remove dirt and oil. The samples were then measured and weighed before being positioned in the tooling assembly with the desired stand-off distance. A final ethyl alcohol rinse followed with compressed air drying was performed just prior to coating deposition to minimize contamination.

Based on the powder formulation, adjustments were made to the powder flow, gas temperature, nitrogen gas pressure, stand-off distance, and traverse speed in an effort to optimize the coating density, and thus the average VHN. Typical cold spray processing parameters for select Cr_3C_2-based coatings are listed in Tables 14.2 and 14.3. Details regarding the powder characteristics (i.e. particle size, morphology, powder size distribution, and density) are discussed in more detail in Section 14.5.

Due to the hardness of the Cr_3C_2 powders, the stainless steel nozzle typically used in aluminum cold spray applications could not be used. Therefore, a proprietary WC–Co nozzle was designed and fabricated to improve the coating process. The particle velocity calculations using the Sandia code,[53,54] suggested a 25% increase in particle velocity based on the nozzle design. The higher particle velocity resulting from the improved nozzle design resulted in increased coating density, and thus increased the average VHN of ~20% discussed later.

The ARL-PSU cold spray unit's nozzle and gas heater are mounted on a six-axis robot enclosed in an acoustic room to minimize noise levels. In addition, the robot uses a multi-axis turntable (capable of holding up to 100 kg) for coating complex geometries. The gas and overspray that flow through the nozzle were collected inside the acoustic room and carried to

Table 14.2 Typical cold spray processing parameters

Gas type	T_g (°C)	P_g (MPa)	Traverse rate (mm/s)	Stand off distance (mm)	Powder flow (g/min)	Coating thickness (μm)
N_2	500	2.41	19–54	12.5–25.4	5.67–7.94	>1000

Table 14.3 Sample series number and powder type description

Sample number	Powder type and supplier
C1	100% nickel (Novamet)
C2	PSU blend #1: Cr_3C_2 (Cerac)–25wt.%Ni (Novamet)
C3	PSU blend #2: Cr_3C_2 (Cerac)–25wt.%Ni (Novamet)
C4	PSU blend #9: Cr_3C_2 (AEE)–18wt.%Ni (Novamet)
C5	PSU blend #3: Cr_3C_2 (AEE)–15wt.%Ni (Novamet)
C6	TAFA 1375V: Cr_3C_2–25wt.%NiCr
C7	Praxair 1375VM: Cr_3C_2–25wt.%NiCr

a dust collector trapping all particulates down to the submicron level. To collect and recycle powders, a cyclone separator could be positioned in-line. Nitrogen gas at pressures up to 1.6 MPa was supplied by vaporizing liquid nitrogen from a large tank. When higher nitrogen pressures were required, a booster pump was used to increase the pressure up to 3.1 MPa.

The coatings were applied to a minimum thickness of 1000 μm. After the desired coating thickness was applied, the cold spray unit was turned off, the samples were allowed to cool, and then measurements and weights were taken. Prior to the tribology testing, the cold spray coatings were machined to approximately 890 μm in thickness with a surface finish of $R_a = 0.61–1.78\,\mu m$.

Lastly, preliminary investigation of combining the cold spray process with laser processing in an effort to further improve the average VHN and thus wear-resistant properties of the coating were investigated. The results strongly suggest that a hybrid technique could substantially improve the quality of cold spray wear-resistant coatings. Selective results will be presented.

14.5 Characterization

The surface finish of the coatings varied for the selected coating due to the type of powder and coating microstructure (i.e. porosity). Using a portable Federal Pocket Surf® III surface roughness instrument, the surface finish of each sample was measured and recorded. One sample from each lot was used for microstructural investigation. Using a LECO diamond wafering saw, sections of the samples were made and mounted in epoxy for coating cross-sectional analysis. Select sample cross-sections were then examined using optical microscopy and scanning electron microscopy. A variety of characterization techniques were used to determine the average VHN, particle size distribution, particle density, select bond strength, and tribological properties (i.e. wear rate and friction coefficients). The Vickers hardness measurements were made using a LECO model M-400-G1 hardness tester with a 300 g load on the polished cross-section of the coatings. A minimum of 10–15 indentations were made for each specimen with the average value being reported. Normal Bragg–Brantano ($\theta/2\theta$) X-ray diffraction continuous scans were performed using $CuK_{\alpha 1}$ radiation = 0.154056 nm, $CuK_{\alpha 2}$ = 0.154056 nm wavelength, over the range of $2\theta = 30°$ to $80°$ at 2 s per step with a step size of 0.030° to determine crystallographic orientation and phase analysis.

14.5.1 Wear resistance

Before proceeding, it is important to describe briefly the various types of wear and to explain why it is so important to design material systems with improved wear resistance. The mechanisms of wear are very complex and the authors refer the readers to the *ASM Handbook*[11] for a more detailed description of the mechanisms of wear. A brief description is presented below. In general, it should be understood that the real area of contact between two solid surfaces compared with the apparent contact area is quite small and is limited to the contact points between surface asperities.

- *Wear*: defined as material loss from a surface due to some mechanical action.
- *Abrasive wear*: defined as the displacement or loss of material from a surface due to hard particles (or protuberances) forced along and moving against a solid surface.
- *Erosion*: material loss (wear) due to a mechanical interaction between the material surface and solid particles, liquid fluid, or mixed combination.
- *Fatigue wear*: wear of a solid surface generally caused by fracture arising from fatigue of the material.
- *Fretting wear*: wear resulting from a small amplitude oscillatory motion between two solid surfaces in contact.
- *Cavitation erosion*: progressive loss of material from a solid surface due to formation and collapse of bubbles or cavities within the liquid that contain gas, vapor, or a mixture of the two.

Typically, under wear applications, the contact area between the two opposing surfaces is very small due to surface asperities. As a result, localized loads in these regions can be very high before transferring the load throughout the contact points. Intrinsic surface properties such as ductility, work hardening, strength, and hardness become very important factors for wear resistance, but equally important and in some cases more important are additional factors such as contact speed, load, corrosion, temperatures, surface finish, etc. (component working environment). In general, increased wear resistance occurs with increasing hardness values, but it is important to consider the working environment, wear characteristics, and the material's microstructure as these will have a profound effect on the coating performance. In the cold spray process, as the particles impinge upon the substrate material, coating build-up relies upon plastic deformation. Hard, ceramic particles such as chromium carbide, titanium carbide, and tungsten carbide will not plastically deform to allow coating build-up. Therefore, a softer binder material is added for coating deposition. The most common Cr_3C_2-based coating formulation is Cr_3C_2–25wt.%NiCr. Various aspects of the cold spray process are discussed – including substrate surface finish, processing parameters (Table 14.2), and powder characteristics – in an effort to better understand the role of various factors on the average VHN of the Cr_3C_2-based coatings and wear-resistance performance. The primary factors that affect particle velocity (and thus coating density and hardness) are: increased gas temperature, increased pressure, type of gas (use helium versus nitrogen), powder characteristics (particle size, morphology, size distribution, and density), nozzle design, stand-off distance, powder feed rate, traverse speed, and type and volume of wear-resistant particles (i.e. Cr_3C_2 versus WC).

Vickers hardness

As previously discussed, the most common parameter used to describe a material's resistance against wear is the hardness of the material. Hardness is defined as a material's resistance against plastic deformation, with ceramic materials generally having a higher theoretical hardness value than metallics. However, the hardness of the material is heavily influenced by variations in composition, crystallographic orientation and phase, and microstructure of the coating often dictated by the coating deposition process. Therefore, since the hardness of a material is defined as the material's resistance to plastic deformation, it is very apparent that the softer materials such as aluminum, copper, and zinc can be applied by the cold spray process with 100% densities whereas applying ceramic–metallic blends is a little more challenging. The hardness of a material has often been correlated with the material's tensile strength whereas the modulus of elasticity affects the material's resistance to deformation. Similarly, the hardness, surface roughness, and the material's chemical affinity are important aspects that influence friction.

Preliminary cold spray trials resulted in average VHN for Cr_3C_2–25wt.%NiCr in the low to high 300 $VHN_{0.300}$ range. The lower than expected hardness values were attributed to the increase in coating porosity (i.e. lower coating density). Since hardness is often correlated with wear resistance, cold spray process conditions and powder characteristics (including composition) were examined in an effort to increase the coating density, and thus VHN. Processing parameters such as stand-off distance, traverse rate, powder feed rate, and nozzle design were investigated for various Cr_3C_2-based powders C3, C4, C5, and C6 and these are shown in Fig. 14.3(a). The average VHN for the uncoated 4140 alloy and pure nickel (used in the powder blends) are included for reference (Fig. 14.3(a)).

Trends were observed as a function of the various processing parameters. In general, the average VHN increased with decreasing stand-off distance, decreased traverse rate, and improved nozzle design. Using the newly designed WC–Co nozzle, a 17% increase in the VHN was obtained and was attributed to an increase in particle velocity resulting in increased density. Using the Sandia code, the Cr_3C_2–25wt.%NiCr particle velocity was

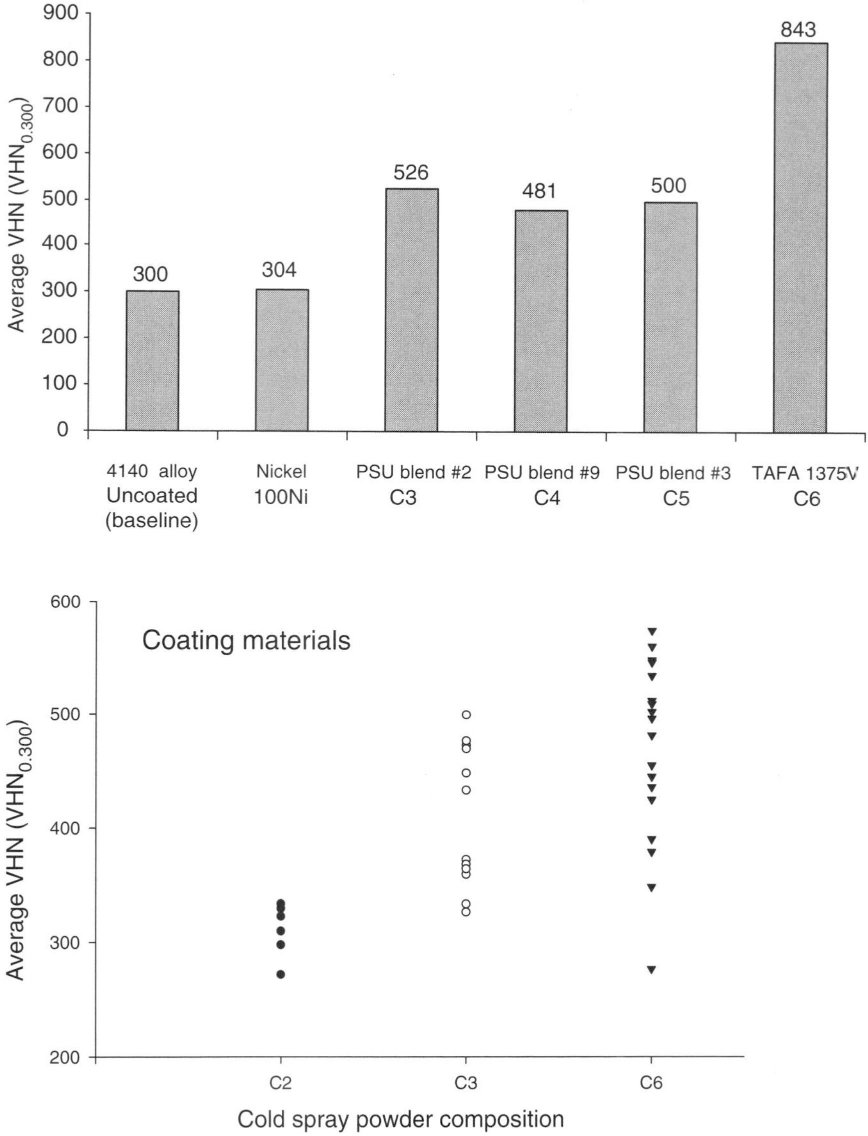

14.3 (a)Average VHN ($VHN_{0.300}$) for uncoated and coated nickel, agglomerated and sintered (C6: TAFA 1375V(Cr_3C_2-NiCr)), and blended (Cr_3C_2-Ni: C3, C4, and C5) coatings applied by the cold spray process. (b) Tailored average VHN ($VHN_{0.300}$) for cold spray coatings C2 (272–334 $VHN_{0.300}$), C3 (327–504 $VHN_{0.300}$), and C6 (277–575 $VHN_{0.300}$) deposited under various conditions showing a wide range of tailored hardness values.

calculated as a function of the nozzle inlet pressure (MPa) for three of the ARL-PSU nozzle designs. As shown in Fig. 14.4, almost a 25% (100 m/s) increase in the particle velocity was obtained with the newly designed nozzle. In addition, a 37% increase in the average VHN was obtained by decreasing the nozzle traverse speed from 54 mm/s to 19 mm/s. Lastly, a 29% improvement in VHN was observed by decreasing the stand-off distance from 25.4 mm to 19 mm. The optimal processing conditions for TAFA 1375V powder using the 3 mm throat/153 mm length WC–Co nozzle were determined to be: gas temperature, 500 °C; gas pressure, 2.41 MPa; stand-off distance, 19 mm; traverse speed, 19 mm/s; powder feed rate, 5.67–7.94 g/min. A 57% improvement in VHN (549 $VHN_{0.500}$) for Cr_3C_2–25wt.%NiCr coatings deposited on 4140 alloy is realized through these optimizations.

In a parallel effort, pure nickel (soft matrix) and Cr_3C_2 (wear resistant) powders were mixed 'in-house' (PSU blend: Cr_3C_2–25wt.%Ni) in order to further improve cold spray coating performance. Two different particle size distributions (PSU blend #1 (C2) and PSU blend #2 (C3)) of Cr_3C_2 powder were mixed in an attempt to increase the volume fraction of hard particles (Cr_3C_2) incorporated within the cold spray coating. The hardness of the PSU blend coating was increased by 29% up to 373 $VHN_{0.500}$. Although the coating density was significantly improved over the pre-alloyed TAFA 1375V powder, the average VHN remained low due to insufficient hard particle incorporation. Comparison of the PSU blend and TAFA 1375V coating cross-sections shows a larger volume percent of Cr_3C_2 in the TAFA 1375V as discussed in Section 14.8.

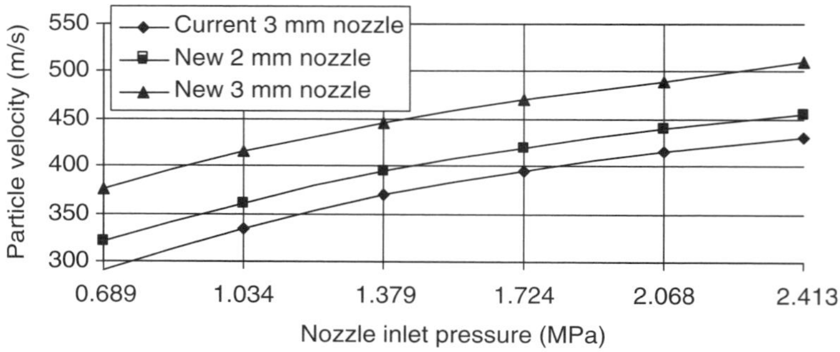

14.4 Particle velocity versus nozzle inlet pressure for Cr_3C_2–25wt.%Ni at 500 °C. Particle velocity predicted by a one-dimensional gas dynamic model. The model predicts a higher velocity for a longer nozzle.

It was determined that the VHN could be tailored to a particular number. Figure 14.3(b) shows selective average VHN values for compositions C2, C3, and C6 for samples deposited under a variety of cold spray conditions (not listed). As seen in Fig. 14.3(b), the VHN can be controlled over a wide range of hardness values which allows tailorability of the coating hardness values for specific applications in which the mating surface may have to be matched in hardness value. Under the parameters studied, the $VHN_{0.300}$ values ranged from 272–334 $VHN_{0.300}$, 327–500 $VHN_{0.300}$, and 277–575 $VHN_{0.300}$ for C2, C3, and C6, respectively. Further refinements were also made in which the $VHN_{0.300}$ was further increased for some of the various powders investigated (i.e. 843 $VHN_{0.300}$ for TAFA 1375V).

14.6 Powder characteristics

Various powder formulations were sprayed in an effort to tailor the average VHN of cold spray coatings and to determine the maximum value for several blended formulations. Table 14.4 briefly describes the chemical composition and production method of the various powders used in this investigation. Several 'off-the-shelf' powders were purchased and investigated for use in applying wear-resistant coatings. These powders were met with mixed success, which led to the preparation of various powder blend formulations in order to improve the wear-resistant properties of the cold spray coatings. Table 14.5 lists the various powders, powder size distributions, densities, and mean particle diameters used in applying the different coatings. Over five different Cr_3C_2–25wt.%NiCr powders were purchased with different particle size distributions and average particle diameters. With the exception of TAFA 1375V, most of the 'off-the-shelf' powders did not build up well. In order to better understand the poor coating build-up, the powders were investigated for density, size distribution, morphology, and average diameter. In addition, in order to achieve a wear-resistant coating for future developmental efforts with and without solid self-lubricating capabilities, several formulations were sprayed using Novamet pure nickel powder and chromium carbide from AEE. Table 14.5 briefly lists the major powder characteristics.

One of the biggest disadvantages to date regarding the cold spray technology is the lack of understanding and characterization of the feedstock (i.e. powders). Several companies have invested heavily in the processing and characterization of select powders for HVOF, high-velocity air fuel (HVAF), and other plasma spray technologies. As a result, significant improvements have been made in coating quality and property performance. As the cold spray process matures, the authors believe that powder vendors will soon develop specialty powders with specific powder characteristics that aid in improving the deposition efficiency of the cold spray

Table 14.4 Chemical composition and production method of cold spray powder feedstock

Powder	Powder supplier	Primary composition	Production method	Chromium	Carbon	Nickel
Ni (C1)	Novamet	100% nickel	Thermal decomposition	0 ppm	430 ppm	Balance
Cr_3C_2 (–325 mesh)	Cerac	100% Cr_3C_2	Crushed grinding and sintered	86.43wt.%	13.16wt.%	NA
Cr_3C_2 (–200/+325 mesh)	Cerac	100% Cr_3C_2	Crushed grinding and sintered	86.53wt.%	12.87wt.%	NA
Cr_3C_2	AEE	100% Cr_3C_2	Sintered	86.45wt.%	13.28wt.%	NA
TAFA 1375V (C6)	TAFA	Cr_3C_2–25wt.%NiCr	Agglomerated and sintered	Balance	ND	ND
Praxair 1375V	Praxair	Cr_3C_2–25wt.%NiCr	Agglomerated and sintered	Balance	10wt.%	19wt.%
PSU blend #1 (C2)	Cerac/Novamet	Cr_3C_2–25wt.%Ni	Blended	Balance	9.96wt.%	25wt.%
PSU blend #2 (C3)	Cerac/Novamet	Cr_3C_2–25wt.%Ni	Blended	Balance	9.96wt.%	25wt.%
PSU blend #3 (C5)	AEE/Novamet	Cr_3C_2–15wt.%Ni	Blended	Balance	11.29wt.%	15wt.%
PSU blend #9 (C4)	AEE/Novamet	Cr_3C_2–18wt.%Ni	Blended	Balance	10.89wt.%	18wt.%

NA, not applicable; ND, not determined.

Table 14.5 Select powder characteristics

Powder	Supplier	Composition	$D_{(v,0.1)}$ (μm)	$D_{(v,0.5)}$ (μm)	$D_{(v,0.9)}$ (μm)	PSD	ρ (g/cm^3)	$\rho_{theoretical}$ (%)
Ni	Novamet	100% nickel	7.46	15.72	37.15	Broad	8.4886	95.6
Cr_3C_2	AEE	100% Cr_3C_2	45.13	65.35	93.15	Narrow	5.5496	82.0
TAFA 1375V	TAFA	Cr_3C_2–25wt.%NiCr	18.83	35.36	59.08	Broad	7.2947	>99.0
Praxair 1375V	Praxair	Cr_3C_2–25wt.%NiCr	21.36	34.22	52.64	Narrow	6.6531	92.3
Praxair 1375VM	Praxair	Cr_3C_2–25wt.%NiCr	21.34	33.25	51.71	Narrow	6.6955	93.4
PSU blend #9	Novamet/AEE	Cr_3C_2–18wt.%Ni	8.71	51.87	91.26	Broad	7.2737	>99.0

PSD, particle size distribution.

process, as well as the coating performance. Therefore, the authors believe a more detailed investigation of the powders is warranted in an effort to help explain the differences in coating build-up and performance.

14.6.1 Nickel powder (C1)

Figure 14.5 shows scanning electron micrographs of the morphology of select powders used for the various coatings applied in this investigation. As shown in Fig. 14.5(a) and (b), the nickel powder (Novamet) prepared by thermal decomposition is categorized as being slightly porous. The nickel powder's measured density (8.4886 g/cm^3) was determined to be 95.6% of its theoretical density and performed well with the cold spray process. The Novamet nickel powder used resulted in nickel coatings that were easy to build up with high coating densities. The high particle velocity of the impinging powders on the substrate surface resulted in the collapse of porosity within the nickel powder due to the plastic deformation, forming a very dense coating. The nickel powder was characterized as having a broad particle size distribution (Fig. 14.6) with a mean particle diameter ($D_{(v,0.5)}$) of 15.72 μm as measured by the laser light scattering technique.

14.6.2 PSU blend #1 and #2: Cr_3C_2–25wt.%Ni (C2–C3)

Figure 14.5(c) and (d) shows the morphology of PSU blend #1: Cr_3C_2–25wt.%Ni. The powder blend is a mixture of nickel (Novamet) and Cr_3C_2 (Cerac, Inc.) powder. The Cr_3C_2 powder is characterized as being very dense, with a sharp faceted morphology. The high density of the Cerac Cr_3C_2 powder is believed to be one of the limiting factors with attempts at increasing the average VHN of this powder blend. The Cr_3C_2 (Cerac) powders were too dense resulting in poor incorporation within the cold spray coating. The high density of the Cr_3C_2 powders resulted in the powders being deflected (bouncing off) from the surface rather than incorporated into the cold sprayed coating. Although this resulted in coating densification, a limited volume fraction of Cr_3C_2 was being incorporated into the coating, resulting in the lower than expected hardness. PSU blend #2 (C3) differed from PSU blend #1 (C2) in that it had a larger particle size (–200/+325) distribution with the same high particle density.

14.6.3 PSU blend #3 (Cr_3C_2–15wt.%Ni (C5)) and #9 (Cr_3C_2–18wt.%Ni (C4))

The powder morphology of PSU blend #3 Cr_3C_2 (AEE)–15wt.%Ni (Novamet) is shown in Fig. 14.5(e) and (f). The Cr_3C_2 powder from AEE is characterized as being slightly more porous than the Cr_3C_2 from Cerac, Inc. with a broad

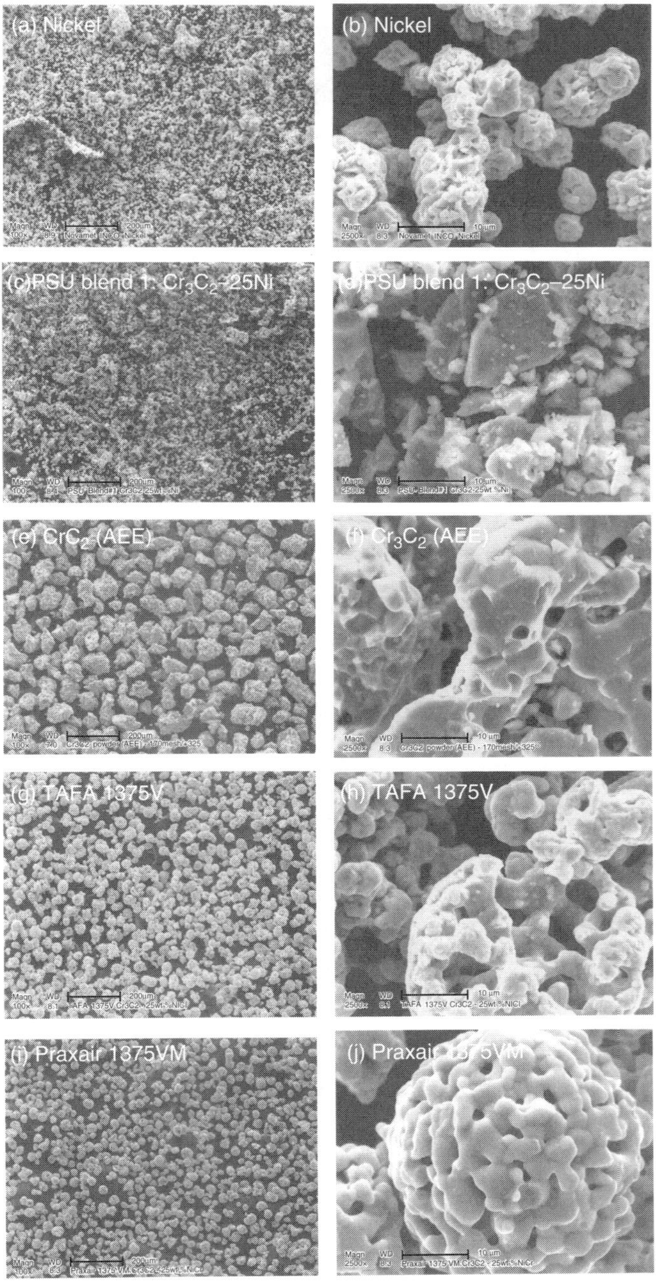

14.5 Scanning electron microscopy showing the morphology of: (a, b) nickel; (c, d) PSU blend #1 – Cr_3C_2 (Cerac)–25wt.%Ni; (e, f) Cr_3C_2 (AEE) powder used in blends (C4, C5); (g, h) TAFA 1375V: Cr_3C_2–25wt.%NiCr; (i, j) Praxair 1375VM: Cr_3C_2–25wt.%NiCr powders used in applying cold spray coatings on 4140 alloy.

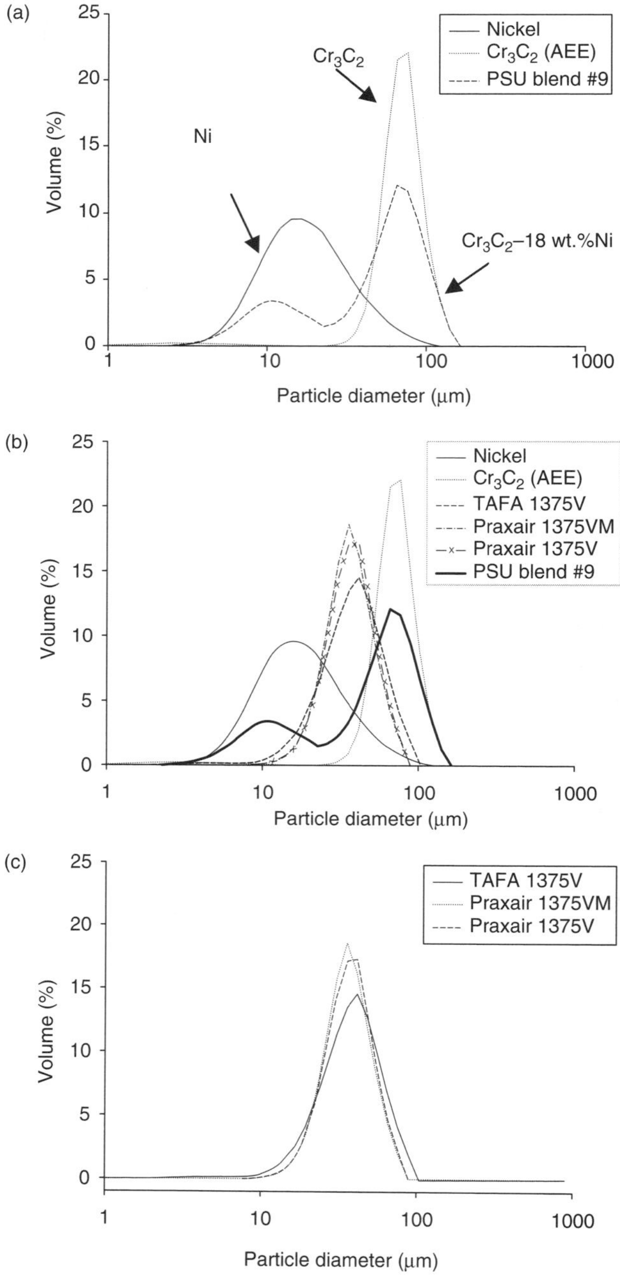

14.6 Particle size distributions for: (a) nickel, Cr_3C_2 (AEE), and Cr_3C_2 (AEE)–18wt.%Ni (PSU blend #9); (b) powders used for blends; (c) TAFA 1375V, Praxair 1375V, and Praxair 1375VM powders.

particle size distribution and $D_{(v,0.5)}$ of 65.35 μm. The AEE Cr_3C_2 was determined to be 5.5496 g/cm^3 which is approximately 82% of the theoretical density for Cr_3C_2. The lower particle density combined with larger particle size resulted in increased volume concentrations of Cr_3C_2 particles incorporated into the various cold spray coating blends and thus slightly higher VHN values. The smaller nickel powder size combined with larger open pores within the AEE Cr_3C_2 powder allowed the nickel powder to become incorporated (penetrate) within the open structure of the Cr_3C_2 where it plastically deformed, resulting in increased Cr_3C_2 deposition efficiency and thus increased Cr_3C_2 volume fractions and VHN values. In addition, upon impact, the porous Cr_3C_2 powder absorbed the impact energy by densifying and fracturing, reducing the number of Cr_3C_2 particles 'bouncing off' the surface of the substrate. However, due to the powders being blends and not chemically reacted, there was still a maximum Cr_3C_2 volume incorporation into the coating that could be obtained regardless of the binder concentration. The maximum Cr_3C_2 volume fraction was obtained with 15wt.%Ni binder (C5). This is primarily the result of the hard chromium carbide powders not being completely surrounded by the soft metallic binder phase (nickel) as is often observed with agglomerated and sintered powders.

14.6.4 TAFA 1375V: Cr_3C_2–25wt.%NiCr (C6)

The highest values for average VHN were obtained for the TAFA 1375V powder, and this is primarily the result of the powder characteristics. The particle size distribution (Fig. 14.6 and Table 14.4) is characterized as being slightly broader than the other 'off-the-shelf' powders with a $D_{(v,0.5)}$ of 35.36 μm. The measured density of the TAFA 1375V was determined to be 7.2947 g/cm^3 which is greater than 99% of the theoretical value of Cr_3C_2–25wt.%NiCr. The morphology of the TAFA 1375V agglomerated and sintered powder is shown in Fig. 14.5(g) and (h). The powders are characterized as being somewhat porous. The observed porosity is not in agreement with the measured density of the powders. Upon further investigation, it was determined that the TAFA 1375V powder was slightly rich in the metallic phase (i.e. NiCr). As a result of this increased softer binder phase, coating consolidation and build-up occurred with the cold spray process. The combination of the porosity and increased binder phase of the agglomerated and sintered powder resulted in particle deformation and particle consolidation, allowing the coating thickness to build up readily.

14.6.5 Praxair 1375VM: Cr_3C_2–25wt.%NiCr (C7)

Lastly, due to the success of the TAFA 1375V powder, additional powder was specially purchased in order to try to reproduce the TAFA 1375V powder performance. Praxair 1375VM was produced closer to the TAFA 1375V design specification with a fairly narrow particle size distribution (Fig. 14.6) and a $D_{(v,0.5)}$ of 33.25 μm. The powder density was 6.6955 g/cm^3 which is approximately 93.4% of the theoretical density value. The agglomerated and sintered powder morphology of Fig. 14.5(i) and (j) again shows that the powder is fairly porous, and was very similar to that of the TAFA 1375V (Fig. 14.5(g) and (h)). However, although this powder was processed to the specification of the TAFA 1375V powder, it did not perform as well as the TAFA 1375V. The results suggested that the Praxair 1375VM powder was closer to the stoichiometric values and not rich in the metallic (NiCr) binder phase (as compared with the TAFA 1375V powder).

Based on the analyzed powder characteristics, a Cr_3C_2-based cold spray feedstock of agglomerated and sintered powders would have a uniform composition (preferably rich in NiCr) and microstructure (open porosity) to allow coating build-up and uniform average VHN values. Based on the powder characterization, it is believed that a slightly higher concentration of the NiCr (approximately 27%) would perform very well with the cold spray process. As previously discussed (Fig. 14.3(a) and (b)), powder blend coatings allow wide average VHN distributions and greater tailorability, and thus wear resistance, especially when mating component wear is a concern.

Once the optimal Cr_3C_2 powder characteristics are identified, it is anticipated that MoS_2, a solid lubricant, could be incorporated into both candidate wear-resistant coatings to reduce the friction coefficient and improve wear resistance. Reducing the friction coefficient of the coating should result in improved wear-resistance properties. In addition to adding solid lubricants, bead peening and laser glazing of the coatings are being explored in order to increase the density of the coating, which will improve the coating hardness, and further improve wear-resistance properties.

14.7 X-ray diffraction and phase analysis

X-ray diffraction (XRD) was performed to determine/confirm whether or not the chromium carbide concentration was altered from the original starting blend composition as compared with the coating material (to help explain the variations in coating hardness) and to determine whether crystallographic phase changes occurred during the cold spray process. In general, little change in the coating composition is observed for cold spray coatings as oxidation and phase transformation that occur due to increased particle or substrate temperatures are not inherent to the cold spray process. However, when mixing powder blends with hard, wear-resistant particles (carbides, nitrides, oxides), solid lubricants (sulfides), and soft metallic powders, variation in the coating composition can be observed due to the hard carbide particles not undergoing plastic deformation and bonding to the substrate or previous coated layer. To show and confirm this, select XRD patterns were performed and compared with the Joint Committee for Powder Diffraction Standards (JCPDS), now known as the International Center for Diffraction Data (ICDD).

Figure 14.7(a) shows the XRD pattern of the pure nickel powder, chromium carbide powder, and Cr_3C_2–25wt.%Ni blend feedstocks prior to applying the cold spray coatings. As observed in Fig. 14.7(a), all of the powders were polycrystalline, with nickel powder being of the cubic crystallographic phase (JCPDS–ICDD #4-850), and the chromium carbide powder being primarily of the orthorhombic crystallographic phase (Cr_3C_2 (JCPDS–ICDD #35-804)) with a small volume fraction of the hexagonal (Cr_7C_3 (JCPDS–ICDD #11-550) crystallographic phase being detected. Due to the complexity of the diffraction patterns and the various overlaps of the planes of diffraction for the various compositional phases, detection of pure chromium (cubic) or delta chromium nickel (cubic) was not confirmed. In order to quantify the various phase mixtures, the internal and external standard methods were used to determine the weight percent of chromium carbide in the powder blend comprising the feedstock for the various cold spray deposition conditions in which the composition of the powder was already known. The intensities of the various crystal structures (i.e. phases) within the coating were then compared with that of the starting raw material with any deviations being attributed to a decrease in chromium carbide weight fraction.

Good agreement was observed between the volume intensities of each of the phases and the known starting composition. For example, for the Cr_3C_2–25wt.%Ni, the weight fraction of Cr_3C_2 was found to be 72.8% which is very close to the 75.0% theoretical value. The small differences were attributed to non-uniform powder mixtures, differences in X-ray absorption coefficients, and variation in the Cr_7C_3 phase content from the raw feedstock. In order to determine the relative amount of chromium carbide and nickel within the cold spray coating, the volume intensities were determined after various background corrections, $K_{\alpha 2}$ stripping and peak smoothing using Philips analytical software was performed. The volume intensities of the cold spray coatings were then compared with the volume intensities of the corresponding known powder blend or raw powder stock material using the following equation to determine the chromium carbide weight fraction:

$$X_{Cr_3C_2} = (I\{111\}_{Ni} / I\{121\}_{Cr_3C_2})(I\{111\}_{Ni} / (I\{111\}_{Ni} + I\{121\}_{Cr_3C_2}))$$

where $X_{Cr_3C_2}$ is the weight fraction of chromium carbide, $I\{121\}_{Cr_3C_2}$ is the volume intensity of the set of (121) chromium carbide diffraction planes,

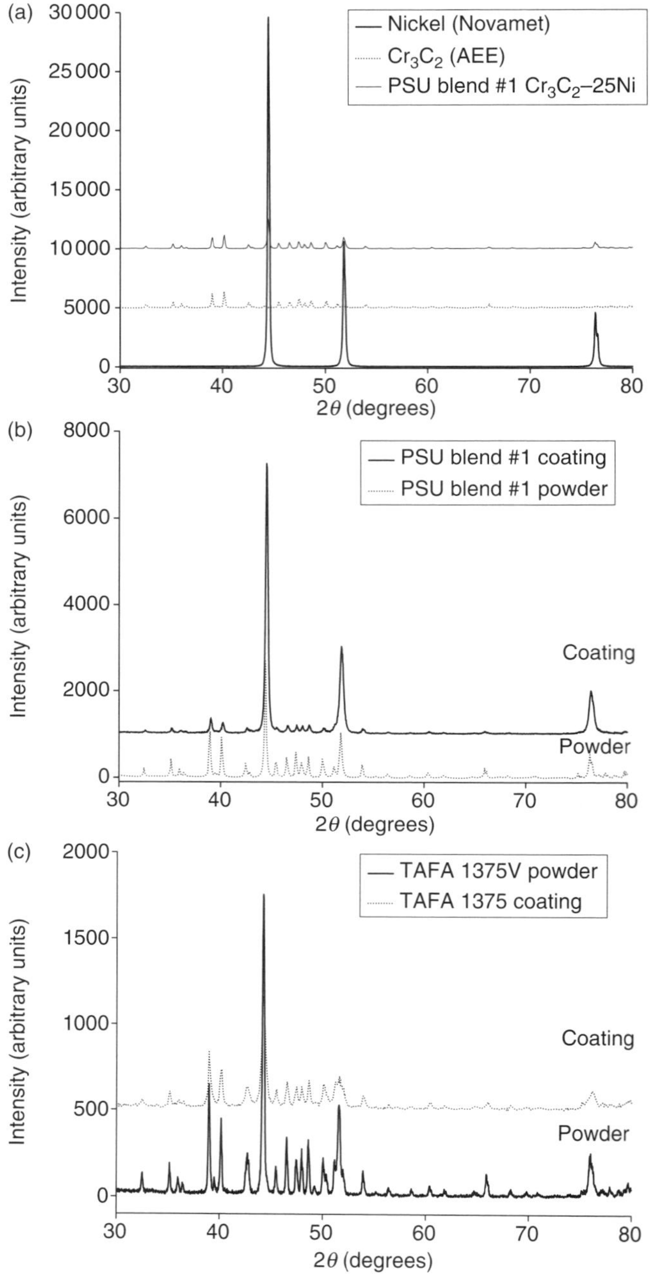

14.7 X-ray diffraction pattern of: (a) nickel, chromium carbide, and PSU blend #1 powders; (b) PSU blend #1 powder and cold spray coating; (c) TAFA 1375V powder and coating. The coatings were all applied on 4140 alloy steel.

and $I\{111\}_{Ni}$ is the volume intensity of the (111) set of nickel diffraction planes. With the exception of the agglomerated and sintered powders, all of the blends showed a reduction in chromium carbide content (discussed below) that was confirmed by the optical and scanning electron microscopy analysis in Section 14.8.

When comparing the diffraction patterns of the feedstock with the cold spray coatings, several differences were observed (Fig. 14.7(b)). In the case of the various coatings using the powder blends, a reduction in the relative intensity of the chromium carbide peak was always observed for the blends containing 15–25wt.%Ni. This confirmed that chromium carbide was not being incorporated into the cold spray blend coating most probably as a result of chromium carbide not plastically deforming, but rather 'bouncing off' the substrate surface. In addition to the lower than expected chromium carbide intensity, the diffraction peak for the nickel (cubic) phase was more broad and often shifted which is indicative of an increase in non-uniform strains within the nickel matrix caused by plastic deformation. As a result of these strains, determination of the various phases present within the coating was subject to some error and, therefore, the volume concentration of the various phases were reconfirmed using analytical software in combination with optical and scanning electron microscopy of polished coating cross-sections. In addition, the peak positions for the nickel phase appear to be shifted from the original powder pattern and strongly suggest residual stresses present in the coating. In general, based on the direction of the peak shifting, the residual stresses were compressive in nature which is supported by the physics of the cold spray process, which results in compressive stresses within the coating due to the nature of particles impinging on the substrate surface acting to make it more dense. In general, the powder blends always showed a lower chromium carbide concentration due to the chromium carbide powders not being surrounded by a soft ductile matrix which would allow the powder to bond to other particles.

In contrast, the TAFA 1375V powder and cold spray coating show almost the same identical diffraction pattern with respect to the relative intensities of the chromium carbide phase and delta nickel chromium (Fig. 14.7(c)). However, it is again observed that non-uniform strains are observed in the NiCr phase as peak broadening was observed. In addition, the crystallinity of the NiCr phase is decreased due to the various degrees of strain induced into the particles resulting from the plastic deformation needed for deposition onto the 4140 alloy, as represented by the peak broadening. However, since the chromium carbide powders of the agglomerated and sintered TAFA 1375V are essentially surrounded by the soft metal matrix (NiCr), coating build-up occurred with little loss of chromium carbide phase being observed, and this was confirmed by the optical and scanning electron microscopy investigation.

However, it should be noted that the hardness (Fig. 14.3(b)) of the coating can be tailored over a wide range of values up to the highest volume concentration of chromium carbide. This could allow a wide range of niche areas for the cold spray process. In addition, it is believed that the volume concentration of the chromium carbide phase could be further increased if the chromium carbide powders were coated with a thin plastically deforming coating that would allow coating build-up. Increasing the volume concentration of the chromium carbide phase in the cold spray coatings would also allow further increases in hardness and increase the range of tailored average VHN values. It is clear that there is substantial potential for the cold spray process for niche areas in tailoring the surface hardness of temperature-sensitive materials that could not be applied by some of the other thermal spray techniques. The authors are in no way suggesting that cold spray will replace other thermal spray processes, but it does allow flexibility in tailoring the wear-resistance properties of the material over a wider range of values. Again, one of the overall objectives of future work is to include solid lubricants into the wear-resistant surface for high temperature applications where liquid lubricants cannot be used.

14.8 Scanning electron microscopy

Select samples were evaluated by optical and scanning electron microscopy in order to determine any microstructural differences attributed to the variations in the average VHN and thus wear resistance. As shown in Fig. 14.8(a), the amount of chromium carbide (Cerac) incorporated into the cold spray coating is much less than the PSU blend #1 (Cerac Cr_3C_2 powder) pre-blended feedstock powder composition. Figure 14.8(b) shows a slightly higher volume concentration of chromium carbide incorporated into the cold spray coating (PSU blend #2). The differences in the amount of chromium carbide incorporation in the cold spray coatings were discussed in Section 14.6 (powder characteristics), and are attributed to the different particle size distributions. For comparison, the cold spray coating using TAFA 1375V powder (Fig. 14.8(c)) shows a much larger volume fraction of chromium carbide being incorporated into the coating. As a result, the average VHN values were greater for TAFA 1375V as compared with the various blended compositions. However, closer observations show that the blended compositions have a higher density than the TAFA 1375V, but the hardness values were lower. Therefore, if increased concentrations of Cr_3C_2 particles could be incorporated into the cold spray coating, higher VHN values would be expected. These results support the XRD results suggesting lower carbide concentrations in the cold spray coatings.

(a) PSU blend #1: Cr_3C_2–25wt.%Ni

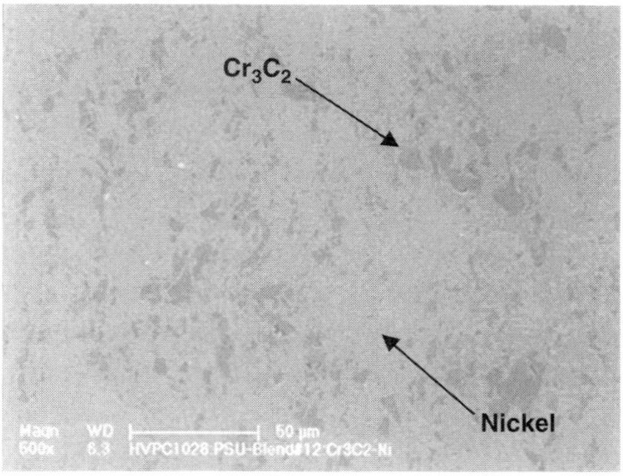

(b) PSU blend #2: Cr_3C_2–25wt.%Ni

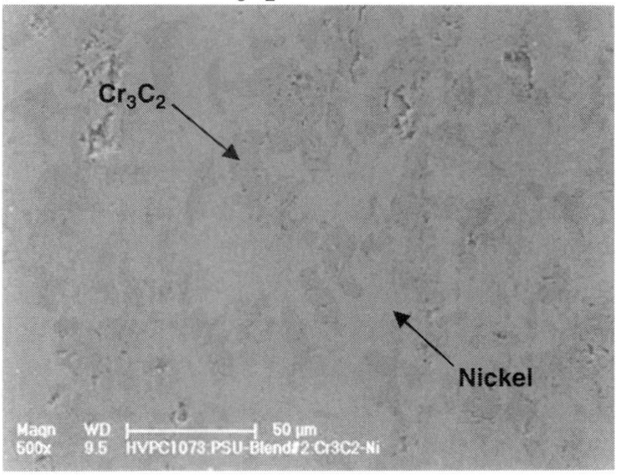

(c) TAFA 1375V: Cr_3C_2–25wt.%NiCr

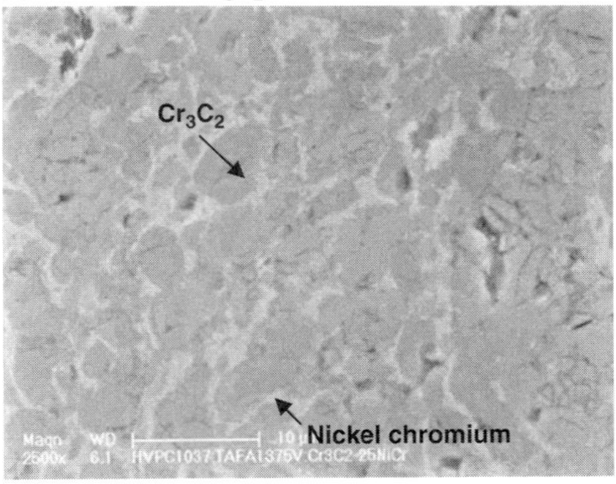

14.8 Scanning electron micrographs of polished cross-sections of cold spray coatings of (a) PSU blend #1, (b) PSU blend #2, and (c) TAFA 1375V powders applied on 4140 alloy steel.

14.9 Tribology

It is often important to understand the wear rates and friction coefficients of mating materials during service life. Therefore, wear tests are often used to better understand a material's performance in contact with itself or the mating material in operation. One of the most popular wear tests for bulk alloys is the ASTM G-65 Dry Sand Rubber Wheel Abrasion Test in which the volume or weight loss is measured in a controlled environment and is often referred to as scratching abrasion since the erosive particles do not fracture during the testing and keep their morphology, shape and size distribution throughout the test. When the abrasive particles fracture, sharp angular surfaces are formed which results in a different wear mechanism. However, one of the biggest challenges often reported is the lack of correlation between wear testing in the laboratory and field wear conditions arising due to the complexity of wear and the various factors contributing to wear.

To investigate the wear resistance of some of the coating materials several cold spray coatings were applied to 4140 alloy, 25.4 mm × 25.4 mm × 6.35 mm (1″ × 1″ × 1/4″), and in addition to the uncoated substrate, were evaluated for wear resistance and friction coefficient determination. The coated and uncoated 4140 alloys were evaluated against 100Cr6 steel (mating material) for the tribology testing under dry conditions. Table 14.6 shows the tribology parameters selected for the evaluation using a CSEM THT tribometer S/N 4-116 instrument.

Two tribology tests were performed for each of the coating materials (Fig. 14.9). The average weight loss of the coating, mating material, and friction coefficient of the various cold spray coatings are listed in Tables 14.7 and 14.8. The average friction coefficient for the uncoated alloy against 100Cr6 mating material was 0.725 ± 0.017. In comparison, the friction coefficients of the nickel, TAFA 1375V, and PSU blend #2 coatings were 0.824, 0.891, and 0.786, respectively. The average mean friction coefficient of the

Table 14.6 Major experimental parameters for tribology experiments

Load	5.00 N
Distance	540.0 × 10^{-3} km
Speed	5.00 cm/s
Radius	3.00 mm
Test time	3 hours
Atmosphere	Air
Humidity	~26% (but varied between 24 and 32%)
Temperature	Ambient (~22–23 °C)
Lubrication	No
Mating material	6 mm diameter 100Cr6 (low carbon steel)

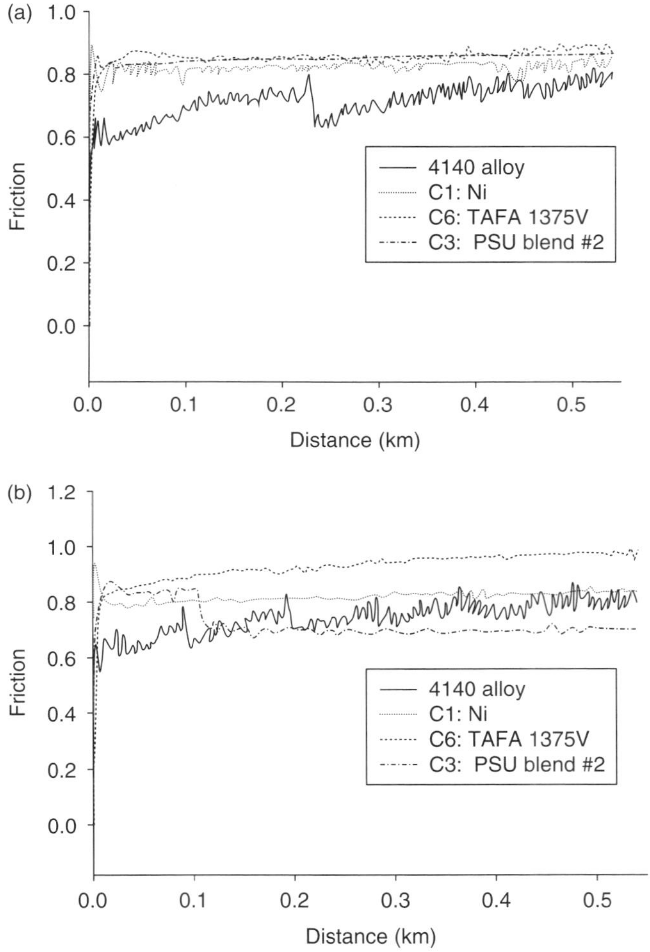

14.9 Friction profiles for 4140 alloy, and C1, C3, and C6 cold spray coatings tested against the 100Cr6 mating surface with (a) 5 N and (b) 10 N loads.

Table 14.7 Tribology results for cold spray coatings tested under the following conditions: 5 N, no lubricant, $R = 3$ mm, 79.6 rpm, and 0.54 km distance traveled

Coating	$VHN_{0.300}$	Average coating loss (g)	Average mating material loss (g)	Average friction coefficient
Uncoated 4140 alloy	444 ± 25	−0.001 78	−0.000 31	0.725
Nickel coating	310 ± 52	−0.004 47	−0.000 11	0.824
PSU blend #2: Cr_3C_2–25wt.%Ni	449 ± 28	−0.000 45	−0.000 07	0.786
TAFA 1375V: Cr_3C_2–25wt.%NiCr	775 ± 127	−0.002 42	−0.000 93	0.891

Table 14.8 Tribology results for cold spray coatings tested under the following conditions: 10 N, no lubricant, $R = 3$ mm, 159.2 rpm, and 0.54 km distance traveled

Coating	$VHN_{0.300}$	Average coating loss (g)	Average mating material loss (g)	Average friction coefficient
Uncoated 4140 alloy	444 ± 25	−0.00337	−0.00007	0.517
Nickel coating	310 ± 52	−0.01997	−0.00034	0.702
PSU blend #2: Cr_3C_2–25wt.%Ni	449 ± 28	−0.00483	−0.00061	0.752
TAFA 1375V: Cr_3C_2–25wt.%NiCr	775 ± 127	−0.00220	−0.00179	0.708
PSU blend #3: Cr_3C_2–15wt.%Ni	483 ± 58	−0.00030	−0.00108	0.663
PSU blend #9: Cr_3C_2–18wt.%Ni	432 ± 98	−0.00033	−0.00121	0.664

various coatings evaluated under dry conditions suggests that the PSU blend coating has the lowest mean friction values. However, in general, the friction increased with increasing time for all coatings tested under dry conditions, as shown by the friction profiles (Fig. 14.9). Preliminary trials were performed with 5 N and 10 N loads and were evaluated for wear and friction coefficient determination. Based on the data in Tables 14.7 and 14.8, the wear rate of the 4140 alloy steel was fairly uniform with linear wear rates. The volume of material was estimated from a computer numerical controlled (CNC) six-axis machine which measured the wear profiles. A minimum of four measurements taken 90° apart were made in order to obtain a statistically weighted average loss of the samples. The wear profile (Fig. 14.10) as measured by the CNC machine appeared to be fairly uniform for the 4140 alloy (smooth surface). However, most of the cold spray Cr_3C_2-based coated samples did not have a smooth uniform surface finish and it was very difficult to determine the amount of wear for the coating accurately. As a result, the degree of wear from the tribology test was based upon the weight loss of the coating using a high-precision balance. The results of the select samples are shown in Tables 14.7 and 14.8. Overall, the chromium carbide coatings showed significant improvements in the wear rates as compared with the uncoated 4140 alloy, but additional testing is required to better understand the wear rates and friction coefficients. In addition, measurements and calculations were performed on the mating material (100Cr6 6 mm diameter ball) to determine the wear of the mating material against the various coatings. However, due to the non-uniform wear patterns of the 6 mm diameter 100Cr6 ball, this was more challenging, as shown by the optical micrographs in Fig. 14.11.

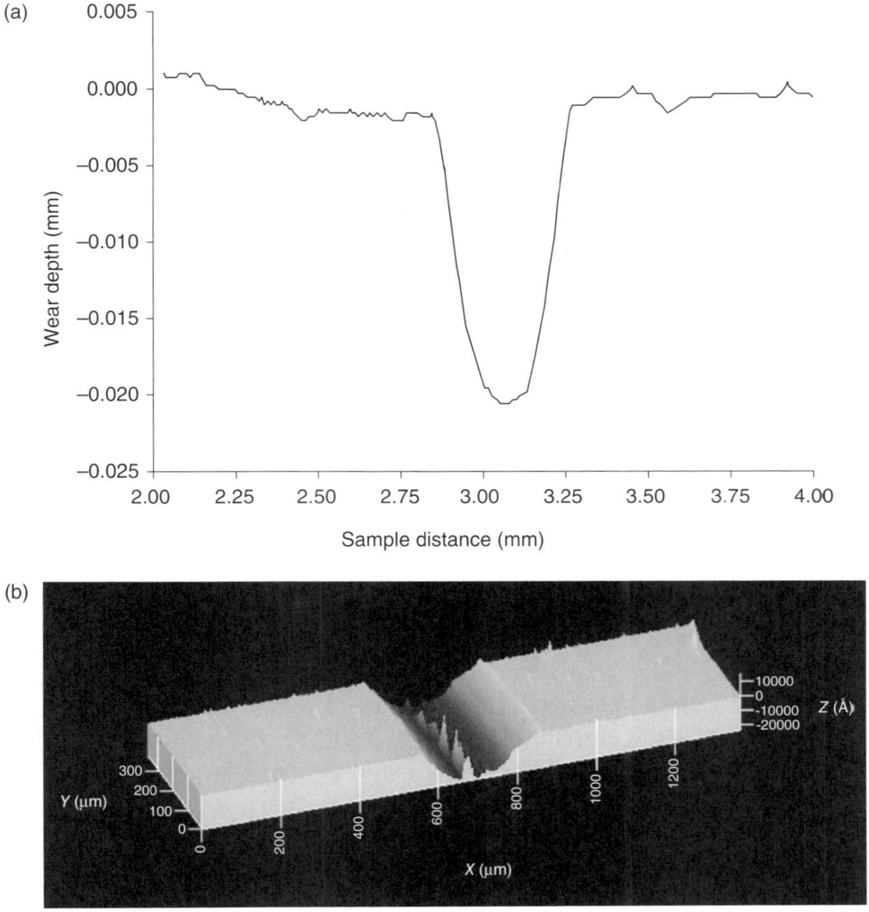

14.10 Wear depth as a function of wear track width.

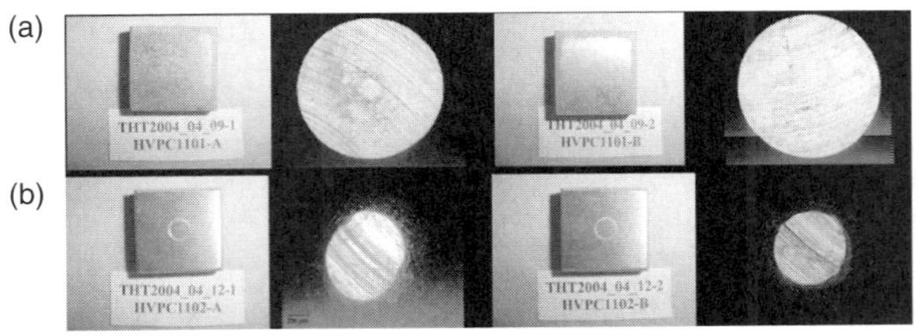

14.11 Digital images of wear block and wear pattern of 6 mm diameter mating material (100Cr6) for (a) TAFA 1375 and (b) PSU blend #2 coatings on 4140 alloy. Tested under the following conditions: 5 N, 3 mm radius, 5 cm/s, and no lubricant.

The non-uniform wear patterns of the mating material (100Cr6) are believed to be associated with the varying degrees of surface finish. It should be noted that prior to the start of the tribology tests, the coated surfaces were ground to a specific surface finish for the coatings. However, due to the varying microstructures and degree of porosity, deviations in the surface roughness/finish occurred that may have influenced the wear patterns and the friction coefficients.

Additional sets of trial experiments were performed with a higher load of 10N and under a sand–oil-lubricated condition. The type of lubrication used was DTE hydraulic oil with 20wt.% Iraq sand (more aggressive environment). The tribology results appeared to show mixed results as to which coating performed the best under lubricated conditions. It appeared that some of the coatings gained weight due to the lubricant being incorporated into the porous coatings. These results will not be discussed. The differences in the wear results from the 5N and 10N loads are believed to be the result of localized changes in friction and coating microstructure. However, depending on the wear application it is not adequate to only look at the amount of wear associated with the coating material. Often it is important to consider both the coating and the mating material wear rates. For this effort, the PSU blend #2 coating showed similar wear results for both the Cr_3C_2-based coating and the 100Cr6 mating material. Similarly, the TAFA 1375V cold spray coating showed the least amount of wear, but showed an increased amount of weight loss for the mating material. Therefore it is important to consider the results of each test and consider the results based on the desired application. For the purpose of this study, the authors were looking for a coating material that had uniform wear against 100Cr6 hardened alloy steel (1000 $VHN_{0.300}$).

14.10 Adhesion and bond strength

In general, factors that affect the bond strength or coating adhesion of the coating are primarily determined by the interfacial phenomena between the coating and the substrate and the microstructure including residual stresses within the coating. Typically, a coating with a high value of tensile residual stresses will have a lower adhesion value than one with compressive stress, given the same degree of interface reactions. Often, at elevated temperatures during deposition, diffusion occurs at the bond coat/substrate interface which results in intermixing of the materials. Depending on the type of interface formed, strong chemical bonding can occur resulting in high values for adhesive strength for chemical vapor deposition (CVD) and particle vapor deposition (PVD) coating processes. Similarly in the thermal spray coating methods, the substrate surface is generally rougher and the coatings are deposited at lower temperatures which reduces the atomic diffusion at the interface generally resulting in lower adhesion. Cold spray and thermal spray processes generally have a rougher interface that relies upon mechanical interlocking for the adhesive strength. However, future trends in cold spray processing have shown that bond strengths in excess of 55 MPa can be obtained with certain materials. In addition, greater adhesion values can be obtained when the carrier gas is helium rather than nitrogen owing to the higher particle velocities. To demonstrate adhesion and to show that the adhesion values for the cold spray process can approach those for thermal spray, bond strength testing was performed.

Using an Instron IX Automated Materials Testing System, cold spray TAFA 1375V and PSU blend #2 coatings were evaluated for bond strength according to ASTM standard C633-01. Special 25.4 mm diameter specimens were prepared for applying both TAFA 1375V and PSU blend #2 cold spray coatings. The coated specimens were bonded to a mating surface using a Loctite Hysol E214HP adhesive bonding agent. Increasing tensile load was applied with a constant cross-head rate of 0.013 mm/s (0.030 in/min) until rupture occurred. The average bond strengths of the TAFA 1375V and PSU blend #2 samples were 27.5 ± 3.5 MPa (3.99 ± 0.716 ksi) and 39.5 ± 2.5 MPa (5.727 ± 0.525 ksi), respectively. The adhesive bond strength of the Cr_3C_2-based coatings applied by the cold spray process in this investigation is comparable with the low to mid range of those values reported in the literature for thermal sprayed Cr_3C_2–NiCr coatings (30–70 MPa).[18–21] The method of coating deposition, the surface condition, surface roughness, etc., all have considerable impact on the degree of coating adhesion. In general, the coatings failed primarily due to adhesive failure and not cohesive failure.

14.11 Potential hybrid process combining cold spray and laser processing

Microscopy of the polished cross-sections in combination with the low hardness measurements of some of the Cr_3C_2–25wt.%NiCr (TAFA 1375V) coatings strongly suggested that the cold spray coatings contained macro-porosity and micro-porosity (Fig. 14.8). This porosity is believed to be the main factor in the relative low average VHN values of the various chromium carbide-based cold spray coatings ranging from 300 $VHN_{0.500}$ to ~575 $VHN_{0.500}$, depending on the deposition parameters. In an effort to increase the VHN, laser glazing trials using a 3 kW YAG laser were performed. Several trials were performed to determine whether laser glazing would further improve the coating quality, density, and microstructure, and thus the cold spray coating properties (particularly VHN). The two primary laser parameters investigated were laser power and traverse speed. Figure 14.12(a) shows a digital image of several laser passes over the surface of a cold sprayed

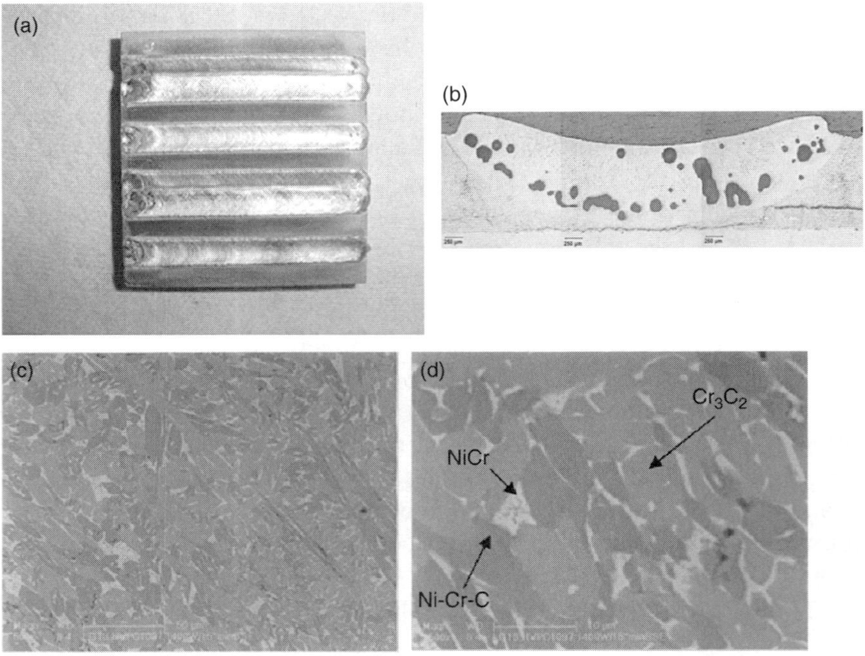

14.12 (a) Digital image of surface; (b) optical micrograph of polished cross-section; (c) and (d) back-scattered electron micrographs of polished cross-sections of a laser glazed cold sprayed TAFA 1375V: Cr_3C_2–25wt.%NiCr coating on 4140 alloy. Laser processing parameters were: 3kW YAG at 1900W with traverse speed of 381mm/min.

Cr_3C_2–25wt.%NiCr coating. Preliminary trials showed densification of the surface material with pore coalescence near the bottom of the laser glaze melt pool. In general, the melt pool depth of penetration can be controlled by adjusting the laser power, beam size, and traverse speed. Also observed in Fig. 14.12(b) is cracking near the interface associated with pore coalescence and shrinking of the coating. Although no cracking was observed in the glazed region, some cracking along the sides was observed. Back-scattered scanning electron micrographs are shown in Fig. 14.12(c) and (d), and show a very dense microstructure. However, due to the high temperatures associated with the laser glazing and localized surface melting, some dissolution of the chromium carbide phase occurred resulting in a phase transition. However, similar phase transitions also occur for other techniques. The preliminary results show a significant improvement in the coating hardness. The average VHN before and after laser glazing was ~450 $VHN_{0.500}$ and 1015 $VHN_{0.500}$, respectively, which is over a 100% increase in the average VHN of the 'as-sprayed' coating. The increase in hardness is

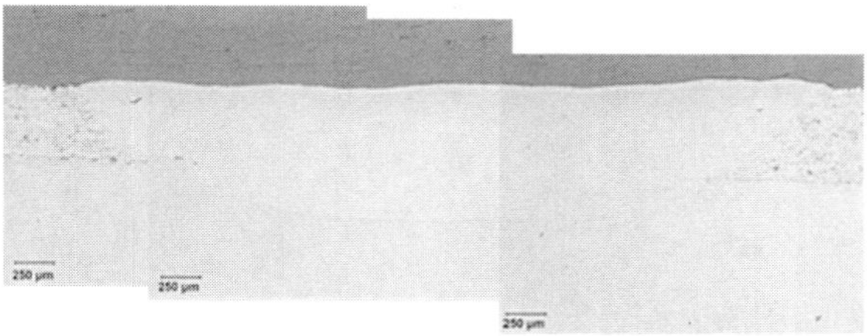

14.13 Optical micrograph of laser glazed surface of PSU blend #2: Cr_3C_2–25wt.%Ni, showing little porosity due to the increased density of the cold spray coating and dissolution with the substrate due to the increased energy. Laser processing parameters were: 3kW YAG at 2300W with traverse speed of 6.35mm/s (381mm/min).

attributed to the densification/solidification of the cold spray coating. However, additional efforts are required as pore coalescence occurred beneath the coating surface, which could act as failure initiation sites for delamination (Fig. 14.12(b)). Defocusing and scanning the laser beam over the surface of the coating should help to minimize the entrapped porosity resulting in a very dense coating with high hardness, as shown in Fig. 14.13. Figure 14.13 shows the cross-section of a PSU blend #2 cold spray coating that was laser glazed, showing increased density and no apparent cracking in the glazed region. Therefore, adding a laser attachment to the cold spray process could result in a new hybrid coating technology addressing some of the limitations of the laser cladding process.

14.12 Conclusions

The cold spray process was successful in applying Cr_3C_2–25wt.%NiCr and Cr_3C_2–25wt.%Ni coatings on 4140 alloy for wear-resistant applications. Average VHN values of the Cr_3C_2-based coatings ranged from 300 to 900 $VHN_{0.300}$ which strongly suggests that the hardness (and thus wear resistance) can be tailored over a wide range of values to meet a variety of hardness specifications. In general, optimized stand-off distance, powder feed rate, traverse speed, and nozzle design lead to over a 57% and 37% increase in average VHN for the agglomerated and sintered powder and powder blends, respectively. The bond strength of the cold spray specimens (24–44MPa) were comparable with the lower end values (30–70MPa) of HVOF sprayed Cr_3C_2-based coatings. Combining the cold spray process with the laser glazing technique resulted in the highest $VHN_{0.500}$ of approximately 1015 $VHN_{0.500}$, and shows promise for refurbishing complex geometries. Tribology tests showed that the wear rates of Cr_3C_2-based coatings are significantly better than those of the 4140 alloy. The presented results strongly suggest that the cold spray process is a new, but versatile process that could be used to tailor the coating composition, friction coefficient, hardness, and wear resistance.

14.13 Future trends

Up until recently, hard wear-resistant coatings have not been explored in depth using cold spray processes. However, with improved helium recovery systems allowing helium carrier gas to be more cost competitive, applying hard and blended materials containing hard ceramic and metallic materials is becoming attractive and competitive. Future advances will continue to be made in the areas of nozzle design and powder development, and optimization for improved coating properties. One of the limitations to date with the cold spray process has been in obtaining the desired volume fraction of hard particles incorporated into the coating material. Advances in coated powders and higher particle velocities should allow increased hard particle incorporation resulting in higher hardness coatings with improved wear performance. Although some work has been done with CrC–Ni, CrC–NiCr and WC–Co compositions, additional efforts are needed in order to define cold spray process composition capabilities and additional opportunities in applying wear- and erosion-resistant coating materials. Future possibilities for applications also exist in applying pure chromium and chromium-rich compounds for wear and corrosion applications. Continued exploration of hybrid processing of cold spray and laser processing will also show promise. Finally, cold spray processing incorporating both solid wear-resistant and lubricant particles should have a strong impact in the coating area.

14.14 References

1 Alknimov A P, Kosarev V F and Apayrin A N, 'Cold gas dynamic spray method for applying a coating,' U.S. Patent 5,302,414, April 12, 1994.

2 Anon. 'Cold gas-dynamic spray method,' *FutureTech*, **224**, August (1998) 1–15.

3 Amateau M F and Eden T, 'High velocity particle consolidation technology,' *iMAST Quarterly*, **2** (2000) 3–6.

4 Kan H K and Kank S B, 'Tungsten/copper composite deposits produced by a cold spray,' *Scripta Materialia*, **49** (2003) 1169–1174.

5 Lee H Y, Yu Y H, Lee Y C, Hong Y P and Ko K H, 'Interfacial studies between cold-sprayed WO_3, Y_2O_3 films and Si substrate,' *Applied Surface Science*, **227** (2004) 244–249.

6 Van Steenkiste T H, Smith J R and Teets R E, 'Aluminum coatings via kinetic spray with relatively large powder particles,' *Surface and Coatings Technology*, **154** (2002) 237–252.

7 Van Steenkiste T H, Smith J R, Teets R E, Moleski J J, Gorkiewicz D W, Tison R P, Marantz D R, Kowalsky K A, Riggs W L, Zajchowski P H, Pilsner B, Mccune

R C and Barnett K J, 'Kinetic spray coatings,' *Surface and Coatings Technology*, **111** (1999) 62–71.

8 Lima R S, Karthikeyan J, Kay C M, Lindemann J and Berndt C C, 'Microstructural characteristics of cold-sprayed nanostructured WC-Co coatings,' *Thin Solid Films*, **416** (2002) 129–135.

9 Mccune R C, Papyrin A N, Hall J N, Riggs W L and Zajchowski P H, 'An exploration of the cold gas-dynamic spray method for several materials systems,' Proceedings of the 8th National Thermal Spray Conference, Houston, TX, eds. Berndt C C and Sampath S, ASM International, Materials Park, OH (1995), pp. 1–6.

10 Li C J and Li W Y, 'Deposition characteristics of titanium coating in cold spraying,' *Surface and Coatings Technology*, **167** (2003) 278–283.

11 Scussel H J, *ASM Handbook* Volume 18: *Friction, Lubrication, and Wear Technology*, ASM International, The Materials Information Society (1992).

12 Toth L, *Transition Metal Carbides and Nitrides*, Academic Press Inc., New York (1971).

13 Storms E K, *The Refractory Carbides*, Academic Press, Inc., New York (1967).

14 Schwarzkopf P, Iieffer R, Leszynski W and Benesovsky F, *Refractory Hard Metals*, The Macmillan Company, New York (1953).

15 Weimer A W ed., *Carbide, Nitride, and Boride Materials Synthesis and Processing*, Chapman & Hall, New York (1997).

16 Bashkirov S I ed., *Hard Metals Production Technology and Research in the USSR*, The Macmillan Company, New York (1964).

17 Pierson H O, *Handbook of Refractory Carbides and Nitrides: Properties, Characteristics, Processing, and Applications*, Noyes Publications, Westwood, NJ (1996).

18 Upadhyaya G S, *Nature and Properties of Refractory Carbides*, Nova Science Publishers, Inc., Hauppauge, NY (1996).

19 Wolfe D E and Singh J, 'Synthesis and characterization of multilayered TiC/TiB_2 coatings deposited by ion beam assisted, electron beam-physical vapor deposition (EB-PVD),' *Journal of Surface and Coatings Technology,* **165** (2003) 8–25.

20 Wolfe D E and Singh J, 'Synthesis of titanium carbide/chromium carbide multilayers by the co-evaporation of multiple ingots by electron beam-physical vapor deposition,' *Journal of Surface and Coatings Technology*, **160** (2002) 206–218.

21 Wolfe D E, Eden T J, Potter J K and Jaroh A P, 'Investigation and characterization of Cr_3C_2-based wear-resistant coatings applied by the cold spray process,' *Journal of Thermal Spray Technology*, **15**(3) (2006) 400–412.

22 Wolfe D E and Singh J, 'Titanium carbide coatings deposited by reactive ion beam-assisted, electron beam-physical vapor deposition,' *Journal of Surface and Coatings Technology*, **124** (2000) 142–153.

23 Cottrell A, *Chemical Bonding in Transition Metal Carbides*, The Institute of Metals, London, UK (1995).

24 Dunand A, Flack H D and Yvon K, 'Bonding study of TiC and TiN. I. High-precision x-ray-diffraction determination of the valence-electron density distribution, Debye-Waller temperature factors, and atomic static displacements in $TiC_{0.94}$ and $TiN_{0.99}$,' *Physical Review B*, **31**(4) (1985) 2299–2315.

25 Didziulis S V, Lince J R and Stewart T B, 'Photoelectron spectroscopic studies of the electronic structure and bonding in TiC and TiN,' *Inorganic Chemistry*, **33** (1994) 1979–1991.

26 Blaha P, Redinger J and Schwarz K, 'Bonding study of TiC and TiN. II. Theory,' *Physical Review B*, **31**(4) (1985) 2316–2325.

27 Kaufherr N, Fenske G R, Busch D E, Lin P, Deshpandey C and Bunshah R F, 'Auger electron spectroscopy and Rutherford backscattering spectroscopy studies of TiN and TiC coatings prepared by the activated reactive evaporation process,' *Thin Solid Films*, **153** (1987) 149–157.

28 Bhaumik S K, Upadhyaya G S and Vaidya M L, 'Design of WC-10Co hard metals with modifications in carbide and binder phases,' *International Journal of Refractory Metals and Hard Materials*, **11** (1992) 9–22.

29 Bordeax F, Moreau C and Saint Jacques R G, 'Acoustic emission study of failure mechanisms in TiC thermal barrier coatings,' *Surface and Coatings Technology*, **54/55** (1992) 70–76.

30 Penlington R, Sarwar M and Lewis D B, 'Application of advanced coatings to narrow neck press and blow plungers in the glass container industry,' *Surface and Coatings Technology*, **76/77** (1995) 81–85.

31 Gotoh Y, Hoven H, Koizlik K, Linke J, Wallura E, Nickel H, Kugel H and Ulrickson M, 'Surface damage in TiC coating layers on PDX wall armor tiles,' *Journal of Nuclear Materials*, **145–147** (1987) 765–769.

32 Cho T, Bahgat D G and Woerner P F, 'Relationship between metal cutting performance and material properties of TiC-coated cemented carbide cutting tools,' *Surface and Coatings Technology*, **29** (1986) 239–246.

33 Dearnley P A, Thompson V and Grearson A N, 'Machining ferrous materials with carbides coated by chemical vapour deposition II: Wear mechanisms,' *Surface and Coatings Technology*, **29** (1986) 179–205.

34 Dearnley P A, Thompson V and Grearson A N, 'Machining ferrous materials with carbides coated by chemical vapour deposition I: Interfacial conditions,' *Surface and Coatings Technology*, **29** (1986) 157–177.

35 Lugscheider E, Barimani C and Lake M, 'Investigation of mechanical properties of Ti(C,N) and TiN thin films deposited on cutting tools,' In *Surface Engineering: Science and Technology I,* eds. A Kumar, Y W Chung, J J Moore, and J E Smugeresky, The Minerals, Metals & Materials Society, (1999), pp. 405–413.

36 Fenske G R, Kaufherr N, Lee R H, Kramer B M, Bunshah R F and Sproul W D, 'Characterization of coating wear phenomena in nitride- and carbide-coated tool inserts,' *Surface and Coatings Technology*, **36** (1988) 791–800.

37 Jamal T, Nimmagadda R and Bunshah R F, 'Friction and adhesive wear of titanium carbide and titanium nitride overlay coatings,' *Thin Solid Films*, **73** (1980) 245–254.

38 Nimmagadda R R, Doerr H J and Bunshah R F, 'Improvement in tool life of coated high speed steel drills using the activated reactive evaporation process,' *Thin Solid Films*, **84** (1981) 303–306.

39 Baker H ed., *ASM Handbook: Alloy Phase Diagrams*, Vol. 3, ASM International, Materials Park, OH (1992).

40 Natesan K and Johnson R N, 'Corrosion resistance of chromium carbide coatings in oxygen-sulfur environments,' *Surface and Coatings Technology*, **33** (1987) 341–351.

41 Gorbatov I N, Ii'chenko N S, Terentyev A E, Martsenyuk I S, Tkachenko E P, Shkiro V M, Shaposhnikova T I and Ivanova N F, *Physics and Chemistry of Materials Treatment*, **25**(3) (1991) 274–277.

42 Vishnevetskaya I A, Golego N N and Solov'ev A V, *Physics and Chemistry of Materials Treatment*, **27**(6) (1991) 589–592.
43 Aubert A and Gillet R, 'Hard chrome coatings deposited by physical vapour deposition,' *Thin Solid Films*, **108** (1983) 165–172.
44 Houck D L and Cheney R F, 'Comparison of properties of Cr_3C_2-Ni-Cr coatings thermally sprayed from pre-alloyed and mechanically mixed powders,' *Thin Solid Films*, **118** (1984) 507–513.
45 Chuanxian D, Bingtang H and Huiling L, 'Plasma-sprayed wear-resistant ceramic and cermet coating materials,' *Thin Solid Films*, **118** (1984) 485–493.
46 Johnson R N, 'Coatings for fast breeder reactor components,' *Thin Solid Films*, **118** (1984) 31–47.
47 Bewilogua K, Heinitz H J, Rau R and Schulze S, 'A chromium carbide phase with B1 structure in thin films prepared by ion plating,' *Thin Solid Films*, **167** (1988) 233–243.
48 Sharma S K and Morlevat J P, 'Structure of reactively sputtered chromium–carbon films,' *Thin Solid Films*, **156** (1988) 307–314.
49 Arai T and Moriyama S, 'Structure of reactively sputtered chromium–carbon films,' *Thin Solid Films*, **259** (1995) 174–180.
50 Aubert A, Danroc J, Gaucher A and Terrat J P, 'Hard chrome and molybdenum coatings produced by physical vapour deposition,' *Thin Solid Films*, **126** (1985) 61–67.
51 Marechal N, Pauleau Y and Paidassi S, 'Sputtered silver and Cr-C films for applications as interconnect materials,' *Surface and Coatings Technology*, **54/55** (1992) 320–323.
52 Kim H-J, Lee C-H and Hwang S-Y, 'Fabrication of WC-Co coatings by cold spray deposition,' *Surface and Coatings Technology*, **191** (2005) 335–340.
53 German J, 'Industry warms up to promises of cold spray,' *Scandia Lab News*, **53**(8) (2001) 1–3.
54 Dykhuizen R C and Smith M F, 'Gas dynamic principles of cold spray,' *Journal of Thermal Spray Technology*, **7**(2) (1998) 205–212.

15
The use of cold spray coating for corrosion protection

R. C. McCUNE, formerly of Ford Motor Company, USA

15.1 Introduction

Surface modifications to materials, of which 'cold spray' technology, in the broadest sense, is a subset, have long been applied to the control of the corrosion of materials. The subject of corrosion control of engineering materials has an extensive history and there are many textbooks and handbooks on the subject.[1–5] The definition of 'corrosion' as the degradative loss of a material or its function due to interaction with its environment, provides hints as to strategic directions for its management. In particular, one either modifies the material, provides some intervention between the material and the environment, or modifies the environment.

The approach specific to the 'material' itself encompasses those aspects of materials science and engineering that treat the design of materials at the microscopic level and their processing to effect durability within a specific environment for the intended service life. The design of material microstructures and composition for corrosion resistance addresses the production of monolithic materials with suitable corrosion resistance for the task at hand. Stainless steels are a quintessential example of the atomic engineering of a base metal (iron) to impart corrosion resistance for a wide range of environments.

The second strategy of 'intervention' encompasses the design and implementation of 'engineered surfaces' as often applied to a less corrosion-resistant base material to impart superior corrosion resistance of critical surfaces or to effect a barrier between the metal and its environment. Galvanizing of steel with zinc or its alloys is a classic embodiment of this approach. The control of corrosion through development of engineered surfaces is probably the most ubiquitous and cost-effective strategy for many practical engineered articles. For example, a modern automobile contains thousands of surface-engineered components, ranging from a plethora of coated fasteners to the entire vehicle body structure. Figure 15.1 illustrates the degree to which various treatments are employed for the corrosion protection and appearance of a typical modern auto body. These treatments can range from modifications to the base metal by heat treatment and diffusion, addition of galvanically protecting layers (e.g. zinc), passivating or adhesion-promoting conversion coatings, electrophoretic layers, top coats, and clear coats.

Lastly, the environment for an engineering structure or material is generally more difficult to control or modify and there are typically fewer options in this regard for corrosion control. In closed systems (e.g. the coolant loop for an automobile engine), there may be the opportunity to introduce corrosion inhibitors that chemically react with internal metal surfaces to reduce the corrosion tendencies and are replenished over the course of time. Thermodynamic control of corrosion for a base metal through externally applied potentials (e.g. cathodic protection) or use of sacrificial anodes is also a system-level approach. In general, however, environmental influences such as temperature, humidity, time-of-wetness, introduction of corrodents (e.g. deicing salts), and formation of poultices are uncontrollable 'noise' factors, and it is incumbent on the materials engineer to develop robust countermeasures perhaps employing any or all of the potential approaches mentioned above.

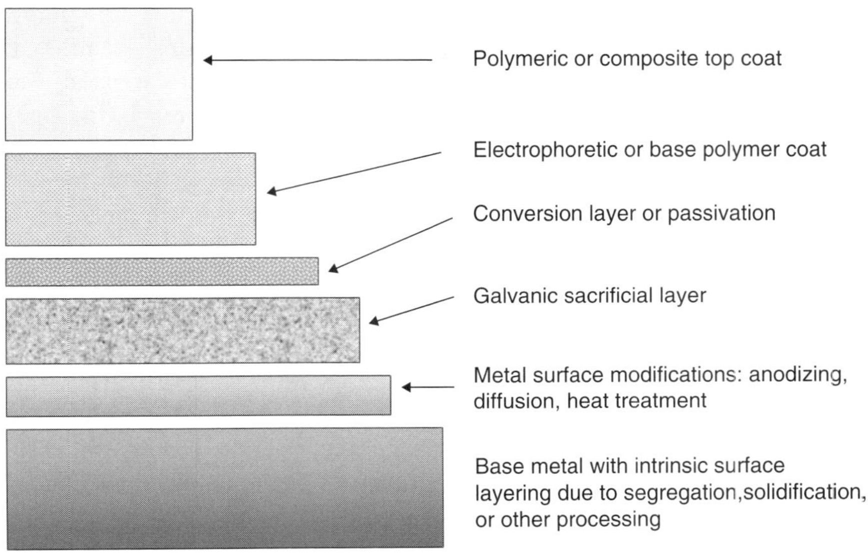

15.1 Schematization of possible layer structure for surface treatments and coatings as applied to automotive structures.

15.2 Cold spraying of metals as a coatings technology for corrosion control

In examining Fig. 15.1, it is apparent that there may be literally hundreds of potential surfacing materials, processes, and combinations thereof to accomplish a particular level of corrosion resistance required for a given application. Entire journals are devoted to corrosion science and protection, organic coatings and engineered surfaces, and to various traditional and novel approaches to corrosion control. In the context of this review, however, the focus will be on the emerging technology, broadly referred to as 'cold spraying'. The following section firstly discusses traditional thermal spray processes as utilized for corrosion protection, and secondly distinguishes those generic aspects of cold spraying that make it particularly suitable for certain corrosion-control applications.

While it is beyond the scope of this chapter to consider thermal spray technology in general, most basic aspects of the various technologies encompassed by the term are discussed in the ASM *Handbook of Thermal Spray Technology*,[6] which also gives an introduction to the cold spray process. Specific treatment of the extensive use of traditional thermal spray processes such as wire arc, plasma spraying, flame spraying, and high-velocity oxy-fuel (HVOF) for corrosion control are provided in other references.[7–9] A primary application of thermal spray technologies for over half a century has been the galvanic protection of steel structures exposed to atmospheric or aqueous corrosion. The portability of thermal spray processes along with the capability for applying highly protective, often aluminum-containing, layers have been crucial to the widespread use of this technology. Traditional thermal spray processes have also been used extensively in providing oxidation and hot-surface corrosion resistance to metals for high-temperature applications including power generation, chemical processing, and aircraft turbines.

Overviews of the fundamentals of cold spray processes are included in this volume, and extensive further discussion is not required here. Details of the process are contained in the ASM handbook[6] and also in Gärtner *et al.*[10] Historically, one major thread for the technology emanated from the work of Alkhimov *et al.*[11] at the Russian Academy of Science, Novosibirsk. An investigative industrial research program, based on this particular methodology, with one of the inventors (A. Papyrin), was conducted through the National Center for Manufacturing Sciences and resulted in several early presentations in the thermal spray literature.[12,13] Some applications of cold spray for corrosion protection, as discussed here, evolved from this earlier program.

Certain attributes of 'cold' spraying as compared with the more-traditional family of thermal spray processes are particularly noteworthy for its use in developing corrosion-protective surface layers. Since the process is conducted with feedstock particles in the solid state, there are effectively no intrinsic defects due to the liquefaction and solidification processes as are characteristic of many other thermal spray processes. This attribute leads to generally very dense metallic coatings, often with limited porosity and mechanical properties approaching those of bulk metals, as well as a tendency to impart residual compressive stresses at the surface of the base metal. On the other hand, because the coating is formed by mechanical compaction, there may be deformation-related ultrastructures (e.g. dislocation arrays, interparticle boundaries, etc.) that are less evident in rapidly solidified coatings. In some cases, it may be possible to recrystallize or anneal such highly deformed cold spray structures through thermal treatments, although the precise impact of these microstructures on corrosion behavior is not known.[14–16] In addition to formation of mechanically dense deposits of metals, cold spray processes have a capability to impart largely compressive residual stresses, much like HVOF or detonation (D)-gun coatings.[13] Such stress patterns generally permit development of thicker deposits than may be possible for other thermal spray processes which typically impart residual tensile stresses and an accompanying tendency for delamination of the coating from the substrate as greater thicknesses are developed. Cold spraying also reduces the tendency for oxidation and volatilization of feedstock constituents as may be encountered in thermal processes that rely on arcs, flames, or plasmas. For production of corrosion-resisting layers, either as galvanic protection or as passivations, this aspect may be significant. From the standpoint of galvanic protection, for instance, it is desirable to have as much of the superficial layer in the form of an oxidizable species, so purity of the deposit is paramount. Additionally, cold spraying may utilize admixtures of deformable and non-deformable powders or coated powders that then permit formulation of specialty layers that might not be achievable from either metal wires or conventionally prepared powders. Because cold spray processes are, by nature, highly collimated, placement control for the coating is achievable, which may be highly desirable for localized corrosion protection, often without the necessity and cost of masking. The 'peening' nature of the process may also limit the need for extensive surface preparation, thereby further advancing the capability for local treatment or restoration.

While the apparent advantages of cold spray technology for deposition of corrosion-resisting layers are evident, there are also some potentially adverse considerations that must be taken into account. One basic hurdle is the general non-portability of many of the early versions of the technology, restricting its use for in-field restoration and primary application processes. Although this has been overcome in yet other versions of the basic process,[17] it remains a limitation for some applications. Since the feedstock material is, of necessity, supplied in powder form, with a generally small particle-size range (typically 10–100 μm), concerns regarding powder handling, inhalable dusts, and pyrophoricity of both incoming feedstocks as well as overspray, cannot be minimized. While many of the feedstock materials of interest (e.g. copper, zinc) do not pose serious handling concerns, powders such as aluminum and titanium need to be carefully handled and controlled to minimize any opportunities for combustion. Storage of powders and prospects for unwanted reactions (e.g. zinc powder + humidity → producing hydrogen) must be considered. As with any spray processing, other safety concerns associated with collection of particulate dusts from overspray, and handling of a particle 'beam' must be taken into account.

A major consideration in the selection of cold spray processing for a given application is its cost relative to alternatives. This may be a significant factor for cost-sensitive situations such as automobile manufacturing; however, it may be less consequential for such industries as aerospace, biomedical, or sporting goods. Blose *et al.*[18] considered the cost implications of cold spray passivation in comparison with alternatives for galvanizing of steel, and concluded that cold spray processing led to cost penalties of at least a factor of two when compared with wire flame spraying. Costs of specialty powders contribute to this, as well as instances where helium is the propellant gas of choice for the application. The use of air or nitrogen gas may contribute to cost reductions; however the stipulation that the feedstock be particulate in nature is not readily overcome at this point. If the costs of masking, unmasking, and surface preparation are considered, then cold spray may offer some advantage for 'selective' coating application.

As of this writing, there are relatively little data in the open literature where the corrosion performance of a cold sprayed coating is directly compared with that for alternatives or some baseline situation. For the purposes of this review, several corrosion situations of interest are identified and discussed with relevant data where generated or located from other sources. These include the selective galvanic protection of steels and the generation of passivating overlayers on more reactive metals.

15.3 Galvanizing of steel structures to prevent corrosion

The baseline for assessing performance of cold spray galvanic protection of steel with zinc–aluminum-based materials is the wealth of practical knowledge of the 'metallizing' thermal spray process dating back practically for almost 50 years. The value of incorporating aluminum into conventional zinc thermal spray coatings was reviewed by Hoar and Radovici,[19] and there have been other treatments focusing on conventional thermal spray processes for galvanic corrosion protection.[20–23] In general, the corrosion protection to steel afforded by sacrificial zinc coating is extended by the incorporation of aluminum into the coating structure. For salt-water corrosion, relatively pure aluminum develops a passivating protection, complementing the sacrificial action of the zinc. The morphology of the zinc and aluminum phases has a significant influence on the protective capability, as shown by Shaw and Moran.[20]

Shaw and co-workers[24,25] reported early results comparing several 'cold' spray coatings to other galvanizing alternatives produced by flame and wire-arc spraying. In one study,[24] a corrosion rate is reported for the free-standing galvanic coatings applied to steel panels at times of 20 and 50 days of exposure to a salt-water environment. A summary of the findings is shown in Fig. 15.2, where it is seen that cold sprayed aluminum (denoted in their work as HVPC (high-velocity powder compaction)) shows measurable improvement over a thermal spray alternative. Results from other comparisons using zinc or zinc–aluminum are mixed, with no overwhelming performance advantage beyond the thermal spray alternative.

In another study of selective galvanizing, McCune and co-workers,[26,27] reported corrosion rates for several coatings and base metals (e.g. zinc 303) using the SAE J-2334 corrosion cycle and ASTM G1 procedures for gravimetric analysis of the corrosion rate of the galvanic layer. Figure 15.3 shows

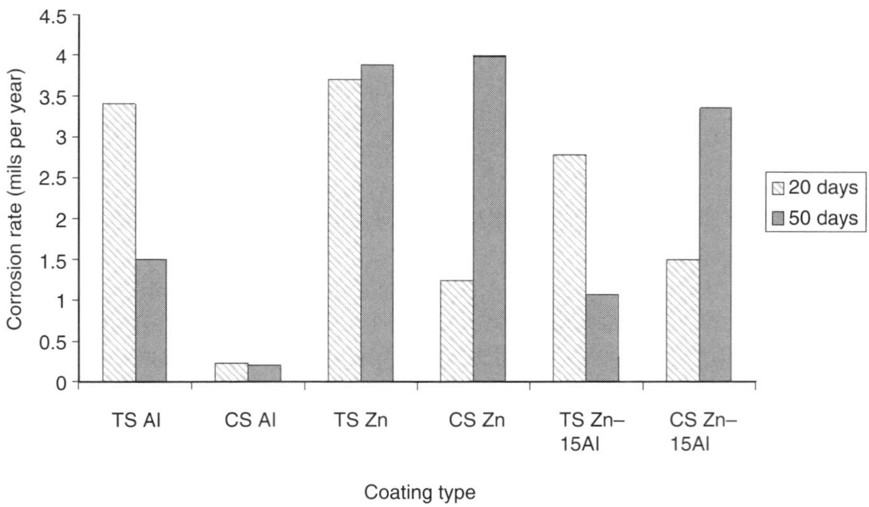

15.2 Comparison of corrosion rates of aluminum, zinc, and zinc–15% aluminum galvanic protection layers on steel panels, for thermal spray (TS) and cold spray (CS) processes at times of 20 and 50 days; after Shaw *et al.*[24]

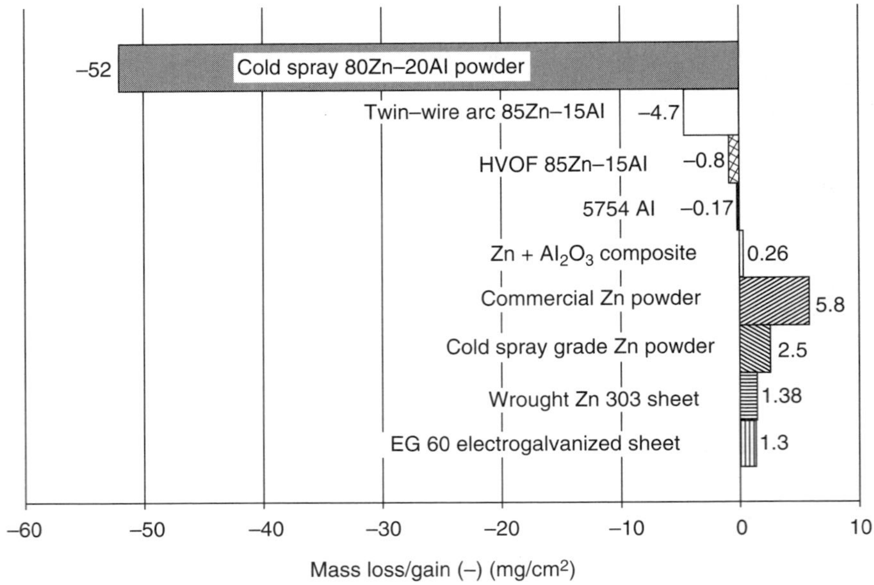

15.3 Summary of weight losses for various Zn and Zn–Al coatings, and reference materials for 20 cycles of the SAE J-2334 corrosion test.

selected outcomes of this study, wherein the cold sprayed zinc was found to corrode at a rate comparable with the pure metal under similar conditions of exposure. Aluminum metal (alloy 5754) and aluminum-containing zinc coatings were observed to effectively gain weight in these experiments, suggesting more stabilized passive protection. Shaw and co-workers[24,25] also measured much reduced corrosion rates for aluminum and aluminum-containing zinc coatings in a salt-water environment. Additionally, polarization resistance studies by Berti *et al.*[25] showed an overall lower rate of corrosion for aluminum applied by a cold spray process, when compared with that applied by either flame or wire-arc spraying. Selective galvanizing has been demonstrated for precision treatment of automotive features including hem flanges and fuel system weldments, where long-term corrosion protection is required.[26–28] Other works by Stehr *et al.*[29] and Ewasyshyn *et al.*[30] address alternative methodologies including cold spraying for protection of detail features – such as hem flanges – on automobiles. Cold spraying has also been shown as a potential process for applying metal as a filler, particularly for cases where applied heat, e.g. welding processes, is unacceptable. Such a technology is expected to be of importance in the restoration of thin-gauge aluminum.[31,32]

Several practical embodiments of the cold spray selective galvanizing process are possible. The first is as a direct substitute for conventional hot-dip or electrogalvanizing on bare steel, with the attributes of cold spray as mentioned in the Introduction, and as an alternative or augmentation to existing thermal spray processes. In a second approach, cold spraying is seen as providing for the prospects of selected increases in galvanic protection at regions of a structure already having a nominal level of galvanic protection from hot-dip or electrogalvanizing. In this instance, the pre-existing galvanic layer provides a readily adhering surface for additional placement of zinc or a selected alloy. The hem flange is an example of such a structure, wherein the nominal galvanic protection of the steel sheet is enhanced in the region of the metal fold, thus providing additional galvanic protection in this region. In a third approach, material of a differing chemical nature (e.g. aluminum or a zinc–aluminum alloy) is deposited selectively at some particular feature, either singularly or atop a pre-existing galvanized layer to provide an alternative mechanism for protection, beyond the sacrificial quality of the base zinc.

In automobile construction, galvanized steel is often painted for reasons of appearance and additional corrosion protection. As suggested by Fig. 15.1, the galvanically protective layer may receive a number of further coating layers including conversion coats, electrophoretically applied polymers, basecoats, top coats, and clear coats. LaLiberte and co-workers[33] have studied the salt-water corrosion performance of a variety of painted galvanically protective coatings including cold sprayed zinc and aluminum, and demonstrated reasonable galvanic protective capability, particularly for cold sprayed zinc with respect to other thermal spray alternatives.

15.4 Preventing aqueous and hot-surface corrosion

In addition to the galvanizing analog presented above, cold spraying has been employed to impart protective layers on more reactive metals, where the objective is not so much to provide sacrificial protection, but to present a generally more passive surface to the environment than the base metal itself. One example utilizes cold sprayed aluminum to protect magnesium articles from salt-water corrosion. It is well-known that aluminum alloying of magnesium creates more passive surfaces than the unalloyed metal, and thus application of cold sprayed aluminum to various magnesium alloys offers a means to impede aqueous corrosion. Two examples are cited; one by McCune and Ricketts,[28] wherein high-purity aluminum was cold sprayed onto AZ91D magnesium die-cast coupons, resulting in corrosion potentials approaching those of commercially pure aluminum. Such polarization behaviors are typified by the plots shown in Fig. 15.4, where the direct current (DC) polarization of the baseline AZ91D magnesium is compared with polarization plots of the same material cold sprayed with high-purity aluminum and a specimen of high-purity aluminum 1199 alloy in a buffered

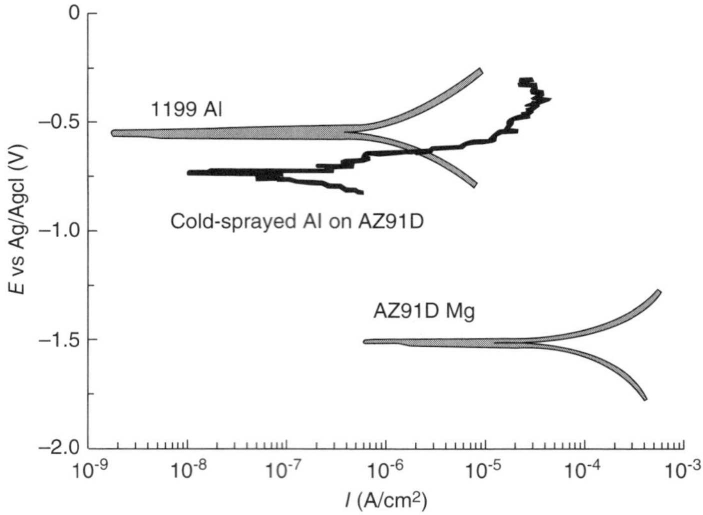

15.4 Direct current polarization behaviors of die-cast AZ91D magnesium, 1199 (99.98% pure Al), and cold-sprayed aluminum-coated AZ91D in a pH 6 buffer solution containing 1000 ppm NaCl.

pH 6, 1000 ppm NaCl electrolyte. Similar results were also reported by Gärtner *et al.*[10] This is an exciting outcome, since there is no effective 'galvanic' protection that is reasonable for magnesium, given its large thermodynamic potential for oxidation. The corrosion resistance of cold sprayed 1100 aluminum has also been studied by Balani and co-workers,[32] further illustrating the protective capability of this coating.

Titanium offers an attractive prospect for aqueous corrosion situations as well as biomedical opportunities; however, despite some level of success in obtaining cold spray deposits by various researchers, it has demonstrated a certain level of difficulty for producing dense coatings, except under very high velocity conditions.[10] The physical basis for apparent difficulties in producing dense titanium coatings was treated by Vlcek *et al.*[34] Specific treatments of titanium cold spraying are also included in references 35 to 37. While high-density coatings may be desirable from the standpoint of surface passivation, the more porous coatings that are typical of cold-gas-deposited layers may be of particular interest in establishing platforms for attachment of biological tissues for such applications as prostheses. Despite the interest in the development of titanium coatings for corrosion-resisting applications, there are very little data in the open literature on the corrosion performance of these materials.

In addition to titanium, there are various literature references to the deposition of nickel-based alloys,[38] stainless steels,[38–40] and tantalum,[41] all of which should be capable of producing passivating surfaces for aqueous corrosion in the form of claddings on more reactive metals such as plain-carbon steels. Again, there are typically little data available on actual corrosion measurements involving these coatings.

15.4.1 Oxidation and hot-surface corrosion

In an introductory paper for the cold spray process,[12] the authors demonstrated an ability to deposit a composite NiCr–Cr_3C_2 coating, typical of that employed in aircraft turbines. This reflected an early and continuing interest by various practitioners in the selected deposition of materials for high-temperature oxidation- and corrosion-resistance applications. As with many other instances in the study of cold spray phenomena, however, there is a general paucity of actual performance data for the applications of interest. Gärtner *et al.*[10] have reported the cold spray deposition of a CoCrAlY alloy which has specific utility as a corrosion-resisting bond layer for turbine engine ceramic thermal barrier coatings. This group has also reported the deposition of 316 stainless steel, which may also be considered for elevated-temperature oxidation- and corrosion-resistance applications. Again, specific performance data for given situations are scarce, although the expectations would be that the advantages of coating density and purity are significant. The mechanical properties of these types of cold spray materials (i.e. those requiring either high particle velocities or some form of particle pre-heating to achieve deformations necessary to effect coating adherence and development) may be less than those of wrought or alternatively produced versions.

15.5 Conclusions

Early investigations of the cold-spray process for selective application of corrosion-resisting coatings have shown encouraging results particularly in the field of sacrificial coatings that extend the capabilities of the zinc or zinc–aluminum coatings presently used. Cold spraying offers both unique opportunities for selective processing and restoration, but also presents challenges of scale-up, cost reduction, powder handling, and safety issues; all of which are well-known to the thermal spray community. Beyond extension of the process to familiar areas such as sacrificial galvanic protection, is the prospect of unique application of unconventional passivations such as deposition of aluminum directly on magnesium alloys. The developing ability to apply stainless steels and certain of the valve metals (e.g titanium and tantalum) suggests a growing capability to produce highly corrosion-resistant surfaces on a variety of substrates. High-temperature oxidation- and corrosion-resistant applications, while envisioned and attempted with stainless steels and certain of the MCrAlY family, are not well-documented in the open literature at this time.

15.6 References

1 Jones D A, *Principles and Prevention of Corrosion, 2nd edn*, Upper Saddle River, NJ, Prentice-Hall, Inc., 1992.

2 Fontana M G, *Corrosion Engineering, 3rd edn*, New York, NY, McGraw-Hill, 1986.

3 Revie R W, ed., *Uhlig's Corrosion Handbook*, New York, NY, John Wiley and Sons, Inc., 2000.

4 Cramer S D and Covino B S, Jr., eds, *Corrosion: Fundamentals, Testing and Protection*, Vol. 13A, *ASM Metals Handbook,* Materials Park, OH, ASM International, 2003.

5 Cramer S D and Covino B S, Jr., eds, *Corrosion: Materials,* Vol. 13B, *ASM Metals Handbook*, Materials Park, OH, ASM International, 2005.

6 Davis J R, ed., *Handbook of Thermal Spray Technology*, Materials Park, OH, ASM International, 2004.

7 Berndt M L and Berndt C C, 'Thermal Spray Coatings,' in Cramer S D and Covino B S, Jr., eds, *Corrosion: Fundamentals, Testing and Protection,* Vol. 13A, *ASM Metals Handbook,* Materials Park, OH, ASM International, p. 802, 2003.

8 Kuroda S and Sturgeon A, 'Thermal Spray Coatings for Corrosion Protection in Atmospheric and Aqueous Environments,' in Cramer S D and Covino B S, Jr., eds, *Corrosion: Materials*, Vol. 13B, *ASM Metals Handbook*, Materials Park, OH, ASM International, p. 422, 2005.

9 Mäntylä T and Uusitalo M, 'Corrosion of Thermal Spray Coatings at High Temperatures,' in Cramer S D and Covino B S, Jr., eds, *Corrosion: Materials,* Vol. 13B, *ASM Metals Handbook*, Materials Park, OH, ASM International, p. 430, 2005.

10 Gärtner F, Stoltenhoff T, Schmidt T and Kreye H, 'The Cold Spray Process and Its Potential for Industrial Applications,' *J. Thermal Spray Technol.* 2006 **15** (2) 223–232.

11 Alkhimov A P, Kosarev V F and Papyrin A N, 'A Method of Cold Gas-Dynamic Deposition,' *Sov. Phys. Dokl.*, 1990 **35** (12) 1047–1049 (Transl. *American Inst. of Phys.*, 1991).

12 Mccune R C, Papyrin A N, Hall J N, Riggs W L and Zajchowski P H, 'An Exploration of the Cold Gas-Dynamic Spray Method for Several Materials Systems,' in Berndt C C and Sampath S, eds, *Advances in Thermal Spray Technology, Proc. Eighth National Thermal Spray Conference*, Materials Park, OH, ASM International, pp. 1–6, 1995.

13 Mccune R C, Donlon W T, Cartwright E L, Papyrin A N, Rybicki E F and Shadley J R, 'Characterization of Copper and Steel Coatings Made by the Cold Gas-Dynamic Spray Method,' in Berndt C C, ed., *Thermal Spray: Practical Solutions for Engineering Problems*, Materials Park, OH, ASM International, pp. 397–403, 1996.

14 Mccune R C, Donlon W T, Popoola O O and Cartwright E L, 'Characterization of Copper Layers Produced by Cold Gas-Dynamic Spraying,' *J. Thermal Spray Technol.* 2000 **9** (1), 73–82.

15 Hall A C, Cook D J, Neiser R A, Roemer T J and Hirschfeld D A, 'The Effect of a Simple Annealing Heat Treatment on the Mechanical Properties of Cold-Sprayed Aluminum,' *J. Thermal Spray Technol.* 2006 **15** (2) 233–238.
16 Call E, Mccartney D G and Shipway P H, 'Effect of Deposition Conditions on the Properties and Annealing Behavior of Cold-Sprayed Copper,' *J. Thermal Spray Technol.* 2006 **15** (2) 255–262.
17 Maev R GR, Strumban E, Leshchynsky V and Beneteau M, 'Supersonically Induced Mechanical Alloy Technology (SIMAT) and Coatings for Automotive and Aerospace Applications,' in *Cold Spray 2004*, Akron, OH, ASM Thermal Spray Society, 2004.
18 Blose R E, Vasquez D L and Kratochvil W R, 'Metal Passivation to Resist Corrosion,' in *Cold Spray 2004*, Akron, OH, ASM Thermal Spray Society, 2004.
19 Hoar T P and Radovici O, 'Zinc-Aluminum Sprayed Coatings,' *Trans. Inst. of Metal Finishing* 1964 **42** 211–218.
20 Shaw B A and Moran P J, 'Characterization of the Corrosion Behavior of Zinc-Aluminum Thermal Spray Coatings,' *Materials Performance* 1985 **24** (11) 22–31.
21 Melzer A, Spriestersbach J, Knepper M and Wisniewski J, 'Thermally Sprayed Coatings of Zinc and Zinc-Aluminium Alloys for Corrosion Protection of Steel Constructions,' in *Euromat 99,* Vol. II, *Surface Engineering*, New York, Wiley and Sons, pp. 53–58, 2000.
22 Nagasaka H and Suzuki R, 'Corrosion Behavior of Various Composition ZnAl Alloys and Coatings Sprayed from them,' in 10th International Thermal Spray Conference, Essen, Germany, DVS Berichte no. 80. *ITSC '83, Proceedings*, pp. 191–194, 1983.
23 Longo F N and Durmann G J, 'Corrosion Prevention with Thermal-Sprayed Zinc and Aluminum Coatings,' in *ASTM STP-646: Atmospheric Factors Affecting the Corrosion of Engineering Metals: Proceedings of the Golden Anniversary Symposium Commemorating 50 Years' Atmospheric Exposure Testing: A Symposium.* Sponsored by Committee A-5 on Metallic-Coated Iron and Steel Products and Committee G-1 on Corrosion of Metals, Philadelphia, PA, American Society for Testing and Materials, 1978.
24 Shaw B A, Berti S, Schmidt D and Shaw B, 'Extension of Vehicle/Component Lifetime Through the Use of Sacrificial Metallic Coatings – PSU Effort,' in *U.S. Army Corrosion Summit*, St. Petersburg, FL, 2002.
25 Berti S, Shaw B A and Shaw W W, 'Corrosion Evaluation of Aluminum Sacrificial Coatings Applied by Various Sprayed Metal Techniques,' in *Corrosion 2002*, Paper 02160, Houston, TX, National Association of Corrosion Engineers, 2002.
26 Mccune R C, Ricketts M S, Gao G, Neiser R A, Puskar J and Roemer T, 'Selective Galvanizing Using Kinetic Spraying,' in *2003 Society of Automotive Engineers World Congress*, Paper 2003-01-1237, Detroit, MI, Society of Automotive Engineers, 2003.
27 Mccune R C, Bomback J L and Gao G, Method for Selective Control of Corrosion Using Kinetic Spraying, U.S. Patent 6,592,947 (B1), July 15, 2003.
28 Mccune R C and Ricketts M S, 'Selective Galvanizing by Cold Spray Processing,' in *Cold Spray 2004*, Akron, OH, ASM Thermal Spray Society, 2004.
29 Stehr J, Borner G and Scherbarth F, 'Verfahrn zur Erzeugung einer Beschichtung, insbesondere Korrosionsschutzschicht' (Transl. 'Process to create a coating,

in particular a layer to protect against corrosion'), German Patent DE 199 18 758 A1, October 26, 2000.

30 Ewasyshyn F J, Kachirine A I, Shkidkin A V, Kliouev O F, Bouzdygar T V, Maev R GR, Paille J M and Strumban E E, 'Method of Forming Dense Coatings by Powder Spraying,' U.S. Patent Application Publication US 2003/0219542 A1, November 27, 2003.

31 Debiccari A, Haynes J D and Rutz D A, 'Structural Repair Using Cold Sprayed Aluminum Materials,' U.S. Patent Application Publication US 2006/0134320 A1, June 22, 2006.

32 Balani K, Laha T, Agarwal A, Karthikeyan J and Munroe N, 'Effect of Carrier Gases on Microstructural and Electrochemical Behavior of Cold-Sprayed 1100 Aluminum Coating,' *Surface and Coatings Technol.* 2005 **195** (2–3) 272–279.

33 Laliberte L H, Miller R G, Shaw B A and Escarsega J A, 'An Evaluation of Sacrifical Metallic Coatings for Service Life Extension of U.S. Army Vehicles,' in *Corrosion 2005*, Houston, TX, National Association of Corrosion Engineers, 2005.

34 Vlcek J, Gimeno L, Huber H and Lugscheider E, 'A Systematic Approach to Material Eligibility for the Cold-Spray Process,' *J. Thermal Spray Technol.* 2005 **14** (1), 125–133.

35 Karthikeyan J, Kay C M, Lindeman J, Lima R S and Berndt C C, 'Cold Spray Processing of Titanium Powder,' in Berndt C C, ed., *Thermal Spray: Surface Engineering via Applied Research*, Materials Park, OH, ASM International, pp. 255–262, 2000.

36 Li C-J and Li W-Y, 'Deposition Characteristics of Titanium Coating in Cold Spray,' *Surface and Coatings Technol.* 2003 **167** 278–283.

37 Marrocco T, Mccartney D G, Shipway P H and Sturgeon A J, 'Production of Titanium Deposits by Cold-Gas Dynamic Spray: Numerical Modeling and Experimental Characterization,' *J. Thermal Spray Technol.* 2006 **15** (2) 263–272.

38 Decker M K, Neiser R A, Gilmore D and Tran H D, 'Microstructure and Properties of Cold-Sprayed Nickel,' in Berndt C C, Khor K A, and Lugscheider E, eds., *Thermal Spray 2001: New Surfaces for a New Millennium*, Materials Park, OH, ASM International, pp. 73–82, 2001.

39 Stoltenhoff T, Kreye H and Richter H J, 'An Analysis of the Cold Spray Process and Its Coatings,' *J. Thermal Spray Technol.* 2002 **11** (4) 542–550.

40 Voyer J, Stoltenhoff T and Kreye H, 'Development of Cold Gas Sprayed Coatings,' in Moreau C and Marple B, eds, *Thermal Spray 2003: Advancing the Science and Applying the Technology*, Materials Park, OH, ASM International, pp. 71–78, 2003.

41 Marx S, Paul A, Kohler A and Huttl G, 'Cold Spray – Innovative Layers for New Applications,' in *Thermal Spray Connects: Explore Its Surfacing Potential. Proceedings of the International Thermal Spray Conference and Exposition (ITSC 2005)*, Basel, Switzerland, DVS-Verlag, for Deutscher Verband fur Schweissen und verwandet Verfahren, Dusseldorf 40010, Germany, 2005.

16
Electromagnetic interference shielding by cold spray particle deposition

D. J. HELFRITCH, US Army Research Laboratory, USA

16.1 Introduction

Electromagnetic interference (EMI) describes the phenomenon that results from allowing conducted and radiated electrical signals to reach destinations where their presence is undesirable. There are basically two purposes for providing EMI shielding in military enclosures. The first is to prevent external EMI sources from penetrating a sensitive environment containing electronic equipment, which is susceptible to the presence of EMI. The second purpose for shielding is to prevent electromagnetic signals generated by equipment within the facility from being transmitted or conducted in sufficient magnitude to be received by sensitive receiving and signal recovery systems.

Shielding effectiveness (SE) is dependent on a number of parameters including electromagnetic frequency, the intrinsic electrical properties of the shielding material, and the number and configuration of discontinuities in the shielding material – such as access doors, hatches, panel joints, piping, conduits, and ventilation duct penetrations. Propagation of EMI may be via radiation and/or conduction, and the required SE must be provided for the total system including all discontinuities and attachments.

SE is defined as 10 times the logarithm of the ratio of the incident electromagnetic power (P_1), to the transmitted power (P_2) with the shielding in place, expressed in decibels (dB) as:[1]

$$SE = 10\log(P_1/P_2) \quad [16.1]$$

For constant wave impedance, the expression can take the more familiar form:

$$SE = 20\log(E_1/E_2) \quad [16.2]$$

where E_1 is the incident electric field and E_2 is the transmitted electric field. The absorption loss (in dB) caused by the shielding material depends on its thickness and electrical properties:[1]

$$SE_a = 131.4d(f\mu_r g_r)^{0.5} \quad [16.3]$$

where:

d = thickness in meters
f = frequency in Hz
μ_r = permeability relative to copper
g_r = conductivity relative to copper

An ideally shielded enclosure would be composed of conductive metal of seamless construction with no openings or discontinuities; however, personnel, powerlines, control cables, and/or ventilation ducts must have access to any practical enclosure. The correct design and construction of these penetrations are critical in order for them to be incorporated without appreciably reducing the shielding effectiveness of the enclosures. EMI leakage (the amount of electromagnetic energy that will leak through a discontinuity) depends mainly on:

- maximum length (not area) of the opening;
- the wavelength of the electromagnetic energy.

Maximum length rather than width of an opening, or slot, is important because the voltage will be highest wherever the detour for the currents is longest. The width has almost no effect on detour length and as a consequence has little effect on the voltage.

Wavelength controls how much the slot radiates. If the slot happens to be 1/4 wavelength or longer, it will be a very efficient radiator; if it is less than 1/100 wavelength, it will be a rather inefficient radiator. Therefore, slots only 0.1 mm wide but 1/4 wavelength or more long can be responsible for large leaks. This is why discontinuities in shields, even if very narrow, can severely reduce the shielding capacity of an enclosure for frequencies above 100 MHz.

Slots are often found at metal panel joints, or seams. Some types of discontinuity, or seams, commonly encountered include:

- seams between two metal surfaces, with the surfaces in intimate contact (such as two sheets of material that are riveted or screwed together);
- seams or openings between two metal surfaces that may be joined using a metallic gasket;
- seams or openings between two metal surfaces that may be joined using an adhesive.

When good shielding characteristics are to be maintained, the permanent mating surfaces of metallic members within an enclosure should be bonded together with conducting materials, such as by welding, brazing, or other metal flow processes. Panels that have been bonded with adhesives present multiple shielding problems, because electrical contact across the seam is lost due to the dielectric nature of organic adhesives. Conductively sealing the seams by means of flowing metal, such as brazing, cannot be done, because of the temperature limitations of adhesives.

The SE required for military communications/control enclosures is typically 24–69 dB in the frequency range of 100 kHz to 100 GHz.[2] The walls of many of these enclosures are of aluminum panels, sandwiching fiber honeycomb for rigidity. The aluminum panels are joined to the honeycomb and to other aluminum panels or terminations by means of adhesives. The seams between aluminum panels and other panels or terminations are a major source of EMI leakage. These seams cannot be sealed through bridging by solder or braze because the required application temperature would degrade the epoxy adhesive. Lower-temperature thermal sprays, such as flame spray, have not produced a metal bridge over the seam that can withstand the vibrations of transit. Cracking occurs during transit, resulting in a failure to meet the 60 dB attenuation. A solution to this seam leakage problem is therefore the identification of a technique that can bridge the aluminum seams with a conducting metal with sufficient integrity to withstand the rigors of battlefield transport, while preserving the integrity of the adhesive.

16.2 Cold spray particle deposition

The cold spray (CS) technique for metal particle deposition has been shown to yield dense, conductive coatings on aluminum surfaces. Since application temperatures are relatively low (<100 °C), nearby adhesives are not damaged. The coatings produced by CS exhibit good bond strength to aluminum,[3] and their density provides high resilience to vibrations. Application of CS to seam closure for EMI shielding requires an identification of the optimum metal sealant and a specification for the optimum spray parameters.

While CS is often carried out with a stationary, robot-controlled system, the large panel size (3 m × 3 m) and military field application favored the use of a hand-held, portable system. A portable CS device, manufactured by the Dymet Corporation, was selected for use on this application.[4] This portable unit uses shop-supplied compressed air as opposed to the significantly higher pressures used by the stationary unit. The lower pressure necessitates that small quantities (~10%) of alumina ceramic are mixed with the feed powder to increase deposition efficiency. The hard alumina particles scour surface oxides, presenting impacting metal particles with a fresh, oxide-free substrate. The Dymet system is shown in Fig. 16.1. The figure shows the complete system and a cut-away view of the hand-held heater–nozzle assembly.

An optimum metal powder for the EMI application would be one that bonds well with the aluminum panel substrate, has high electrical

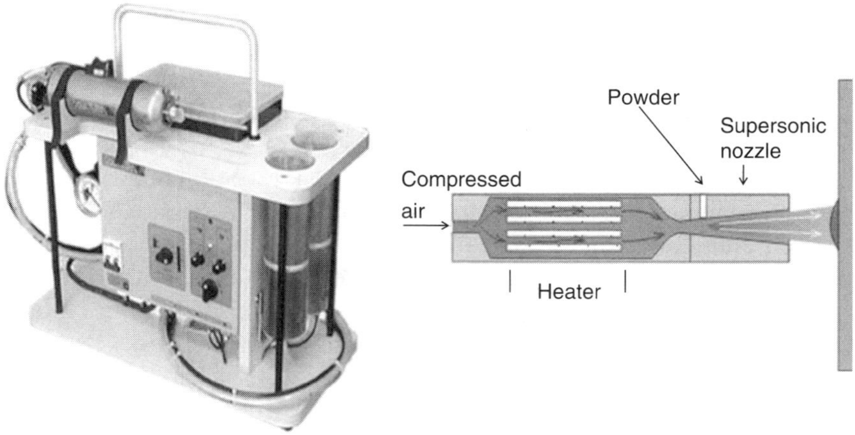

16.1 The Dymet™ portable cold spray system.

conductivity, and is rugged enough to withstand the stresses encountered in field transport of the enclosure.

16.3 Screening tests

The parameters that needed to be varied in order to optimize the CS process of aluminum seam-sealing for EMI shielding were as follows.

1 Powder characteristics:
   - metal and/or ceramic components;
   - particle morphologies;
   - component percentages.

2 Surface treatment:
   - untreated;
   - grit blast.

The characteristics of the resulting CS seal to be evaluated were:

- coating bond strength with the substrate;
- density (or porosity) of the coating;
- electrical conductivity of the coating;
- seam coverage uniformity.

The powders chosen for trial all have proven CS applications, and all have high wrought metal conductivity, greater than 15% International Annealed Copper Standard (IACS). The powders include aluminum, zinc, and tin. Trials were made with the pure metal powders, and with powder mixtures of the metals and alumina additive. The powders selected for evaluation were:

- pure aluminum;
- 90%Al–10%Al_2O_3 powder blend;
- 90%Al–10%Zn powder blend;
- pure tin;
- 90%tin–10%Al_2O_3 powder blend;
- pure zinc;
- 80%Zn–20%Al alloyed.

The micrographs of aluminum and zinc powders in Fig. 16.2 show the morphology of the <325 mesh commercial powders used in this program. The individual particles are roughly spherical, with diameters less than 50 μm. These characteristics make the powders suitable for CS.

The optimum powder for this shielding application was determined through screening tests of deposits on coupons. A 6062 aluminum alloy was used as the test coupon substrate in all cases. The CS nozzle traverse motion was robot-controlled for all test coupon runs for consistency of application. Gas for particle acceleration is heated prior to insertion into the nozzle in order to achieve high velocity, but expansion through the nozzle cools the gas substantially before it exits the nozzle. CS operating conditions for all runs are given in Table 16.1. The tests used to characterize the CS coatings are described below.

16.3.1 Bond strength

The bond strength, or adhesion, of the CS coating to the substrate indicates the ability of the coating to withstand impact and vibrational stresses. Adhesion of the coatings was measured according to ASTM Standard C633-79. Bond strengths of the coatings to an aluminum substrate were made by cold

16.2 Zinc (left) and aluminum (right) powders.

Table 16.1 Cold spray operating parameters

Parameter	Condition
Compressed air	0.6 MPa
Gas feed temperature	>300 °C
Gas exit temperature	<100 °C
Standoff distance	10 mm
Traverse speed	25 mm/s
Line-to-line increment	1 mm

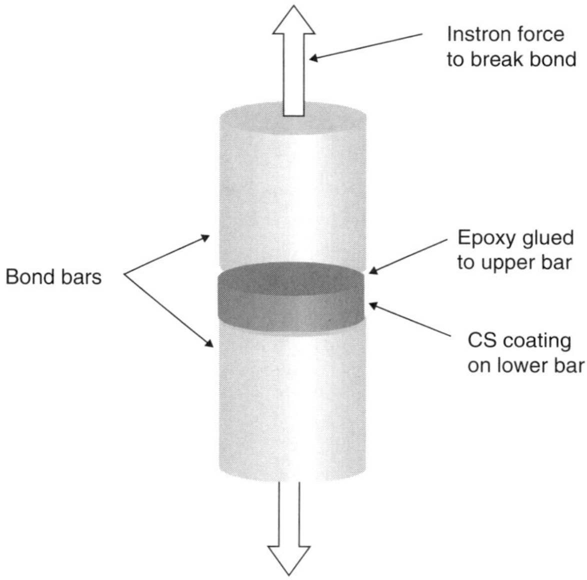

16.3 Adhesion strength test arrangement.

spraying the coating onto the flat end of a 1.6 cm aluminum rod, threaded at the opposite end. The sprayed coating was then glued to a second rod by means of epoxy glue, FM-1000, and subsequently cured at 300 °F. The adhesively bonded bars were then threaded into the cross-heads of an Instron tensile test machine and pulled apart. The loads were measured and converted to tensile strength. When the failure occurs at the coating–substrate interface, adhesion strength is then the applied force at failure, divided by the 1.6 cm circular area. A glue failure indicates that the coating–substrate bond is greater than the measured stress at glue failure. Figure 16.3 shows the bond bar arrangement.

16.3.2 Density

A dense, non-porous, conductive coating provides good electromagnetic shielding and has rugged structural properties. The coating density was determined by optically measuring the pore/solid ratio in a magnified cross-section of the joint. The cross-section was digitally scanned and the ratio determined by AnalySIS™optical software.

16.3.3 Electrical conductivity

The electrical conductivity was measured by a Jentek GridStation™ Meandering Winding Magnetometer (MWM).[5] MWM methods use an advanced, conformable eddy current sensor combined with model-based measurement grids for applications such as coatings characterization, residual stress measurements, crack detection and sizing, fatigue monitoring, and corrosion detection. The MWM sensor is geometrically designed so that its response fits a known magnetic field model. This enables the sensor response to be accurately predicted, permits model-based simulations to be performed for sensor optimization, and enables the measurement of material properties in real-time with minimal calibration requirements. The MWM determines material conductivity by measuring the modification of an applied magnetic field, caused by the presence of the material. The depth of measurement is given by the magnetic skin depth for a given driving frequency, by the equation:

$$\delta = (\pi f \mu \sigma)^{-1/2} \qquad [16.4]$$

where:

δ = depth of penetration (mm)
f = frequency (Hz)
μ = magnetic permeability (H/mm)
σ = electrical conductivity (% IACS)

Conductivity as a function of depth into the material is then obtained when the driving frequency is varied. This method also measures the depth of the coating by giving the depth of the measurement when the conductivity changes from that of the coating to that of the substrate.

16.4 Test results

The need for substrate conditioning (grit blasting) was evaluated before coupons were sprayed for other testing. The grit blast of test sections was carried out with 60 grit alumina media. The effects of grit blasting on the deposition of tin onto aluminum are shown in Table 16.2. It is seen that grit blasting of the aluminum substrate does not affect coating porosity or conductivity (which are not substrate related), but significantly affects the build-up thickness and the coating adhesion to the substrate. These results demonstrate the advantage of grit blast to the quality of the coating, and all coupon substrates used in the subsequent tests were grit blasted.

Typical coating cross-sections are shown in Fig. 16.4. Pure aluminum and aluminum/alumina blend powder coatings are shown. The alumina particles appear black in a white aluminum matrix for the case of the aluminum/alumina blend. The coatings are quite dense, with porosities of <1%. A tight bond line can be observed between the coating and the aluminum substrate, with bond strengths greater than 6 and 12 MPa for the pure aluminum and the aluminum/alumina blend, respectively.

The conductivities of zinc coatings measured by MWM are shown in Fig. 16.5. This figure shows the conductivity of coatings as a function of the skin depth of the applied magnetic field. It is thus a plot of conductivity versus depth into the coating. As the depth of measurement increases, the conductivities transition from that of the coating material to that of the aluminum substrate. As reference, the conductivities of the pure materials

Table 16.2 Surface conditioning effects on tin deposition

Surface preparation	None	Grit blasted
Thickness (μm)	80	230
Adhesion (MPa)	0.86	3.96
Porosity (%)	0.2	0.2
Conductivity (% IACS)	13	13

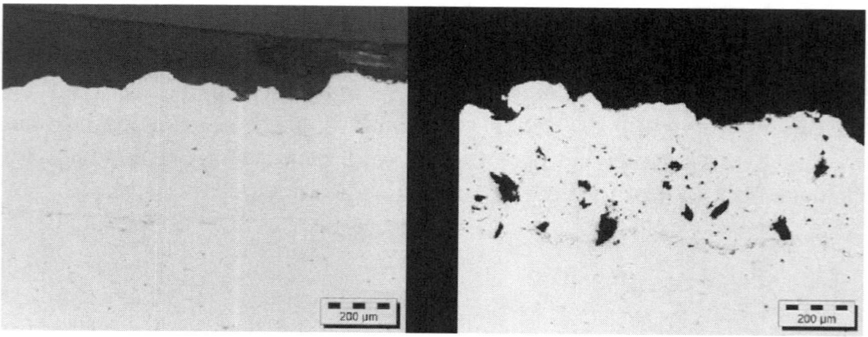

16.4 Cross-sections of CS-deposited Al (left) and Al/Al_2O_3 (right) on aluminum.

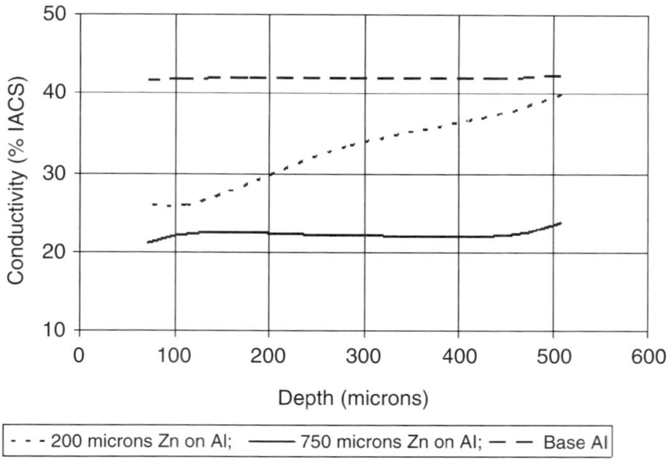

16.5 Zinc conductivity as a function of depth over an aluminum substrate.

Table 16.3 Comparative performances of coatings

Coating powder	Thickness (μm)	Adhesion (MPa)	Conductivity (% IACS)
Aluminum	175	5.9	43
90%Al–10%Al_2O_3	300	12.2	44
90%Al–10%Zn	500	9.6	21
Tin	380	4.0	13
90%tin–10%Al_2O_3	75	4.2	12
Zinc	150	6.2	19
80%Zn–20%Al (Alloy)	300	11.4	24

are 29% IACS for zinc and 42% for the 6062 aluminum alloy base. The 750 μm coating shows a constant conductivity of 23% IACS for all measurement depths down to 500 μm, the MWM penetration limit. A 200 μm coating approaches 23% IACS wrought metal value near the surface and approaches the 42% IACS of 6062 aluminum, as measurement penetration increases toward the depth of the aluminum substrate.

A comparison of the measured properties of the metal coatings is given in Table 16.3. The coating thickness is an indication of the deposition efficiency of the process, since the powder feed rate was equal for all runs. Significant differences in adhesion and conductivity can be seen. It was determined from these results that the aluminum–alumina mixture gave the best combined performance and was therefore chosen to be used for trial on a full-scale enclosure.

16.5 Application

A lightweight, transportable, (EMI) protected, rigid wall, tactical shelter was used for a demonstration of this seam shielding technique.

The shelter is constructed from 3 cm thick panels consisting of aluminum skins that are bonded to a fibrous honeycomb core. These panels are assembled as the walls of the enclosure. A typical seam at the junction of an aluminum skin panel and a panel termination strip is shown in Fig. 16.6. This seam between the panels forming a lap-joint is one of many such seams comprising the shelter which must be sealed with a conducting fill in order for the structure to achieve the necessary shielding.

Based upon the test results reported in Section 16.4, a 90% Al–10% Al_2O_3 powder blend was selected to seal these seams. This powder was applied to a shelter enclosure by means of the hand-held CS system, using the following operating parameters: compressed air, 0.6 MPa; gas inlet temperature, >300 °C; standoff distance, 10 mm. The CS application of the aluminum powder mix to the shelter enclosure is shown in Fig. 16.7. Approximately 30 cm of seam per minute was EMI shielded in this fashion. An example of this fill seal is shown by the cross-section presented in Fig. 16.8. The seam is completely filled with the aluminum–alumina blend. The interface between the fill and the substrate aluminum panel is tight, with no gaps or delaminations.

Except for powder cost, all material and utilities costs for this application are negligible. At a spray rate of 30 cm of seam per minute, a labor cost of $40 per hour, a powder cost of $30 per kg, and 50% deposition efficiency, the operating cost for this technology is $7 per meter of seam.

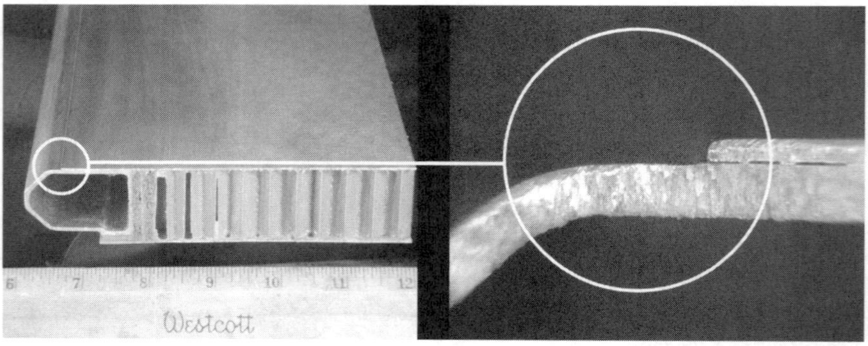

16.6 A shelter wall panel seam.

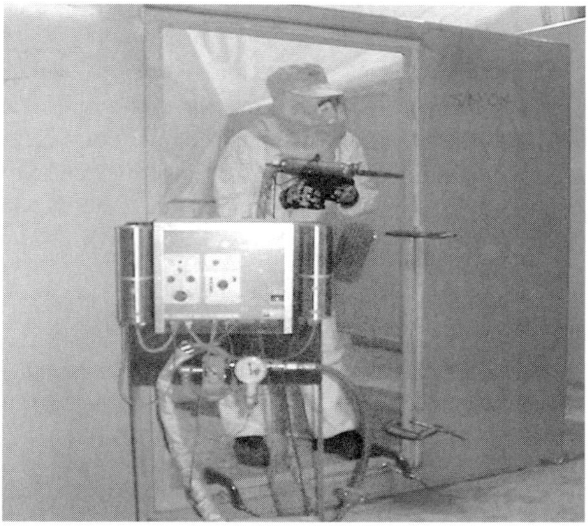

16.7 Cold spray application to the shelter.

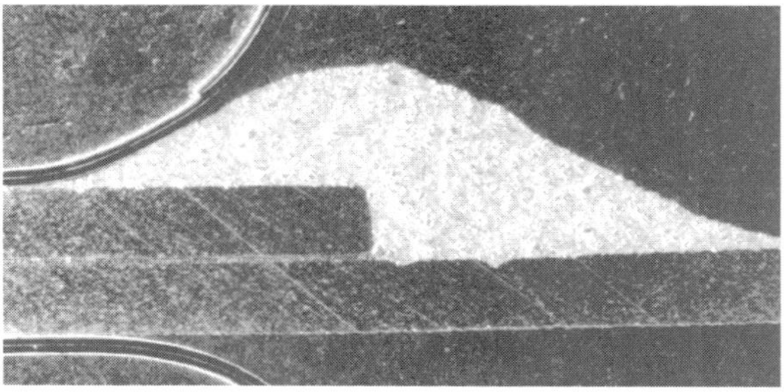

16.8 Filled seam cross-section.

16.6 Conclusions

A low-temperature EMI sealing method for panel seams has been demonstrated. The CS method has produced a well-adhering, high-conductivity, seam filler for aluminum panel joints that are glued together. It has been shown that a portable CS system can seal seams with consolidated aluminum powder having an electrical conductivity equivalent to the panels. Adhesion tests have shown that the deposited aluminum has bond strength with the aluminum substrate greater than 12 MPa. The seams of a complete personnel shelter were sealed in this manner in a few hours. With reasonable ventilation, this method of seam-sealing can be accomplished on the manufacturing shop floor. The hand-held CS system is compact, and requires only shop compressed air and a plug-in electrical supply.

16.7 Acknowledgments

The author wishes to thank Phillip Leyman and Robert Lempicki of the ARL Materials Applications Branch for their help in the operation of the cold spray system. The author would also like to thank Youn Lee of ARL Sensors and Electron Devices Directorate, for his introduction to this application and his suggestions with respect to military requirements and parameter ranges.

16.8 References

1 MIL-HDBK-1195, 'Radio Frequency Shielded Enclosures', September, 1988.
2 MIL-STD-461E, 'Requirements for the Control Electromagnetic Interference Characteristics of Subsystems and Equipment', August, 1999.
3 Champagne V, Helfritch D, Leyman P, Grendahl S and Klotz B (2005), 'Interface Material Mixing Formed by the Deposition of Copper on Aluminum by Means of the Cold Spray Process', *J. Thermal Spray*, **14**(3), 330–334.
4 Maev R, Strumban E, Leshchynsky V and Beneteau M (2004), 'Supersonically Induced Mechanical Alloy Technology (SIMAT) and Coatings for Automotive and Aerospace Applications', *Proceedings of Cold Spray 2004*, 2004.
5 Goldfine N, Washabaugh A, Zilberstein V, Schlicker D, Shay I, Grundy D and Windoloski M (2002), 'Absolute Electrical Property Imaging Using High Resolution Inductive, Magnetoresistive and Capacitive Sensor Arrays for Materials Characterization', *Proceedings of the 11th International Symposium on Nondestructive Characterization of Materials*, 2002.

17
Repair of magnesium components by cold spray techniques

V. K. CHAMPAGNE, US Army Research Laboratory, USA

17.1 Introduction

The US Army and Navy have experienced significant corrosion problems with magnesium alloys that are used to fabricate aircraft components. The most severe of these are associated with large and expensive transmission and gearbox housings for rotorcraft, which have to be removed prematurely because of corrosion, adversely affecting fleet readiness and safety. Many of the parts cannot be reclaimed because there is not an existing technology that can restore them adequately for service. The replacement of these parts is very expensive, ranging in the millions of dollars every year. One common repair technique, used for some of those parts that can be salvaged, involves the use of aluminum shims, which are adhesively bonded over areas where the corrosion has been ground down and dimensional restoration is required.

The US Army Research Laboratory (ARL) Center for Cold Spray Technology? has developed a cold spray process to reclaim magnesium components that shows significant improvement over existing methods and is in the process of qualification for use on rotorcraft. The cold spray repair has been shown to have superior performance in the tests conducted to date, is inexpensive, can be incorporated into production, and has been modified for field repair, making it a feasible alternative over competing technologies. The work presented in this chapter represents the first 2 years of a 3 year effort, which will result in the establishment of a demonstration cold spray facility at a naval air depot where the overhaul and repair of Navy rotorcraft is accomplished. The program involved the participation of all branches of the US Department of Defense (DOD), major US helicopter companies, academia, and international groups.

17.2 Problems in using magnesium components

The widespread use of magnesium in aircraft occurred during the Vietnam era to reduce weight and increase performance.[1] Magnesium is approximately 35% lighter than aluminum and has exceptional stiffness and damping capacity.[2] Magnesium alloys, therefore, are used for the fabrication of many components on US DOD aircraft, especially for complex components such as transmission and gearbox housings on helicopters, and gearbox housings on fixed-wing aircraft.

Magnesium is a very active metal electrochemically and is anodic to all other structural metals. Therefore, it must be protected against galvanic corrosion in mixed-metal systems because it will corrode preferentially when coupled with virtually any other metal in the presence of an electrolyte or corrosive medium.[3] Many of the corrosion problems associated with magnesium helicopter components occur at the contact points between inserts or mating parts, where ferrous metals are located, creating galvanic couples.[4] In addition, magnesium alloys are also very susceptible to surface damage due to impact, which occurs frequently during manufacture and/or overhaul and repair. Scratches from improper handling or tool marks can result in preferential corrosion sites. The DOD and the aerospace industry have expended much effort over the last two decades to develop specific surface treatments to prevent corrosion, to increase surface hardness, and to combat impact damage for magnesium alloys in order to prolong equipment service life;[5] however, the means to provide dimensional restoration to large areas on components where deep corrosion has occurred remains a challenge.

17.2.1 Environmental and safety concerns

The current surface protection methods employed by many original equipment manufacturers (OEMs) pose special environmental and safety concerns. Much of the industry still relies on hard anodizing using a process designated Dow 17 or a chromate conversion process designated MIL-M-3171, which is used by repair depots. All involve the use of hexavalent chromium. The electrolyte used to apply the coating comprises sodium dichromate, ammonium acid fluoride, and phosphoric acid which are hazardous to the environment and difficult to recycle. These operations also produce harmful vapors and hazardous waste in the form of contaminated wastewaters and solid waste, which can be costly to dispose of. There are also safety issues to contend with, such as the formation of hazardous vapors that contain metal salts and carcinogenic substances. These rise up from heated bath solutions and require additional worker protection measures.

These surface treatment processes are followed with the application of a phenolic resin sealer and then a chromated primer and topcoat, on most surfaces. Robinson[6] has reported that this surface treatment regimen, which incorporates the use of sealers, can increase the resistance of magnesium to salt spray corrosion. However, regardless of the progress made to improve the corrosion resistance of magnesium alloys, millions of dollars are expended each year on repair and replacement of magnesium alloy components that have corroded or have been damaged. In addition, emerging legislation will be reducing the hexavalent chromium permissible exposure limit from 52 to 1 $\mu g/m^3$ making the replacement of processes using this compound mandatory. The qualification of the cold spray repair process for magnesium alloy components would represent a significant achievement in the elimination of hexavalent chromium-containing compounds currently used to provide surface protection.

17.2.2 The cost of corrosion

The Army and Navy helicopter fleet comprises more than 3000 aircraft, each having several components manufactured from magnesium alloys, generally consisting of transmission and gearbox housings. In addition to helicopters, magnesium alloy gearboxes are used extensively in fixed-wing aircraft. This represents thousands of magnesium components currently in service within the DOD. The cost of each component ranges from several thousand to over four hundred thousand dollars. The total cost of all magnesium alloy components on all DOD helicopters easily exceeds several hundred million dollars. The susceptibility of these components to corrosion and damage, even with the current surface protection schemes, is significant, resulting in very high repair/overhaul costs, to the extent that the Navy has even considered different alloys to replace magnesium.

In 2001, the US Navy conducted an extensive review of the cost of corrosion on the main transmission of one type of helicopter and reported that from 1991 to 2000, the total estimated cost for both unscheduled maintenance and module replacement was $41 million. The US Army has reported costs of $8 million annually. Most of the corrosion occurs at attachment points where a dissimilar metal is in contact with the coated magnesium component. This includes flanges, mounting pads, tie rods, lugs, and mounting bolts. Figure 17.1 is a schematic of the main transmission housing for a helicopter showing the areas most susceptible to corrosion. Figure 17.2 shows the extent of corrosion damage on one of the mounting pads after one overhaul cycle. Because of the localized nature of the corrosion, surface treatments intended to mitigate the problem would only have to be applied in these specific areas.

17.3 Limitations of current technologies

It had been determined through extensive testing and evaluation by several major US aerospace manufacturers and the DOD that the deposition of

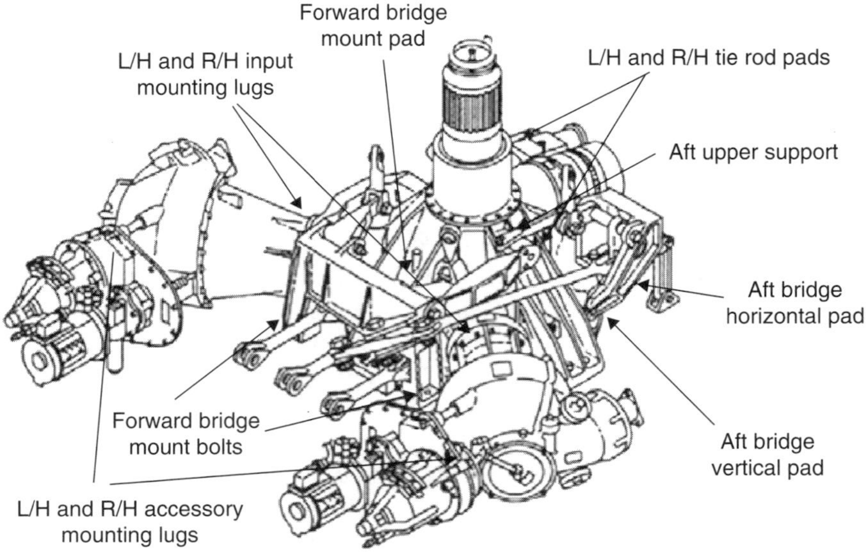

17.1 Schematic of a main transmission housing showing areas most susceptible to corrosion. L/H and R/H, left-hand and right-hand, respectively.

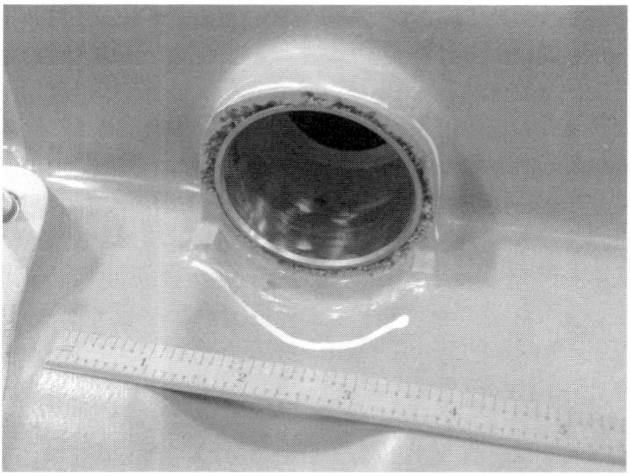

17.2 Corroded area on a magnesium transmission housing.

aluminum and certain aluminum alloys followed by the phenolic resin and a sealant would have the highest probability of success in reducing corrosion and impact damage. The technical challenge was to identify a method to deposit the aluminum alloys onto the magnesium and meet all service requirements for bond strength and corrosion resistance without sacrificing the structural integrity of the substrate. The processing of production parts as well as field repair capability was an important attribute in the selection of a viable process. Initially, a review of the current state-of-the-art was conducted and critiqued before cold spray was down-selected as the technology of choice for this application.

17.3.1 Ion-vapor deposition (IVD)

Ion-vapor-deposited aluminum is a method of applying pure aluminum onto metallic, as well as non-metallic substrates, through metal vaporization. The process is conducted in a vacuum chamber and requires stringent controls, proper air conditioning, dehumidifying, and avoidance of contaminates; otherwise, the coating will have poor integrity. Small parts to be coated are loaded onto rotating stainless steel barrels, while larger parts are attached by hooks or wire onto a rack prior to charging the IVD vacuum chamber. The vacuum chamber is evacuated to purge the system, subsequent to backfilling with argon, which ionizes and creates a glow discharge around the parts, bombarding them with positively charged ions. This plasma operation serves to clean the substrate by removing surface contamination. The anode is able to accept aluminum wire, which is fed into the system and melted through the use of a high current before it passes through the glow discharge where it combines with the ionized argon and is attracted to and deposited onto the substrate, providing a uniform, dense, and adherent coating.

IVD aluminum would not be viable because it would be difficult to place large components in a vacuum chamber, the coatings are too porous, and it is not possible to deposit the aluminum in the required localized areas. In addition, rework parts would not be clean enough for this process and field repair would be impossible.

17.3.2 AlumiPlate®

AlumiPlate® is a process for electroplating high-purity aluminum onto various conductive substrates in a completely closed system. However, it has been found that AlumiPlate® does not adhere well to magnesium and would require a bond layer. It is also difficult to deposit in localized areas. In addition, AlumiPlate® cannot deposit coatings with sufficient thickness for most repair operations requiring rebuild to restore original dimensions.

17.3.3 Plasma and high-velocity oxy-fuel (HVOF) thermal spray

Plasma and HVOF thermal spray have been investigated for deposition of aluminum, but the results have generally been unsatisfactory due to inconsistent coating integrity. Poor adhesion and delamination of the coatings are typically the cause for high rejection rates. Both of these processes involve the use of high thermal energy to melt or partially melt the coating material before it is accelerated onto the surface of the substrate. The tiny molten or partially melted particles rapidly solidify upon impact with the substrate and contract forming tensile residual stresses in plasma spray coatings. This is not always the case with the HVOF process because the particles are accelerated at high velocity and have been known to form coatings that are in compression. Regardless, failures occur because the plasma and HVOF processes can generate excessive heat causing the formation of an oxide on the magnesium that is detrimental to adhesion. Besides, the thermal spray pattern is very wide so that it would be difficult to apply the coatings to localized areas requiring repair or rebuild.

17.4 Key issues in using the cold spray process

Because of the previous and ongoing work being conducted by the US ARL, it was clear that commercially pure aluminum (CP-Al) and 5056 Al coatings of various thicknesses could be deposited onto Mg alloys using the cold spray process. The coatings have been shown to be very dense and have high adhesion values. Therefore, the technical risk associated with the actual deposition process was very low. The principal technical risk was related to the extent to which the application of the aluminum cold spray coatings would improve the corrosion performance of the Mg alloys and the effect on fatigue properties of the substrate. It was anticipated that a significant amount of parametric optimization work would have to be performed in order to optimize the corrosion resistance of the coatings. Other risks were associated with the cost of the process as well. Higher particle velocities and, in many instances, superior coatings are obtained using helium gas which is very expensive and might be cost-prohibitive. An issue was raised as to whether acceptable coatings properties could be achieved using less-expensive nitrogen or whether a recycling system was required with helium. If the desired level of corrosion performance could be achieved using nitrogen or by using a helium recovery system, then there would be little risk associated with application of the coatings onto actual components and the implementation of the technology into a depot environment because of the similarity of the cold spray process to other thermal spray processes, which are used on components of similar geometry and complexity.

The development and qualification of the cold spray process to deposit CP-Al was recommended by the ARL Center for Cold Spray for the purpose of dimensional restoration and protection to magnesium. The cold spray process was viewed as the best possible method for depositing the aluminum coatings and would be viewed as part of an overall strategy of replacement of the chromate processes such as Dow 17 and MIL-M-3171 in use today, eliminating environmental and worker safety issues, while significantly improving performance and reducing life-cycle costs.

17.5 Developing and testing cold spray coating of magnesium

The critical tasks associated with the qualification of the cold spray coatings on magnesium components was established in a document known as the Materials Joint Test Protocol (JTP) developed by representatives of the DOD, OEMs, industry, and academia. The JTP comprised mechanical, physical, and chemical tests designed specifically for this application. The most important qualification tests were identified as adhesion (uniaxial tension), salt fog exposure (ASTM-B117), and microstructural analysis, which will be the primary focus of this discussion. These screening tests would form the basis of the data required to recommend the use of cold spray to repair magnesium aircraft components. Additional tests would later be conducted to satisfy specific application requirements for structural repair.

17.5.1 Nonstructural repair test plan

The optimal cold spray process parameters to deposit CP-Al onto ZE-41A magnesium for nonstructural applications were developed by executing the following test plan and leveraging off of previous work:

- *Cold spray systems*: ARL developed a portable high-pressure cold spray system specifically for this application. In addition, a new nozzle was designed and fabricated to prevent clogging which often occurs while spraying aluminum.
- *Selection of gases*: initial spray trials using helium gas were very successful in achieving high adhesion and corrosion resistance. Cold spray process parameters were also developed with nitrogen to reduce costs while maintaining satisfactory coating performance.
- *Powders*: various commercially available CP-Al powders were chosen for their flow characteristics, size, geometry, and cost to improve flow and increase deposit efficiency. Aluminum alloys Al-125i and 5056 were also recommended for use.
- *Predictive modeling*: the predictive model developed at ARL was used to determine starting values of gas temperature and pressure, which resulted in an appropriate particle velocity and surface temperature for optimal coating deposition and minimal damage to the substrate.

- *Cold spray trials*: a series of cold spray trials were performed to optimize coverage, adhesion, and coating integrity. Adjustments of the spray parameters were recorded and correlated to test results to improve the overall cold spray process.
- *Coating characterization*: test coupons of the cold spray coating were subjected to adhesion, hardness, and corrosion testing according to the requirements established by the JTP. Optical and electron microscopy were used to characterize the coating/substrate interface and to access the integrity of the deposit. The density, microstructure, and inherent material features of the coating were characterized.

17.5.2 Structural repair test plan

Additional mechanical tests were recommended by the JTP as part of a test plan to qualify the cold spray process for structural repair of magnesium components, which is not the focus of this work but is included for future reference for those wishing to consider this repair for structural components. The following tests would be performed to determine whether the cold spray coating adversely affected the fatigue resistance of the substrate and to test the impact resistance of the coating.

- *Fretting fatigue testing.* This would be performed using an apparatus similar to the United Technologies Research Center (UTRC) equipment which utilizes standard axial fatigue specimens except with a rectangular cross-section in the gage area, where the coatings to be evaluated are applied. Fret pins bear against the center of the gage surface with a constant load and slip against the specimen at an amplitude controlled by an actuator. The fatigue specimens are subjected to cyclical loads that would be expected to result in specimen failure in approximately 105 cycles in the absence of the fretting. The number of cycles to failure is measured and compared with a baseline material.
- *Impact testing.* Because much of the damage observed on magnesium alloy components in service is due to impact from foreign objects, evaluation of cold spray-coated specimens would be appropriate. Two types of impact testing would be performed: the first being high-velocity, small-particle impact (evaluated by gravelometry) and the second low-velocity, large-item impact simulating dropped tools, handling damage, etc. The gravelometry would follow the ASTM D310 protocol, which involves feeding a stream of gravel into an air jet and allowing it to strike the coated specimen at a defined angle. The large-item impact would involve dropping a steel ball bearing from a specific height onto a coated specimen and evaluating the surface damage.

17.5.3 Additional corrosion tests

Corrosion tests would later be performed on selected Mg alloy samples having only the cold spray CP-Al alloy coating and on Mg alloy samples containing a 'complete' finishing operation that would include the cold spray Al coating, Rockhard resin, and sealant once the screening tests were completed (Section 17.5.1).

- Potentiodynamic polarization measurements in accordance with ASTM G5 and G59 in different electrolyte solutions to characterize performance of the cold spray Al alloy coatings in terms of open circuit potential, passivation behavior, and pitting potentials.
- ASTM G85, Annex 4 SO_2 salt-fog corrosion tests on coated panels.
- Crevice corrosion testing.
- Field corrosion testing. The Navy maintains corrosion test racks on selected aircraft carriers for the purpose of conducting field trials on materials and coatings. Mg alloy panels containing the complete surface finishing of cold spray CP-Al alloy, Rockhard resin, and sealant would be evaluated in comparison to the current surface finishing process. An assessment would be made of the extent to which the corrosion life has been increased through the use of the new coating process.

17.5.4 Flight testing

In addition to development and execution of the Materials JTP and in conjunction with the demonstration of the cold spray coatings on actual components, lead-the-fleet flight testing was included for structural repair of magnesium components. For this type of application, component rig tests are generally not applicable since they do not simulate the corrosive environment encountered in service, so flight testing is the best way of evaluating the airworthiness of the cold spray coatings.

17.6 Cold spray technology for coating magnesium

The ARL Center for Cold Spray currently maintains two high-pressure stationary cold spray systems, one manufactured by Ktech Corporation and the other by CGT Incorporated, as well as four portable cold spray systems. Two of the portable cold spray systems were manufactured by Dymet Inc., one by Centerline, and the remaining system was produced at ARL. All of the portable systems are considered low pressure with the exception of the ARL system which can operate at low or high pressures. The work described in this chapter was conducted with the Ktech system and the portable ARL system.

17.6.1 Cold spray nozzle design

The conventional cold spray nozzle that is used with the Ktech system is normally fabricated from stainless steel or tungsten carbide. Various nozzle configurations have been designed and tested at ARL and it is not the intent of this chapter to repeat the research of others but to relate those aspects of nozzle design specific to this application. The primary concern was clogging of the nozzle while spraying CP-Al. Clogging can occur in the throat of the de Laval-type nozzle, where higher temperatures are employed. Aluminum particles tend to stick to the sides of the nozzle interfering with proper gas flow, which adversely affects coating deposition. To mitigate clogging, a plastic nozzle was used with absolute success. ARL and others have conducted studies of various nozzle configurations and materials. Comparisons have been made of nozzles fabricated from ceramic, plastic, carbide, and other metallic materials. ARL has designed a nozzle fabricated from a high-temperature-resistant plastic that can be operated at 400 °C with satisfactory results. There are several plastics on the commercial market that can be successfully machined and used as a nozzle material. Such plastics can maintain their properties at high operating temperatures (400 °C) and are adequately wear resistant for use with CP-Al for extended periods of time. These plastics tend to be proprietary in nature but can be obtained from the cold spray equipment manufacturers.

17.6.2 Selection of gases

The decision to use helium over nitrogen relies primarily on the added costs involved. However, from a technical standpoint the velocities that can be achieved by helium and the resultant density of the coatings may well be worth the extra cost, especially for components that are valued at \$400 000 each. During operation at ARL, the Ktech cold spray system uses 67 m^3/hour of nitrogen at a cost of \$0.1/m^3 for a gas operating cost of \$0.116/ minute. With helium, it uses 118 m^3/hour at a cost of \$4.10/m^3 for a gas operating cost of \$8.05/ minute. Therefore, the costs are substantially lower when using nitrogen. To put the gas costs into perspective, the labor rate for operating a cold spray system is about \$1.00/minute and the powder costs are about \$2.00/ minute. Helium recycling systems have been designed and put into use that are able to recover approximately 90% of the helium, thereby greatly lowering the operating cost.[7] For a research and development cold spray facility that operates intermittently, the cost of the recycling system cannot be justified. However, for a production cold spray facility, the payback time on the cost of the recycling system would be fairly short for these expensive components. Regardless, the results of this work demonstrated that nitrogen can be used as a carrier gas to produce satisfactory cold spray coatings of CP-Al.

17.6.3 Powders

Vlcek *et al.*[8] has provided extensive information related to the physical processes that occur during the cold spray deposition of different materials and why some work better than others. Impact heating, equations of state during impact, dynamic yielding of the particles, and impact pressures were examined and it was concluded that the materials that are most amenable to cold spray are those with a face-centered cubic structure, which includes aluminum.

The powders under consideration for the cold spray repair of magnesium transmission housings were CP-Al, Al-12Si, and Al 5056. The criteria were based on the requirements associated with this application where galvanic corrosion and corrosion pitting were the primary causes for removing the components from service. In addition, the repair was confined to nonstructural areas of the transmission and gearbox housings. There were advantages and disadvantages associated with each material.

The 5056 Al (composition Al–5Mg–0.1Mn–0.1Cr) was considered because it is compatible with magnesium and has better tensile, yield, elongation, and fatigue strength than any magnesium alloy. The corrosion resistance of 5056 is also among the best of any aluminum alloy. The presence of the manganese is important because it serves to tie up any residual iron contamination which has been shown to degrade corrosion performance. The Al–12Si was chosen as a candidate based upon its excellent mechanical properties and resistance to wear and corrosion. It is also used extensively with thermal spray and the powders are commercially available.

Initially, concern was expressed over CP-Al as the material of choice for the cold spray repair of magnesium ZE-41A because it was lower in hardness and strength than other aluminum alloys that were also being considered for this application. It was later selected since the primary reason that the magnesium components were removed from service was due to corrosion and not wear, and the cold spray repair was to be performed on nonstructural areas of the component, making the strength requirement less of an issue. Additionally, the CP-Al powder could be doped with a certain percentage of hardened particles to impart wear resistance if required. The cost of CP-Al was attractive at approximately $5/kg and it was commercially available, while many other alternative alloys originally considered were much more expensive and/or had to be produced as a specialty item. The Al 5056 was an example of a powder that was not commercially available. The stock material used in the melt of the Al 5056 powder was also not available forcing the powder manufacturer to purchase and subsequently mix the raw materials to produce the alloy prior to atomization.

17.3 Particle shape effects on deposition.

Finally, the spherical particle shape of the CP powder was preferred over that of the Al 5056 flake. Spherical particles form a more densely packed structure when deposited by the cold spray process resulting in better protection for the magnesium substrate. Figure 17.3 shows micrographs of the two powder types.

17.7 Predictive modeling for process optimization

The nature of the bond created during particle consolidation and the properties of the material produced by the cold spray process have been modeled at ARL. These predictive models are used to establish and optimize cold spray process parameters. Modeling efforts predict the amount of mixing at the interface between the particles and the substrate with concomitant high coating adhesion when the particle velocity reaches a certain minimum value. Compressible, isokinetic flow equations are used to model gas flow within the nozzle. Modified drag and heat transfer coefficients are then used to iteratively calculate the resulting particle velocities and temperatures. An example of this calculation for the aluminum/helium nozzle acceleration is shown in Fig. 17.4. The particle size for this calculation is 20 μm, and the initial gas pressure and temperature are 2.75 MPa and 20 °C, respectively. Calculated particle velocities are verified experimentally by means of a dual-slit, laser velocimeter.

An empirical relationship of the penetration of micrometeorites into spacecraft skin is used to model the interface between the deposit and the substrate, and an empirical relationship between particle characteristics and critical velocity is used to model deposition efficiency. The cold spray material is evaluated for shear strength at the interface and for hardness. Magnified cross-sections are examined for density and interface locking.

The demonstrated prediction accuracy of these results allows the definition of operating parameter values and expected coating characteristics

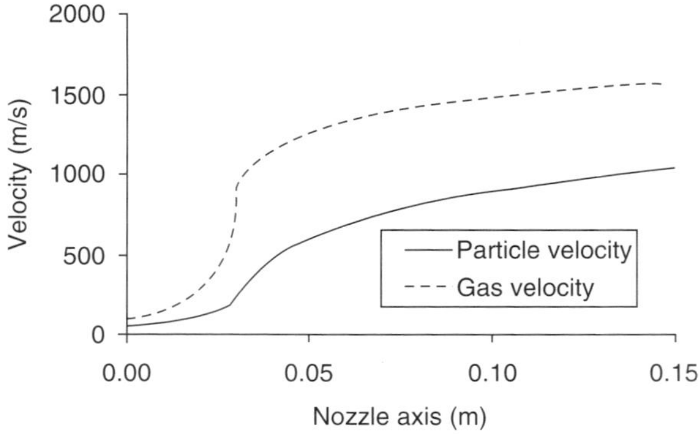

17.4 Aluminum particles (20 μm) accelerated through a 150 mm nozzle by helium.

prior to cold spray operation. This prediction algorithm was subsequently used on multiple coating/substrate combinations with favorable results.

17.8 Cold spray trials

A series of cold spray trials was performed to optimize coverage, adhesion, and coating integrity. Each trial run required the specification of the powder, powder feed rate, gas, gas pressure, and gas temperature. Gas pressure and temperature were specified by model calculations to produce sufficient particle velocity for good deposition. Subsequent coating evaluation (see Section 17.9 below) prescribed changes to the operating conditions if coating characteristics needed improvement.

A stationary and a portable cold spray system were used in these tests. A schematic of each system is shown in Fig. 17.5. The stationary system, supplied by Ktech allows for the range of operation parameters given by Table 17.1. Because the portable system does not have a gas heater, nitrogen gas cannot be used, as unheated nitrogen gas cannot achieve the required particle acceleration. The operating limits for the portable system are also given in Table 17.1.

Besides the limitations imposed by maximum allowable temperature and pressure, two operational restrictions narrowed the available parameter range – these were nozzle fouling and ambient temperature gas (for the portable unit). Low melting point powders can stick to nozzle walls, eventually plugging the nozzle. This is especially true for aluminum powders. Limiting gas heating to lower temperatures can prevent nozzle fouling.

A series of cold spray trials were carried out within the parameter limits listed in Table 17.1, in order to determine optimum spray conditions for the

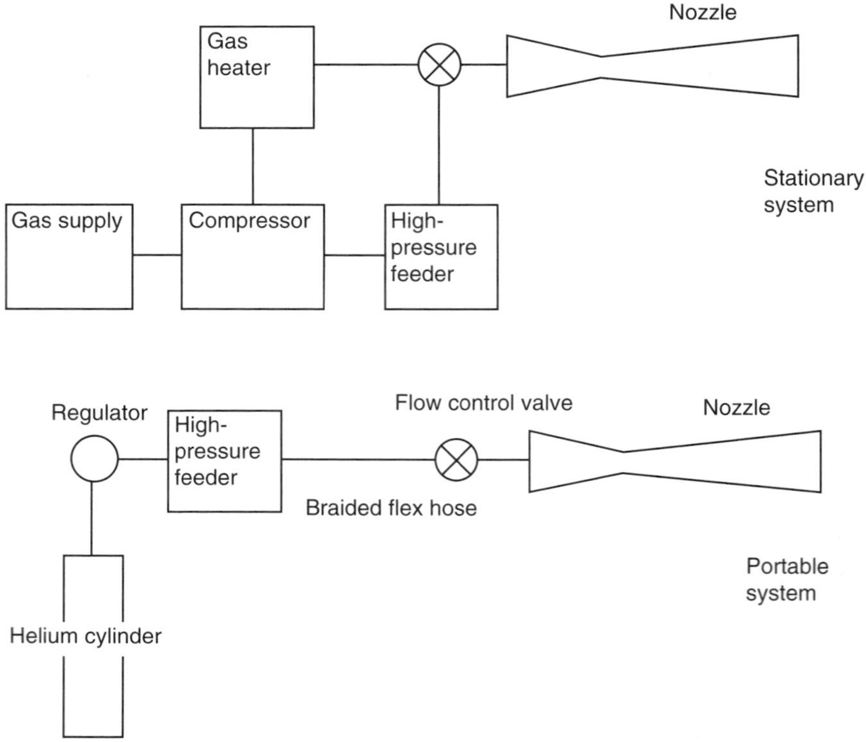

17.5 Stationary and portable cold spray system schematics.

Table 17.1 Cold spray operating parameters

Parameter	Stationary system	Portable system
Gas	Nitrogen, helium	Helium
Gas pressure	1–2.74 MPa	1–2.74 MPa
Gas temperature	20–500 °C	20 °C
Powder feed rate	1–50 g/min	1–5 g/min

two systems. As expected, maximum pressure and temperature produced the best coatings, where the coatings were evaluated by the methods described in Section 17.9 below. In addition, it was found that helium gas produced superior coatings as compared with nitrogen when using a conventional nozzle fabricated from stainless steel or tungsten carbide. However, fouling occurred when temperatures in excess of 250 °C were used with conventional nozzles. It was therefore necessary to conduct

Table 17.2 Optimum conditions

Parameter	Stationary system (standard nozzle)	Portable system (standard nozzle)	Stationary system (plastic nozzle)
Gas	Helium	Helium	Nitrogen
Pressure	2.74 MPa	2.74 MPa	2.74 MPa
Temperature	250 °C	20 °C	400 °C
Gas flow rate	122 m^3/hour	41 m^3/hour	39 m^3/hour
Powder feed rate	3 g/min	1 g/min	3 g/min
Stand-off distance	25 mm	10 mm	25 mm
Particle MMD*	20 μm	20 μm	15 μm

* Mass mean diameter.

similar studies incorporating a plastic nozzle. Higher gas temperatures could be attained without nozzle fouling through the use of plastic nozzles, consequently denser deposits with significantly increased bond strength were achieved. Optimum conditions determined by these tests are given in Table 17.2.

17.9 Coating characterization

The mechanical properties, corrosion resistance, and microstructural features of the CP-Al cold spray coatings were analyzed. The results obtained are provided in this section, as well as a discussion of the test methods employed, when they are determined to be of significant importance to the data obtained or are necessary in order that test results may be duplicated by other researchers. Special surface preparation techniques, test fixtures, specimen geometry, or other procedures that affect test results are worthy of discussion and will provide valuable insight to those wishing to gain a better understanding of the evaluation of cold spray coatings. Bond strength and corrosion resistance were used as the primary screening tests to qualify the use of cold spray for this application.

17.9.1 Bond strength

The adhesive strength (strength of the bond between the coating and the substrate) and cohesive strength (strength of the bond between particles) of the cold spray coatings was determined to be an important factor in the qualification of the cold spray process for use on magnesium by the aerospace community. A major problem associated with the application of thermal spray coatings onto magnesium is the formation of oxides on the surface of the substrate, which adversely affects adhesion.

Conventional thermal spray processes involve preheating the powders to a semi-molten state prior to being deposited onto the substrate. Magnesium is a highly reactive material and is very susceptible to oxidation. An advantage of the cold spray process for this application is that the powders are not heated sufficiently to cause the formation of a deleterious oxide layer.

Adhesion of the cold spray coatings was measured according to ASTM Standard C633-79. This test required the use of two bars fabricated from the substrate material (ZE-41A magnesium). The bars each had a diameter of 2.5 cm and were threaded on one end, in order that they could be inserted into the test fixtures of a tensile test machine. The bond bar assembly is shown in Fig. 17.6. In the first step of this test, a cold spray coating was deposited onto the nonthreaded portion of one of the bars that had been lightly abraded with 60 grit aluminum oxide. It was essential that the grit should not be contaminated with extraneous particulate matter, such as iron, which could become imbedded into the substrate and affect adhesion or corrosion test results. The main purpose of the surface cleaning operation was to remove the magnesium oxide layer and expose the fresh metal beneath in order to facilitate the bonding mechanism. The cold spray coating was applied immediately afterward. Further process development negated the requirement of grit blasting but the fact remains that the surface of the substrate should be free from oxide and debris for optimal adhesion.

Next, the second bar was bonded to the coated end of the first bar with an adhesive agent, and subsequently cured at 150 °C. The adhesively bonded bars were then threaded into the cross-heads of a tensile test machine and

17.6 Bond bar test procedure.

pulled apart. The loads were measured and converted to tensile strength. This test is preferred over the conventional 'dolly' or the 'Patti' test where 2.5 cm diameter plugs are bonded to a coated test panel because the bond bar test conducted with the use of a tensile test machine best ensures that uniaxial tension is maintained throughout the duration of the test and also it eliminates the risk of excess adhesive seeping out from around the edges of the plugs, which in turn can bias the test results by increasing the cross-sectional area being tested, thus yielding erroneously higher adhesion values.

The test results shown in Table 17.3 revealed that in all instances where helium was used as the accelerating gas the glue failed before the coating. Therefore, the reported adhesive strengths represent that of the glue and those of the coating are higher. In the test trial where nitrogen was used with the standard nozzle the values were significantly lower as a result of the lower particle velocities achieved at low gas temperatures. In order to achieve higher gas temperatures without clogging of the nozzle a plastic nozzle was incorporated with satisfactory results as Table 17.3 indicates. The results show that increased adhesion values were achieved with nitrogen at 400 °C and by means of applying a thinner coating per pass.

17.9.2 Hardness

The hardness of the cold spray coatings of CP-Al as measured from metallographic cross-sections with a Wilson Tukon Micro-Hardness Tester utilizing a load of 500 g and a diamond indentor. The values obtained were converted to Brinell hardness numbers for comparative purposes. Figure 17.7 shows a comparison between annealed and fully worked wrought aluminum, and a CP-Al cold spray coating.[9] It is important to note that the values of the cold spray coating were significantly higher than values achieved by conventional materials processing. The cold spray coatings had an average value of 57 Brinell while that of fully hardened wrought aluminum was 45 Brinell. The tremendous plastic deformation of each particle as it impacts the surface of the substrate during the cold spray process results in microstructural changes that increase the hardness. It has been

Table 17.3 Adhesion values of CP-Al deposited by the cold spray process

Conditions	Adhesion (MPa)	Failure mode	Microns/pass
N_2, 2.74 MPa, 250 °C	18.9	Cohesive	114
He, 2.74 MPa, 20 °C	>45.0	Glue failure	64
He, 2.74 MPa, 20 °C	>58.6	Glue failure	64
N_2, 2.74 MPa, 400 °C	>71.4	Glue failure	64

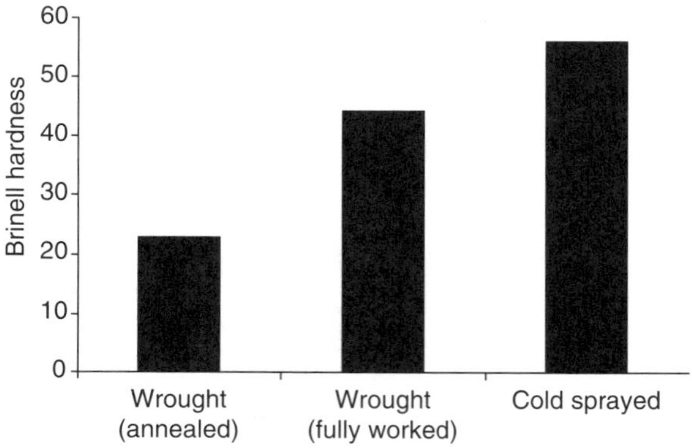

17.7 Hardness of CP-Al cold spray coatings compared with fully worked and annealed wrought material.

well established that cold spray is considered to be a powder shock compaction and consolidation process resulting in high localized strain and substantial grain refinement via fracturing or the formation of sub-grain structures.[10–12] Therefore, an increase in hardness, commensurate with the amount of plastic deformation of each particle during consolidation, was anticipated. Even though this consolidation theory has been associated with the deposition of powder mixtures it is apparent that as a result of the high localized strain that occurs within each particle during impact the conditions were satisfied for significant grain refinement.

17.9.3 Microstructural examination

The microstructural features of the CP-Al cold spray coatings were examined utilizing optical and electron microscopy. Cold spray coatings were produced with the high-pressure Ktech system and the high-pressure ARL system using both helium and nitrogen for comparative purposes. Figure 17.8 shows a representative micrograph of the CP-Al coating produced using helium gas at room temperature and at a gas pressure of 2.74 MPa with the ARL portable system. The coating is very dense and shows no evidence of significant inherent material defects. The coating/substrate interface is free from voids, entrapped grit, or areas of delamination. The coating material is in intimate contact with the substrate forming a metallurgical bond as a result of the severe plastic deformation of the accelerated particle impact.[13]

Figure 17.9 shows a higher magnification of Fig. 17.8 but subsequent to etching with Keller's reagent. There was evidence of plastic deformation of the consolidated particles and significant grain refinement that was the

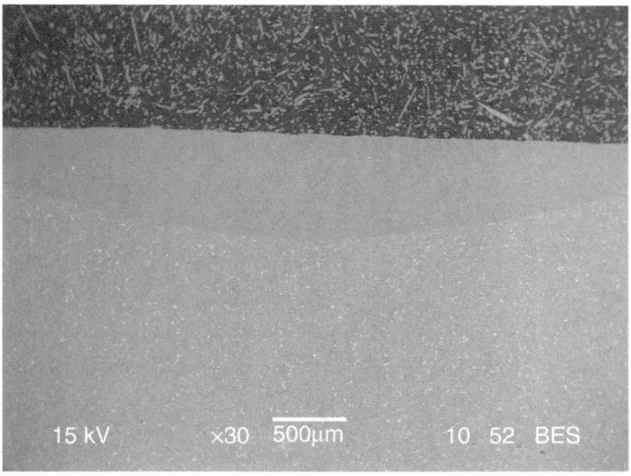

17.8 CP-Al deposited by the ARL portable cold spray system on ZE-41A magnesium interface.

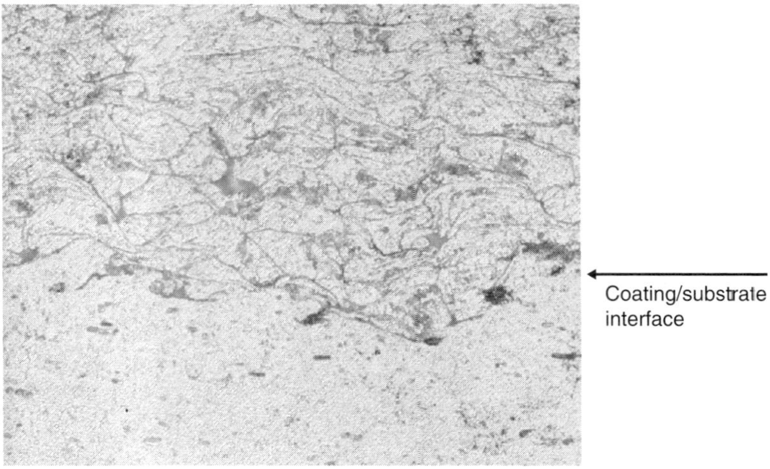

17.9 Etched sample using Keller's reagent showing grain refinement and plastic deformation.

result of the shock compression that occurs during particle impact causing grain size reduction.

Figure 17.10 is an example of a coating of Al 5056 produced with the Ktech high-pressure stationary system also using helium gas at a temperature of 400 °C and a pressure of 2.74 MPa. It was necessary to incorporate the use of helium at high gas temperatures in order to attain a dense deposit. The figure clearly illustrates the potential of cold spray to produce

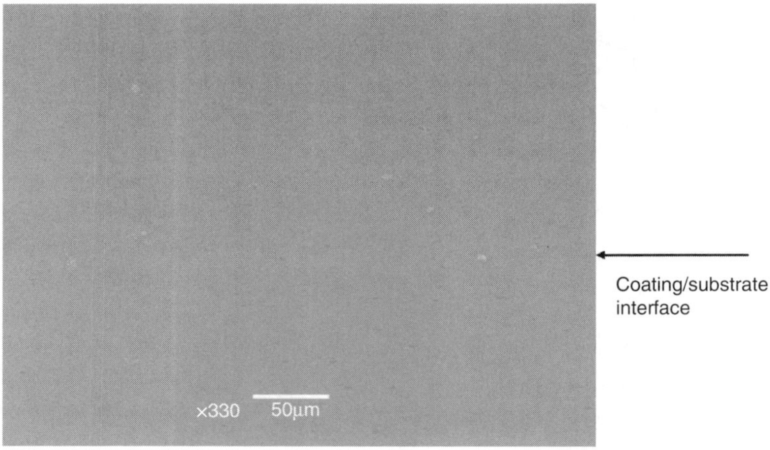

17.10 Al 5056 deposited by the stationary Ktech system using helium as the carrier gas. Note the high level of density.

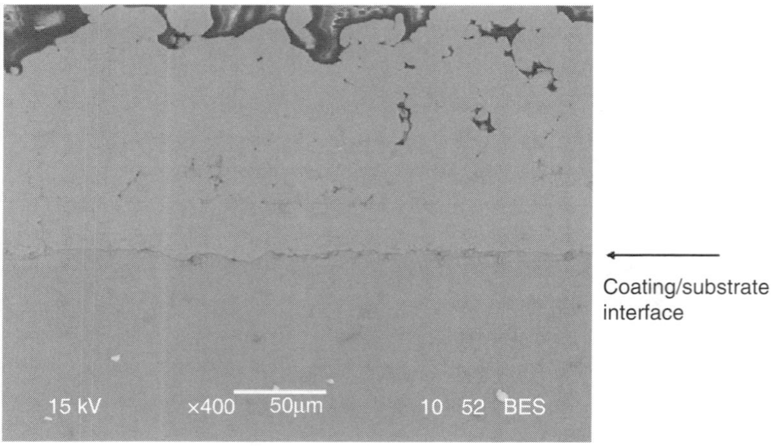

17.11 CP-Al deposited by the stationary Ktech system using nitrogen as the carrier gas at 250 °C . Note the high level of porosity.

an adherent and dense coating. The microstructure is free from voids and porosity and the interface between the coating and the substrate is barely discernible.

Figure 17.11 shows a representative example of a cold spray CP-Al coating deposited with the use of nitrogen as the carrier gas at a temperature of 250 °C and at a gas pressure of 2.74 MPa. The resultant coating was porous and had low bond strength because the gas temperature could not be increased until the incorporation of a plastic nozzle that prevented nozzle fouling. Figures 17.12 and 17.13 show significant improvement of the coating, when the gas temperature was increased to 400 °C. The density had improved as a result of the higher particle velocity as had the interface between the coating and the substrate.

17.9.4 Corrosion testing

The requirement for salt fog exposure established by the Army and Navy Aircraft Commands, was 336 hours according to the procedure of ASTM B117, 'Standard Practice for Operating Salt Spray (Fog) Apparatus'. A total of six test panels of ZE-41A magnesium (7.62 cm × 15.24 cm × 0.64 cm) were machined from a plate, degreased, and then grit blasted lightly with 60 grit Al_2O_3 that was free of any contamination. This is important as any iron contamination can adversely affect corrosion test results. The panels were

17.12 CP-Al deposited by the stationary Ktech system using nitrogen as the carrier gas. Slight evidence of porosity.

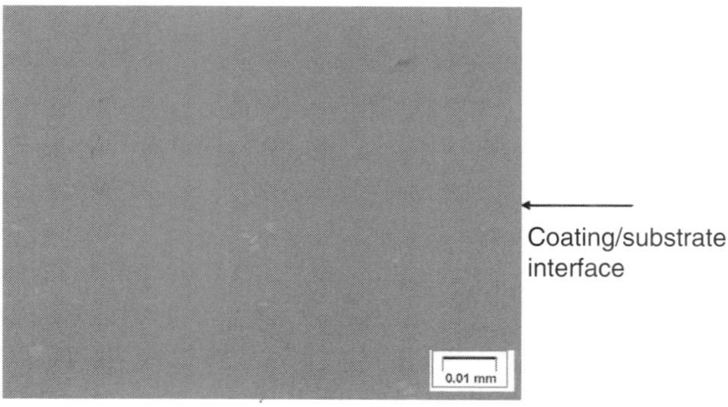

17.13 High-magnification scanning electron micrograph of CP-Al deposited by the stationary Ktech system using nitrogen as the carrier gas.

subsequently coated with 0.30 mm of CP-Al via cold spray. The carrier gas was helium, the pressure was 2.74 MPa, the traverse rate was 100 mm/s, the feed rate was 2 g/min, and the standoff distance was 25 mm. In addition, for comparative purposes, test panels of Al AA2024-T3 and AA7075-T6, and 4130 steel, were also prepared in a similar fashion. The back side and the edges of each panel that were uncoated were lacquered to prevent any infiltration of the environment. The test panels were placed into a salt fog chamber operating at 35 °C with a 5% salt solution and periodically examined every 4 hours. The lacquer applied to the test panels of magnesium failed prematurely allowing corrosion to take place from a few areas on the edges of the panels. When this occurred, the magnesium test panels were removed from the salt fog chamber and the edges were cleaned, recoated with lacquer, and corrosion testing was resumed. The magnesium panels had to be repaired in this manner several times during corrosion testing and after 610 hours, the repair procedure could not be attempted again since too much of the magnesium substrate had corroded away from the edges and back side of the test panels. However, the minimum requirement of 336 hours had been achieved. The test panels of aluminum and steel had been exposed for 7000 hours before some visual evidence of rust spots appeared on the steel panels. The aluminum test panels never experienced any substrate corrosion after 7000 hours but the test was stopped at this point.

A new series of six corrosion test panels were prepared as previously described, but the test panels were machined to round off sharp edges and the entire coupon surface was coated with cold sprayed CP-Al. The process parameters were adjusted to test coatings produced using nitrogen as the carrier gas at the preheat temperature of 400 °C and a pressure of 2.74 MPa. A plastic nozzle was used for these tests. These panels remained intact over 1000 hours in the salt fog environment and exposure continues at this writing.

17.9.5 Oxide analysis

Inert gas fusion according to ASTM E1019 was used to analyze the oxygen content of a CP-Al deposit produced by cold spray. These results were subsequently compared with the oxygen content of the starting powder. In this test, the starting powder consisted of CP-Al, –325 mesh (Fig. 17.14). The results revealed that the cold spray deposit did contain less oxygen (0.25%) as compared with the powder (0.34%). The reason for this decrease is that the brittle oxide layer on each particulate fractures during deposition as a result of the extreme amount of plastic deformation, and a portion of it falls away, never being incorporated into the deposit. The oxygen content of the CP-Al cold spray coating is largely determined by the oxygen content of the original powder, not the process.

17.14 Oxygen content measured by inert gas fusion.

17.10 Cost savings and implementation

The service life of the magnesium components is currently limited due principally to corrosion at mating and mounting surfaces, plus they are especially susceptible to damage resulting from handling and during repair operations. In many cases, the extent of the corrosion or damage is such that there are no repairs available and the component must be replaced. In addition, the corrosion or damage leads to the performance of unscheduled maintenance in the field which is very expensive and impacts on mission readiness. The Navy performed an analysis of these costs and determined that in the year 2000, 23 main transmission modules were replaced on one type of helicopter and 23 unscheduled maintenance actions were performed on this component at a total cost of $8.5 million. Similar costs are incurred by the Navy for overhaul of magnesium components on the two other types of helicopters.

The Army fleet of helicopters exceeds 1400. Overhaul of the main or tail rotor transmission costs up to $400 000, a substantial portion of which is due to corrosion and damage on magnesium housings. Increasing their lifetime through use of cold spray coatings to improve corrosion performance and increase resistance to impact damage would significantly reduce these repair costs. One major aerospace OEM has performed an analysis related to the main and tail rotor transmission housings of their fielded helicopters. In 2003, 21 housings had to be removed from service at a cost of $610 000, which did not include the cost of labor for removal and reinstallation. Based on an analysis of the reasons for removal and the anticipated increased service life associated with use of the cold spray aluminum alloy coatings, it was calculated that 40% of the components would not have had to be removed from service. The cost of application of the cold spray coatings on these components was estimated to be approximately $25 000, so the net annual benefit would be in the range of $250 000 just on the one type of component at one repair facility. Therefore, the payback on the investment in cold spray equipment would be very rapid. The cost of acquisition of new magnesium housings throughout the Army is approximately $8 million annually. The projected potential savings to the entire fleet of helicopters in the Army and Navy results in annual savings of several million dollars through use of the cold spray technology.

In a similar fashion, the Navy will transition the technology, through the Materials Review Board (MRB) which has a similar mission to the SAFR Program, and will be directly involved in the project to provide guidance on implementation requirements.

17.11 Conclusions

In recent years, experimental and computational studies at universities such as Helmut Schmidt University in Hamburg, Germany, have led to a better understanding of the cold spray process.[14] Modeling of the particle impact and bonding mechanisms has been performed as well as measurements of the effect of process variables on particle temperatures and velocities. For the former, a number of researchers have likened the bonding mechanisms in cold spray with those identified in explosive welding, where bond formation relies on deformation under high pressures.[15] With respect to process variables, helium or nitrogen are the most commonly used gases with cold spray, with higher particle velocities obtained with helium but nitrogen is generally preferred because it is much less expensive. The results of this study have shown this to be the case but that adequate adhesion, density, and corrosion resistance can be achieved with the use of nitrogen for CP-Al.

The feasibility of using the cold spray process to repair nonstructural magnesium aircraft components has been demonstrated by the satisfactory results obtained from adhesion, corrosion testing, and microstructural analysis. Adhesion of the coating was considered to be the most important evaluation criteria by AMCOM. The cold spray process yielded adhesion values in excess of 58.6 MPa when helium was used as the carrier gas and 71.3 MPa when nitrogen was employed. These values represent the strength of the adhesive used, as all failures occurred at the interface between the adhesive and the coating. The cold spray coating was not pulled off the substrate and the coating did not fail cohesively.

The cold spray coating was required to withstand a minimum of 336 hours of salt fog exposure according to the requirements of ASTM B117. The CP-Al coating applied by cold spray on ZE-41A magnesium test panels to a thickness of 0.305–0.381 mm using helium as the carrier gas lasted at least 610 hours before the lacquer used to seal the edges and back face of the test panel failed. The coating was intact and did not fail from the front face of the test panel. Corrosion test panels fabricated from aluminum and steel that had also been coated with CP-Al by cold spray at the same time as the ZE-41A magnesium panels using the identical process parameters, lasted 7000 hours in the salt fog chamber. These results serve as a testimony that adherent and dense coatings can be achieved by the cold spray process for this application.

The cold spray process has matured from an emerging technology to a viable alternative to thermal spray for selected applications.[16] This study has shown it to be a promising cost-effective and environmentally acceptable technology to impart surface protection and restore dimensional tolerances to magnesium alloy components on helicopters and fixed-wing aircraft.

17.12 References

1 Levy M, Bombard R, Huie R and Lei K, 'Assessment of Some Corrosion Protection Schemes For Magnesium Alloy ZE41A-T5', *Tri-Service Corrosion Conference*, Atlantic City, 1989.

2 Yamauchi M, Seki J, Sakita E, Miyata Y and Arita K, 'Corrosion Resistant Composite Layer on Magnesium Alloys', *Advanced Composite Materials*, 1991, **1** (1), 3–10.

3 Gonzalez-nunez M, Nunez-lopez C, Skeldon P, Thompson G, Karimzadeh H, Lyon P and Wilks T, 'A Non-Chromate Conversion Coating for Magnesium Alloys and Magnesium-Based Metal Matrix Composites', *Corrosion Science*, 1995, **37** (11), 1763–1772.

4 Gorman W and Woolsey E, 'Selective Anodization Process for Repair of Magnesium Helicopter Components', *Tri Service Corrosion Conference*, Las Vegas, 2003.

5 Griffin R and Zuniga D, 'Evaluation of Coatings on Mg Alloy ZE41A Used in Helicopter Rotor Gearboxes', *Tri Service Corrosion Conference*, Orlando, 2005.

6 Robinson A, 'Evaluation of Various Magnesium Finishing Systems', *Proceedings of International Magnesium Association, 42nd World Conference*, New York, 1985.

7 Johnson A, 'Helium Recycle – A Viable Industrial Option for Cold Spray', *Cold Spray Conference*, Akron, OH, ASM International, 2004.

8 Vlcek J, Gimeno L, Huber H and Lugscheider E (2005) 'A Systematic Approach to Material Eligibility for the Cold Spray Process', *Journal of Thermal Spray Technology*, **14** (1), 125–133.

9 Rooy E, 'Introduction to Aluminum and Aluminum Alloys', *ASM Handbook*, Tenth Edition, Volume 2, p. 49.

10 Stoltenhoff T, Kreye H, Krommer W and Richter H, 'Cold Spraying – From Thermal Spraying to High Kinetic Energy Spraying', *HVOF Colloquium 2000*, Gemeinschaft Thermisches Spritzen e.V., 2000, pp. 29–38.

11 Stoltenhoff T, Kreye H, Richter H and Assadi H, 'Optimization of the Cold Spray Process', *Proceedings of the 2001 International Thermal Spray Conference*, Eds C.C. Berndt and E.F. Lugscheider, ASM International, Materials Park, OH, 2001, pp. 409–416.

12 McCune R, Donlon W, Popoola O and Cartwight E, 'Characterization of Copper Layers Produced by Cold Gas-Dynamic Spraying', *Journal of Thermal Spray Technology*, 2000, **9** (1), 73–82.
13 Maev R, Leshchynsky V and Papyrin A, 'Structure Formation of Ni-based Composite Coatings During Low Pressure Gas Dynamic Spraying', *Proceedings of the 2006 International Thermal Spray Conference*, Seattle, Washington, USA, ASM International, Materials Park, OH, 2006.
14 Kreye H, 'The Cold Spray Process and Its Potential for Industrial Applications', *Cold Spray 2004*, Akron, OH, The Thermal Spray Society of ASM International, Materials Park, OH, 2004.
15 Zhang D, Shipway P and Mccartney D, 'Cold Gas Dynamic Spraying of Aluminum: The Role of Substrate Characteristics in Deposit Formation', *Journal of Thermal Spray Technology*, 2005, **14** (1), 109–116.
16 Papyrin A, 'Cold Spray Technology', *Advanced Materials and Processes*, 2001, **159** (9), 49–51.

Index